Surfaces and Interfaces
of Liquid Crystals

Th. Rasing I. Muševič

Surfaces and Interfaces of Liquid Crystals

Springer

Prof. Theo Rasing
NSRIM
University of Nijmegen
Toernooiveld 1
6525 ED Nijmegen
The Netherlands

Prof. Igor Muševič
Jožef Stefan Institute
Jamova 39
1000 Ljubljana
Slovenia
and
Faculty of Mathematics and Physics
University of Ljubljana
Jadranska 19
1000 Ljubljana
Slovenia

ISBN 978-3-642-05868-4

Springer is a part of Springer Science+Business Media

springeronline.com

Printed in Germany

Final processing by Frank Herweg, Leuthershausen
Cover design: *design & production* GmbH, Heidelberg

Printed on acid-free paper 56/3141/tr 5 4 3 2 1 0

Contents

List of Contributors

Riccardo Barberi
LiCryL, INFM Regional Liquid
Crystal Laboratory
Physics Department
University of Calabria
via Pietro Bucci - Cubo 31C
87036 Rende (CS), Italy
barberi@fis.unical.it

Roberto Bartolino
LiCryL, INFM Regional Liquid
Crystal Laboratory
Physics Department
University of Calabria
via Pietro Bucci - Cubo 31C
87036 Rende (CS), Italy
bartolino@fis.unical.it

Lev M. Blinov
LiCryL, INFM Regional Liquid
Crystal Laboratory
Physics Department
University of Calabria
via Pietro Bucci - Cubo 31C
87036 Rende (CS), Italy
blinov@fis.unical.it

Giovanni Carbone
LiCryL, INFM Regional Liquid
Crystal Laboratory
Physics Department
University of Calabria
via Pietro Bucci - Cubo 31C
87036 Rende (CS), Italy
gcarbone@fis.unical.it

Martin Čopič
[1]Faculty of Mathematics and Physics
University of Ljubljana
Jadranska 19, 1000 Ljubljana
Slovenia and
[2]Jožef Stefan Institute
Jamova 39, 1000 Ljubljana
Slovenia
martin.copic@ijs.si

Gregory P. Crawford
Division of Engineering
Brown University, Providence
Rhode Island 02912
USA
gregory_crawford@brown.edu

Irena Drevenšek-Olenik
[1]Faculty of Mathematics and Physics
University of Ljubljana
Jadranska 19, 1000 Ljubljana
Slovenia and
[2]Jožef Stefan Institute
Jamova 39, 1000 Ljubljana
Slovenia
irena.drevensek@ijs.si

Daniele Finotello
Dept. of Physics, Kent State
University
Kent, Ohio 44242
USA
dfinotel@kent.edu

Jan Gerritsen
University of Nijmegen
NSRIM
Toernooiveld 1
6525 ED Nijmegen
The Netherlands
jang@sci.kun.nl

Klemen Kočevar
Jožef Stefan Institute
Jamova 39, 1000 Ljubljana
Slovenia
klemen.kocevar@ijs.si

Jan Lüning
Stanford Synchrotron Radiation Laboratory
2575 Sand Hill Rd
Menlo Park, CA 94025
USA
J.Luning@Stanford.edu

Gerd Marowsky
Laser-Laboratorium Goettingen e.V.
P.O. Box 2619
37016 Goettingen
Germany
gmarows@gwdg.de

Igor Muševič
[1]Jožef Stefan Institute
Jamova 39, 1000 Ljubljana
Slovenia and
[2]Faculty of Mathematics and Physics
University of Ljubljana
Jadranska 19, 1000 Ljubljana
Slovenia
igor.musevic@ijs.si

Cindy Nieuwkerk
Philips Research
Prof. Holstlaan 4
5656 AA Eindhoven
The Netherlands
cindy.nieuwkerk@philips.com

Theo Rasing
University of Nijmegen
NSRIM
Toernooiveld 1
6525 ED Nijmegen
The Netherlands
th.rasing@sci.kun.nl

Philippe Richetti
Centre de Recherche "Paul Pascal"
Pessac - France
Avenue Albert Schweitzer
33600 Pessac
France
richetti@crpp-bordeaux.cnrs.fr

Mahesh G. Samant
IBM Research Division, Almaden Research Center
650 Harry Road, San Jose
California 95120
USA
mahesh@almaden.ibm.com

Maria P. De Santo
INFM Research Unit of Calabria
Physics Department
University of Calabria
Via Pietro Bucci, Cubo 33B
87036 Rende (CS)
Italy
desanto@fis.unical.it

Andreja Šarlah
[1]Hahn-Meitner Institut
Abteilung Theorie
Glienicker Strasse 10
14109 Berlin
Deutschland and
[2]Faculty of Mathematics and Physics
University of Ljubljana
Jadranska 19, 1000 Ljubljana
Slovenia
sarlah@hmi.de

Miha Škarabot
Jožef Stefan Institute
Jamova 39, 1000 Ljubljana
Slovenia
miha.skarabot@ijs.si

Silvia Soria
ICFO-Institut de Ciencies
Fotoniques
Jordi Girona 29-1D
Edifici Nexus II
08034 Barcelona
Spain
Silvia.Soria@upc.es

Marija Vilfan
Jožef Stefan Institute
Jamova 39, 1000 Ljubljana
Slovenia
mika.vilfan@ijs.si

Mojca Vilfan
Jožef Stefan Institute
Jamova 39, 1000 Ljubljana
Slovenia
mojca.vilfan@ijs.si

Bianca van der Zande
Philips Research
Prof. Holstlaan 4
5656 AA Eindhoven
The Netherlands
bianca.van.der.zande@philips.com

Boštjan Zalar
Jožef Stefan Institute
Jamova 39
1000 Ljubljana
Slovenia
bostjan.zalar@ijs.si

Bruno Zappone
LiCryL, INFM Regional Liquid
Crystal Laboratory
Physics Department
University of Calabria
via Pietro Bucci - Cubo 31C
87036 Rende (CS)
Italy
zappone@fis.unical.it

Slobodan Žumer
[1]Faculty of Mathematics and Physics
University of Ljubljana
Jadranska 19
1000 Ljubljana and
[2] Jožef Stefan Institute
Jamova 39
1000 Ljublijana
Slovenia
slobodan.zumer@fiz.uni-lj.si

Prologue

In 1998 we started a collaborative effort between several European Laboratories to study liquid crystal interfaces with novel, surface and interface-sensitive experimental techniques, such as Atomic Force Microscopy, Scanning Tunneling Microscopy and Second Harmonic Generation, together with more conventional though very sensitive techniques like Optical Birefringence, Ellipsometry, Dynamic Light Scattering and Nuclear Magnetic Resonance. These experimental techniques were complemented with computer simulations and theoretical modeling.

The main goals of the project were:

- to obtain a microscopic understanding of the liquid crystal-wall interactions on a molecular level.
- to understand how these interface properties can determine the long range ordering and properties of LC devices.
- to study how this understanding can be exploited to manipulate the relevant interface properties, in order to control LC-based devices.

The resulting European Network "Surfaces and Interfaces of Liquid Crystals (SILC)" that originated from a discussion between Ljubljana and Nijmegen, further involved groups from Italy, Germany and France, whereas also the industry expressed their interest through the involvement of Philips. At its concept, about 7 years ago, we could start our network proposal summary by stating:"The microscopic understanding of the interaction between a macromolecule (liquid crystal, polymer) and a wall is a major scientific problem that, despite its technological importance and thirty years of research, is still poorly understood".

The main reason for this was the lack of surface sensitive techniques that could probe interfaces at the proper molecular level. Therefore most attention was directed to understand the bulk properties whereas interface properties were only addressed by macroscopic, empirical studies like wetting and spreading.

Now, 6 years and around 100 joint publications later, we can state that we more than fulfilled our original goals. In particular the development of the Scanning Probe methods and the parallel progress in theory have largely contributed to our deeper understanding of the liquid crystal wall interactions on a molecular level.

At our final meeting in Sardinia, spring 2002, it was therefore suggested we should combine our joint expertise and results in a book that would present the state of the art of this field that is both scientifically and technologically very challenging. We are extremely happy that, apart from those who were actually present at its original conception, like Gerd Marowsky, Cindy Nieuwkerk, Slobodan Žumer and Riccardo Barberi, we also got Jan Lüning interested in contributing a very interesting and well fitting chapter on NEXAFS. Together with the already mentioned approaches, this allowed us to present indeed a fairly complete overview and detailed insight in the mechanisms that are responsible for the alignment of liquid crystal molecules on solid and soft-matter interfaces and the possible applications of this knowledge in many LC based devices.

We thank all of them, plus all the other contributors, Marija Vilfan, Boštjan Zalar, Gregory P. Crawford, Daniele Finotello, Klemen Kočevar, Giovanni Carbone, Bruno Zappone, Silvia Soria, Maria De Santo, Lev M. Blinov, Miha Škarabot, Andrea Šarlah, Bianca van der Zande, Philippe Richetti, Roberto Bartolino, Irena Drevenšek Olenik, Mojca Vilfan and Martin Čopič for their willingness to prepare their manuscripts in exactly the way and style we wanted it. And last but not least, Marilou van Breemen from Nijmegen who put a tremendous amount of energy and time in fixing all those cases where this was not completely succesful and who basically corrected all the manuscripts.

The work described in this book was for a large part suported by the EU through the European Network SILC (ERBFMRX CT98-0209).

May 2004

Theo Rasing, Nijmegen
Igor Muševič, Ljubljana

1 Introduction

Igor Muševič, Cindy Nieuwkerk and Theo Rasing

Since the pioneering work on surface-induced alignment of liquid crystals, performed by Lehmann [1], Grandjean [2], Mauguin [3], Chatelain [4], and others [5], scientist have been looking for the answer to the question: *why do certain surfaces align liquid crystals and others not?*

The answer to this question has become even more important with the advent of modern liquid crystal display technologies, that are based on reliable and technologically controllable surface alignment of liquid crystals, used in a variety of electrooptic devices, such as liquid crystal displays, light modulators, optical shutters, switches, holographic systems, etc.

During the last decade, the progress in the technology of liquid crystal devices, as well as the discovery of a variety of novel liquid crystalline phases have triggered a considerable and intense scientific interest in the microscopic origin of surface alignment. Fortunately, this renewed scientific and technological interest was accompanied by the advent of modern, surface sensitive experimental techniques, that have been successfully used in the study of liquid crystal interfaces. Whereas a decade ago the mechanisms of surface alignment were "poorly understood", nowadays we can claim that we do understand most of the "mysteries" of the surface alignment of liquid crystals. This understanding of the fundamental physics and chemistry of liquid crystal interfaces has also resulted in a variety of new and exciting techniques for producing and patterning of high quality surface alignment layers. This has subsequently led to the development of well controlled structures where the liquid crystal molecules are confined in thin (micrometer) cells with corresponding optical properties. Thin in this respect refers to dimensions comparable or smaller than typical length scales, which, for photonic applications, is the wavelength of light.

Apart from the significant technological advancement in recent years, confined liquid crystals are by themselves a very interesting subject from a fundamental point of view. During the last decade, the scientific curiosity about the fundamental aspects of interfacial phenomena has resulted in extensive studies of confined liquid crystals (for some reviews see [5–9]).

In this Chapter we want to introduce some basic concepts and present an overview of the mechanisms, that are in our opinion responsible for the alignment of liquid crystals on solid and soft-matter interfaces. In addition we will present and discuss recent developments and results in surface sensitive techniques, that have allowed us to obtain a molecular scale picture of liquid crystal alignment.

1.1 Surface Alignment, Length-scales, Symmetry and Microscopic Interactions

The influence of confining surfaces on the liquid crystal alignment can be seen in a simple experiment, such as the observation under a polarizing microscope of a thin layer of a nematic liquid crystal between two glass slides that have been unidirectionally rubbed with a piece of paper. This simple experiment has in fact led to the idea that nematic liquid crystals are aligned on "microgrooves", that have been created during the rubbing process. The resulting model is based on the elastic continuum theory of the nematic phase [10] and the liquid crystal alignment is a result of the minimum of the free energy of the system. For parallel grooves, the nematic liquid crystal aligns with the director along the grooves, which is most favourable from the point of view of the elastic distortion and the corresponding elastic energy.

The rubbing grooves can indeed be observed using nanoscale imaging techniques, such as Atomic Force Microscopy (AFM). Moreover, well defined microgrooves can be created, using laser ablation [11] or electron beam litography. As an example, Fig. 1.1 shows an AFM image of a periodic array of parallel microgrooves, created in the surface of a thin polymer layer by laser ablation.

Using these microgrooved surfaces as an aligning layer, the nematic liquid crystal is indeed aligned with the optical axis (and the director) along the grooves, as shown on the polarizing micrograph in Fig. 1.1, that shows a number of pixels created by single shot laser ablation. The multidomain structure that is visible in some of the pixels in Fig. 1.1b indicates that these grooved substrates do not have a pre-tilt angle. This simple experiment confirms that liquid crystal molecules spontaneously order along a macroscopic direction in space. Although the structure of the "orienting" substrate is disordered on the microscopic scale, it is the overal macroscopic order, that induces the alignment of the liquid crystal.

In fact, this is not surprising, if we consider that a liquid crystal phase corresponds to a spontaneously orientationally ordered macroscopic system of a broken continuous rotational symmetry. In this picture, the orientational order of a nematic liquid crystal is established by a condensation of a spatially homogeneous mode. The effect of a confining surface can be considered as a coupling of a spatially homogeneous potential to the excitation modes in the nematic. The modes that are unstable in the nematic (positional and orientational), decay exponentially from the surface and give rise to a surface ordered (or disordered) layer with a thickness of the order of the correlation length. There is only one mode that couples to a homogeneous surface and that is correlated over macroscopic distances: the Goldstone mode, which represents

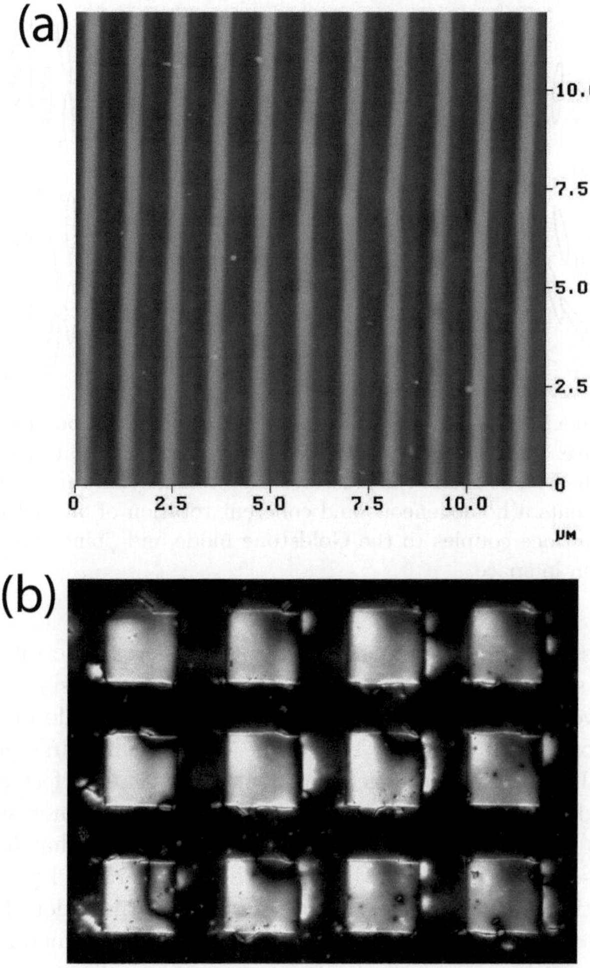

Fig. 1.1. (a) AFM image of a periodic array of about 25nm deep grooves, created by laser ablation of a thin polymethylmethacrylate layer using one single pulse at $350 J/cm^2$. (b) crossed polarizer micrograph of a twisted nematic cell, assembled from a rubbed substrate and a substrate with a series of grating pixels, each created by single shot laser ablation (from [11]).

a homogeneous rotation of a nematic. An even more striking manifestation of this "orientation" pinning by the macroscopic direction at the surface, is the case of chemically patterned isotropic surfaces [12]. We can therefore summarize that in a macroscopic picture, a confining homogeneous surface aligns a nematic liuid crystal because it represents a state of spontaneously broken continuous rotational symmetry, as shown in Fig. 1.2.

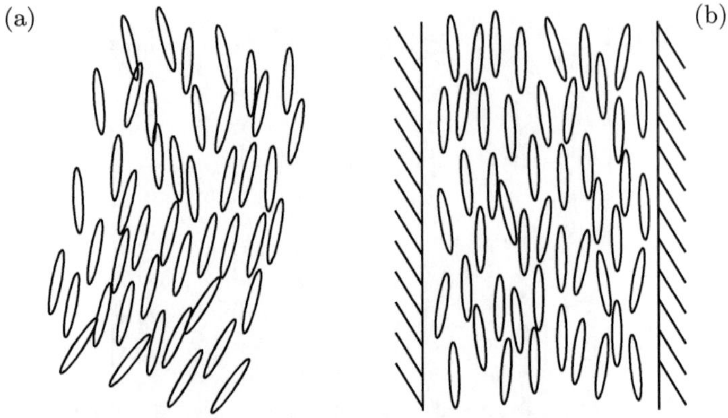

Fig. 1.2. (a) in a bulk nematic liquid crystal, the director can point in an arbitrary direction in space. This is a signature of a broken continuous rotational symmetry of the isotropic phase. The mode that restores the broken symmetry is the Goldstone mode. It represents a homogeneous and coherent rotation of all molecules. (b) the homogeneous surface couples to the Goldstone mode and "pins" the director in a certain direction in space.

When crossing from the macroscopic to the microscopic (atomic, molecular) view of surface alignment, we note that structures, similar to the microscale grooves, can be produced on a nano (or meso) scale by scanning an AFM tip across a thin polymer layer. The experiments, first performed by Ruetschi et al. [13] and Pidduck et al. [14] show that perfect alignment of nematic liquid crystals can be obtained on grooved polymer surfaces. Figure 1.3 shows a tapping mode AFM image of a polymer that has first been "rubbed" by scanning the AFM tip across the polymer in a litography mode. The cross section shows relatively narrow grooves of the order of several tens of nanometers width and depth of the order of several nanometers. The corresponding polarized micrograph of a twisted nematic cell, assembled from an AFM patterned substrate, demonstrates perfect alignment. This raises the question whether it is only the geometrical-macroscopic confinement that is responsible for the alignment or whether there is something on the nanoscale that does it as well.

This question can be answered by considering the experiments, where liquid crystals have been spread across solid crystal surfaces [15]. Macroscopic observations under a polarizing microscope reveal that *crystal surfaces do orient nematic liquid crystals*, which leads to the conjecture that there may be microscopic positional and orientational order at a crystal-liquid crystal interface. Indeed, this local interfacial order can be revealed in experiments with a Scanning Tunneling Microscope (STM), first performed by Foster and Frommer [16]. An example of an STM image of a crystal-nematic liquid crystal interface is shown in Fig. 1.4 for the case of 8CB on graphite. One can

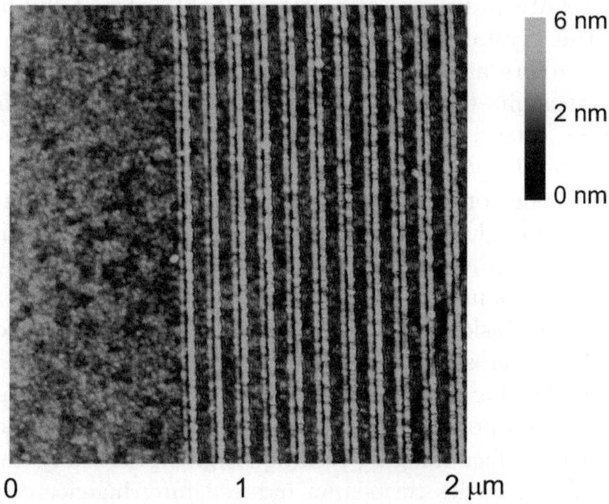

Fig. 1.3. A polyimide surface partly "rubbed" by scanning the AFM tip in a regular manner across the surface. This creates tiny grooves-corrugations that have been imaged using a tapping mode AFM. This is a much gentler mode that does not significantly disturb the soft polymer surface.

Fig. 1.4. STM image of the first molecular layer of 8CB molecules on graphite. The image has been taken at a tunneling current of 20pA and a tunneling voltage of 870mV. The image was taken by spreading a drop of 8CB on graphite and immersing the STM tip into the LC.

clearly observe both positional and orientational order of the liquid crystal molecules on this crystalline surface.

There is therefore an *epitaxially ordered* first molecular layer at the interface, that transmits the aligning action of the surface into the bulk. The positional order, that is unstable in the bulk nematic, decays within the *smectic correlation length* (several nm) from the surface, whereas the surface induced orientational order is of long range (hundreds of micrometers). It is important to note, that in this case, the microscopic interactions (i.e. the molecule-molecule and atom-atom interactions) are responsible for inducing a long range orientational order.

The overall conclusion from these considerations is, that both types of interactions (i.e. macrosopic-geometrical and microscopic) are likely to be effective in a particular alignment technique. As a typical example, we can consider the case of a polymer alignment layer that has been unidirectionally rubbed. Whereas surface imaging techniques give practically no evidence of microgrooves, linear optics reveals an induced birefringence of the aligning layer, which is of the order of $0.01 - 0.03$ [17]. This is a clear evidence of the rubbing-induced order on a molecular scale, that can be clearly observed, for example, using Near Edge X-ray Absorption Fine Structure (NEXAFS) spectroscopy [18]. The scattering cross section of this technique is tensorial and is sensitive to the orientational order of the molecular clusters that absorb X-ray photons. These can be, for example, phenyl rings, so that NEXAFS can measure the degree of molecular (not collective) order of particular molecular species on a surface. Indeed, recent NEXAFS experiments on rubbed polymer films clearly indicate that rubbing induces a significant degree of order of particular segments of a polymer.

1.2 Methods and Techniques of Surface Alignment

Rubbing

The most conventional and widely used method of surface alignment is rubbing [17,20–22] a thin (50-100 nm thick) organic film, such as polyimide (PI), polyvinyl acohol (PVA), nylon, etc., by moving a cloth over the substrate (see Fig. 1.5). Due to mechanical action, the rubbing induces a certain degree of orientational order in the molecular segments of the polymer, which is very likely induced by the thermally assisted plastic deformation due to a mechanical shearing force [23,24]. The resulting order in the segments of the polymer interface can be observed either by the thus induced linear optical birefringence [17] or by NEXAFS [18]. The aligning action of a partially aligned polymer layer proceeds to the "first molecular layer" via the anisotropic van der Waals interaction and excluded volume effects due to the topology of the interface. Further into the bulk, the orientational order is transmitted via

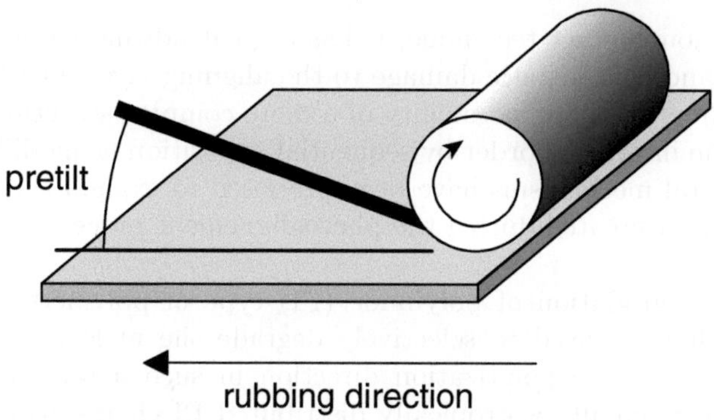

Fig. 1.5. Principle of rubbing. A rotating drum moves over the surface inducing an LC orientation as indicated by the rod.

intermolecular forces that are responsible for the orientational order in liquid crystals.

The rubbing method has its advantages and drawbacks. The advantage is its relative simplicity and technological robustness, whereas the disadvantages are the generation of dust particles, mechanical damage (rubbing stripes) and possible electrostatic hazard to the thin film transistor circuitry in complex screens. Therefore other alignment technologies like e.g. photoalignment [25–27], oblique evaporation [7,28] and ion beam alignment [19,29,30] are actively studied for mass manufacturing.

Photoalignment

A very promising method for reducing the above mentioned disadvantages of mechanical rubbing is a photoalignment technique [25–27]. As shown in Fig. 1.6, the process uses UV linearly polarized light and selective UV induced chemical reactions to induce a certain degree of surface ordering in a photosensitive polymer layer, deposited on a substrate.

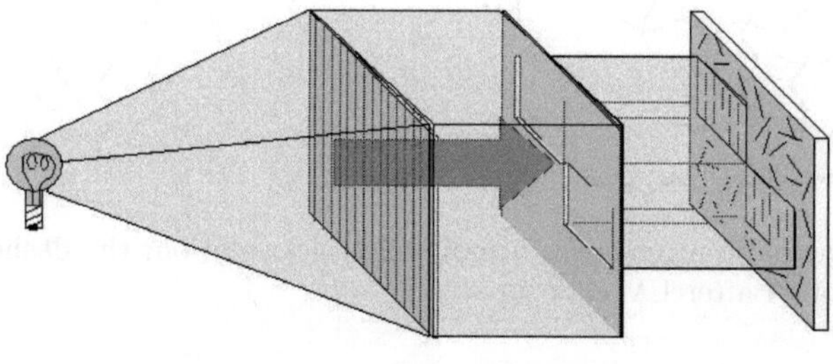

Fig. 1.6. Principle of photoalignment.

Being a non-contact technique, it has a great advantage of imposing no mechanical and electrostatic damage to the aligning layer. Furthermore, the technique offers a unique possibility of a more complex structuring and patterning of the interfacial order by sequential exposition using different masks [26, 27]. Several mechanisms have been proposed to explain the onset of orientational order created during the photoalignement process:

(i) selective degradation of polyimide (PI) type of polymers. Linearly polarized UV light is used to selectively degrade the molecules [31, 32] that are aligned along the polarization direction in such a way that after UV exposure the originally isotropically distributed PI chains attain a prefered direction (see Fig. 1.7). The degradation process is initiated in the UV sensitive groups, which are subjected to decomposition by different photochemical reactions, such as photo-oxidation, deimidization, and even ablative photodecomposition. The UV light causes bond scission in the polyimide material, resulting in the creation of free radicals [33, 34]. These may further react and various gaseous products may be released from the exposed surface. If the number of simultaneously broken bonds is large, photo-ablation may occur. The type and rate of the photo reactions depend on the energy dose (i.e. incident power per unit surface), and the length of the light pulse [36]. One of the disadvantages of this process is the creation of electric charges during the photochemical reaction, that results in charge trapping on the alignment surface and image sticking [35, 36]. The second disadvantage is possible overdosing, where the bonds that are directed in the preferred direction are progressively broken due to an overdose. Some examples of photodegradable polyimides are shown in Fig. 1.8.

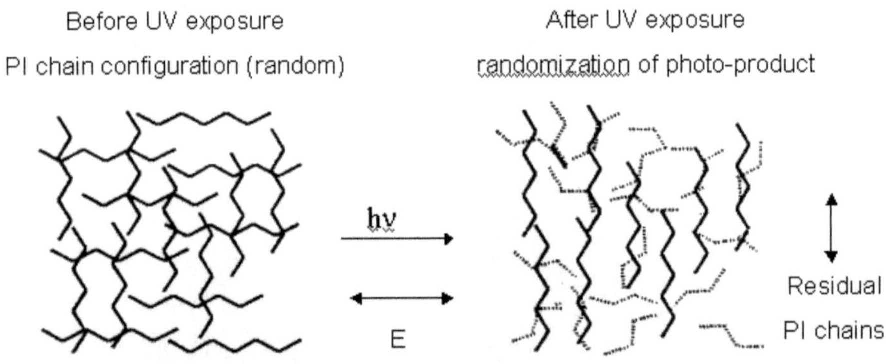

Fig. 1.7. Schematic mechanism of polyimide degradation; the dashed lines show by-products of PI after UV exposure.

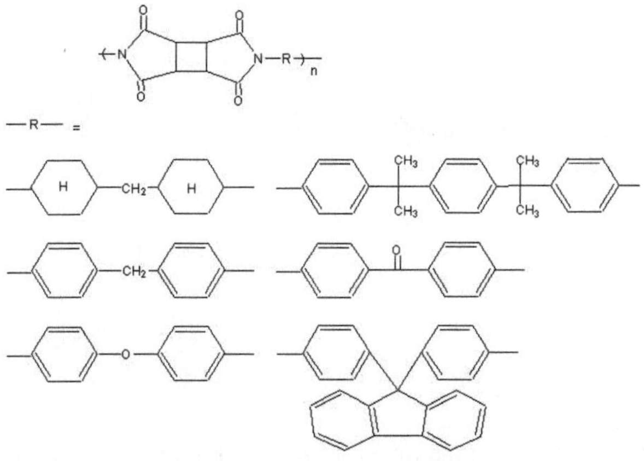

Fig. 1.8. Examples of photodegradable polyimides.

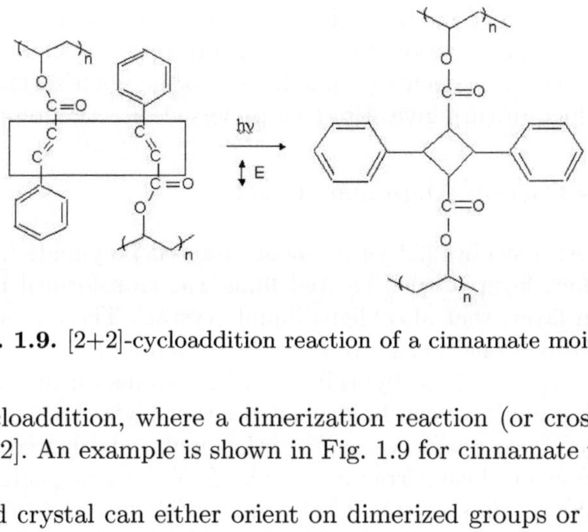

Fig. 1.9. [2+2]-cycloaddition reaction of a cinnamate moiety.

(ii) [2+2]-cycloaddition, where a dimerization reaction (or crosslinking) occurs [26, 37–42]. An example is shown in Fig. 1.9 for cinnamate type of polymers.

The liquid crystal can either orient on dimerized groups or on the unreacted moities. Depending on the type of dimerisable polymer, either a parallel or a perpendicular aligning direction can be obtained.

(iii) isomerization mechanism, shown in Fig. 1.10. The mechanism is typical of materials, containing azobenzene and stilbene units. These moities have two possible molecular conformations, e.g. a trans- and cis-configuration [43]. Under the influence of UV light, these molecular parts can transform from one to the other state. Usually, the cis-isomer is less stable due to inherent "molecular strain" and relaxes to the trans configuration. Under the influence of UV light, the molecules with their transition moments along the polarization direction will absorb light and transform into this cis-isomer. As

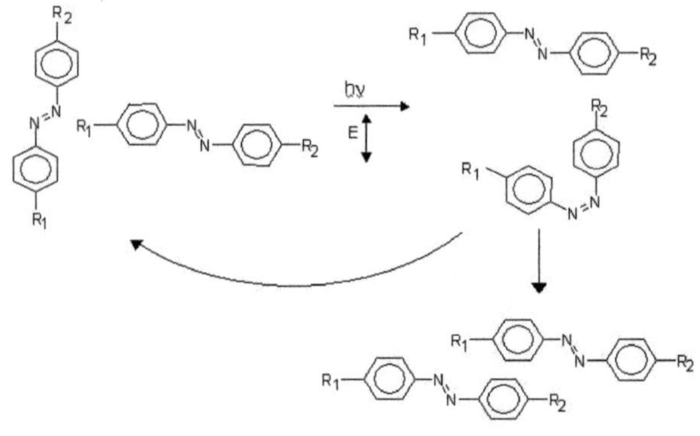

Fig. 1.10. Example of the isomerisation of an azobenzene.

the configuration is unstable, the molecules would relax back to the trans configuration, but now the transition moments can be either parallel or perpendicular to the polarization direction. As a result, the number of molecules with their transition moments perpendicular to the polarization will increase with time, which in turn gives rise to a preferred direction on the surface.

Amorphous Carbon Alignment Layers

Recent NEXAFS studies [29] of ion-beam treated polyimide films indicated that the surface layer of such treated films was transformed into an amorphous carbon layer, that also aligns liquid crystals. The reason for this was found in the directional nature of the carbon double (or triple) bonds, that are formed by sp_2 (and sp) hybridized carbon atoms during ion bombardement. The aligning layer can be deposited either by sputtering or CVD [29]. Besides carbon, also a wide variety of other materials showed liquid crystal alignment upon ion beam irradiation, like SiN_x, hydrogenated amorphous silicon, SiC, SiO_2, glass, Al_2O_3, CeO_2, SnO_2, $ZnTiO_2$, and $InTiO_2$. In principle any material with oriented bonds, which can be preferentially broken by bombardement, can introduce a surface orientational anisotropy.

Alignment by Nanostructuring Polymer Surfaces

Recently, Ruetschi et al. [13] showed that the AFM in the contact mode can produce relatively smooth aligning surfaces with controllable anchoring strength. This alignment method has already been used to obtain specific surface configurations of polymer layers, controllable grey scale and liquid crystal gratings [44, 45], bistable [46–48], and multistable surfaces [49]. As one can control the scanning parameters very well, the method was also used to study the mechanism of liquid crystal alignment [50, 51]. Of particular importance seem to be patterned surfaces that cause frustration to the alignment of liquid crystal molecules. This can result in bistable and multistable [49] surfaces, i.e. the surfaces that have several well defined "easy" directions, as well as extremely large values of surface pretilts, of the order of 40 degrees [52].

Other contact-free alternatives of surface alignment include alignment by electric [53] or magnetic [54] fields. In particular the last approach seems quite promissing, as it appears to be possible to obtain very good anchoring properties by using moderate magnetic fields of less than 2 Tesla, that can be applied to very large surfaces in a very homogeneous way. Another contact-free alternative is using laser abblation [11], which is, similar to the AFM patterning, limited to rather small surface areas.

We have not considered here alignment techniques, such as SiO_x deposition or alignement by monolayer (Langmuir) deposition on surfaces. Details on these techniques of surface alignment can be found elsewhere [5–9]. We also do not discuss recently introduced active (command) surfaces, where the anchoring conditions are controlled by light [55], electrochemistry [56] and electric field [57].

1.3 Overview of the Contents of the Book

The purpose of this book is two-fold. On the one hand, we want to present a selection of recently developed experimental methods, their potential and selectivity in the studies of surface and interfacial phenomena of liquid crystals. This includes Atomic Force Microscopy (AFM), Near Edge X-ray Absorption Fine Structure (NEXAFS) spectroscopy, Scanning Tunneling Microscopy (STM) and Second Harmonic Generation (SHG) from surfaces. In combination with already mature, but still powerful spectroscopic techniques, such as Nuclear Magnetic Resonance (NMR), Dynamic Light Scattering (DLS) and Brewster Angle Ellipsometry (BAE), this forms the solid experimental backbone of this book, that in combination with a Theory Chapter gives to our opinion a solid foundation for our present understanding of the surface alignment of liquid crystals. On the other hand, we also want to present in a transparent way the relevant physics, from the nanometer to the micrometer scale, that plays an essential role at the various liquid crystal interfaces, as well as how this understanding is translated in a large variety of liquid crystal applications.

The book starts with an NMR Chapter, that in our opinion presents both the static and dynamic aspects of liquid crystal interfaces in a most complete manner. The next Chapter is an excursion into the nanoscale world, where the surface and intermolecular forces at liquid crystal interfaces can be measured with unprecedented precission. Chapters 4 and 5 are optical chapters, where the combination of linear and non-linear optics is applied to the study of the structure of the liquid crystal interfaces. Chapter 6 introduces a very powerful new technique to the field of liquid crystal interfaces, Near

Edge X-ray Absorption Fine Structure (NEXAFS), and gives a beautiful example of its surface sensitivity. Chapter 7 brings us to the STM images at the molecular level and discusses and demonstrates the presence of surface charges on alignment layers. A comprehensive review of the theoretical tools (mean field and simulation), that are conveniently applied to the analysis of soft matter interfaces, is given in Chapter 8. Last, but not least, we present in Chapter 9 the important reason, why we are interested in the interfacial physics of liquid crystals. Applications and implications of the physics and technology of liquid crystal interfaces are considered in this Chapter. Finally, we give at the end of this book an epilogue, which may be considered as our condensed view of interfacial phenomena in liquid crystals, as well as some future aspects.

References

1. O. Lehmann, Handbuch der biologischen Arbeitsmeithoden, ed. E. Abderhalden, Berlin, 1922.
2. F. Grandjean, Bull.Soc.Fr.Mineral.Cristallogr.**39**, 164 (1916).
3. C. Mauguin, Bull.Soc.Fr.Mineral.Cristallogr.**34**, 71 (1911).
4. P. Chatelain, Bull.Soc.Fr.Mineral.Cristallogr.**66**, 105 (1943).
5. H. Yokoyama, Chapter 6 in Handbook of Liquid Crystal Research, edts. P. J. Collings and J. S. Patel, Oxford University Press, Oxford, (1997).
6. B. Jerôme, Rep.Prog.Phys. **54**, 391 (1991).
7. J. Cognard, Mol.Cryst.Liq.Cryst. **78**, Supplement**1**, 1 (1982).
8. A. A. Sonin, *The Surface Physics of Liquid Crystals*, Gordon and Breach Publishers, Amsterdam B.V., (1995).
9. L. M. Blinov, A. Yu. Kabayenkov, A. A. Sonin, Liq.Cryst. **5**, 645 (1989).
10. D. W. Berreman, Phys.Rev.Lett. **28**, 1683 (1972).
11. M. Behdani, S. H. Keshmiri, S. Soria, M. A. Bader, J. Ihlemann, G. Marowsky, Th. Rasing, Appl.Phys.Lett. **82**, 2553 (2003).
12. Baek-woon Lee, N.A. Clark, Science **291**, 2576 (2001).
13. M. Ruetschi, P. Grutter, J. Funfschilling, H.J.Guentherod, Science **265**, 512 (1994).
14. A. J. Pidduck, S. D. Haslam, G. P. Bryan-Brown, R. Bannister, I. D. Kitely, Appl.Phys.Lett. **71**, 2907 (1997).
15. M. Glogarova, J.Physique **42**, 1569 (1981).
16. J. Foster, and J. Frommer, Nature **60**, 1418 (1988).
17. J. M. Geary, J. W. Goodby, A. R. Kmetz, J. S. Patel, J.Appl.Phys. **62**, 4100 (1987).
18. M.G. Samant, J. Stöhr, H.R. Brown, T.P. Russell, J. M. Sands and S.K. Kumar, Macromolecules **29**, 8334 (1996).
19. J. van Haaren, Nature **411**, 29 (2001).
20. J. Stöhr, M. G. Samant, J.Elec.Spectrosc.Relat.Phenom. **98-99**, 189 (1999).
21. N. Ito, K. Sakamoto, R. Arafune, S. Ushioda, J.Appl.Phys. **88**, 3235 (2000).
22. N. F. A. van der Vegt, F. Müller-Plathe, A. Geleßus, D. Johannsmann, J.Chem.Phys. **115**, 9935 (2001).

23. A. J. Pidduck, G. P. Bryan-Brown, S. Haslam, R. Bannister, I. Kitely, T. J. McMaster, L. Boogaard, J.Vac.Sci.Technol. **A14**, 1723 (1996).
24. A. J. Pidduck, G. P. Bryan-Brown, S. D. Haslam, R. Bannister, Liq.Cryst. **21**, 759 (1996).
25. W. M. Gibbons, P. J. Shannon, S.-T. Sun, B. J. Swetlin, Nature **351**, 49 (1991).
26. M. Schadt, K. Schmitt, V. Kozenkov, V. Chigrinov, Jpn.J.Appl.Phys. Part 1 **31**, 2155 (1992).
27. M. O'Neill, S. M. Kelly, J.Phys.D:Appl.Phys. **33**, R67 (2000).
28. M. Lu, K. H. Yang, T. Nakasogi, S. J. Chen, SID 00 digest **2000** , page 446.
29. P. Chaudhari, J. Lacey, J. Doyle, E. Galligan, S. C. A. Lien, A. Callegari, G. Hougham, N. D. Lang, P. S. Andry, R. John, K. H. Yang, M. H. Lu, C. Cai, J. Speidell, S. Purushothman, J. Ritsko, M. Samant, J.Stöhr, Y. Nakagawa, Y. Katoh, Y. Saitoh, K. Sakai, H. Satoh, S. Odahara, H. Nakano, J. Nakagaki, Nature **411**, 56 (2001).
30. J. Stöhr, M. G. Samant, J. Luning, A. Callegari, P. Chaudhari, J. P. Doyle, J. A. Lacey, S. A. Lien, S. Purushothaman, J. L. Speidell, Science **292**, 2299 (2001).
31. M. Nishikawa, J. L. West, IDW'**98**, 327 (1998).
32. M. Nishikawa, J. L. West, Y. Reznikov, Liq. Cryst. **26**, 575 (1999).
33. J. Lu, S. Deshpande, J. Kanicki, A. Lien, R. A. John, W. L. Warren, AMLCD conference digest **1995**, page 97.
34. Y. Wang, C. Xu, A. Kanazawa, T. Shiono, T. Ikeda, Y. Matsuki, Y. Takeuchi, J. Appl. Phys.**84**, 4573 (1998).
35. K. H. Yang, K. Tajima, A. Takenaka, H. Takano, Jpn.J.Appl.Phys.Part 2 **35**, L561 (1996).
36. C. J. Newsome, M. O'Neill, J.Appl.Phys. **88**, 7328 (2000).
37. O. Yaroshchuk, G. Pelzl, G. Pitwitz, Y. Reznikov, H. Zaschke, J.-H. Kim, S. B. Kwon, Jpn.J.Appl.Phys. Part 1 **36**, 5693 (1997).
38. Y. Makita, T. Natsui, S. Kimura, S. Nakata, M. Kimura, Y. Matsuki, Y. Takeuchi, J.Photopol.Sci.Techn. **11**, 187 (1998).
39. S. Nakata, M. Kimura, A. Kumano, Y. Takeuchi, SID 01 digest **2001**, page 802.
40. M. Schadt, H. Seiberle, A. Schuster, Nature **381**, 212 (1996).
41. M. Obi, S. Morino, K. Ichimura, Chem.Mater. **11**, 656 (1999).
42. P. O. Jackson, R. Karapinar, M. O'Neill, P. Hindmarsh, G. J. Owen, S. M. Kelly, IS&T/SPIE Conference on Liquid Crystal Materials, Devices and Applications VII **1999**, 38.
43. T. Ikeda, A. Kanazawa, Bull.Chem.Soc. **73**, 1715 (2000)
44. B. Wen, R. G. Petschek, Ch. Rosenblatt, Appl.Opt. **41**, 1246 (2002).
45. B. Wen, M. P. Mahajan, Ch. Rosenblatt, Appl.Phys.Lett. **76**, 1240 (2000).
46. J. H. Kim, M. Yoneya, J. Yamamoto, H. Yokoyama, Appl.Phys.Lett. **78**, 3055 (2001).
47. M. Yoneya, J. H. Kim, H. Yokoyama, Appl.Phys.Lett. **80**, 374 (2002).
48. J. H. Kim, M. Yoneya, H. Yokoyama, Appl.Phys.Lett. **83**, 3602 (2003).
49. J. H. Kim, M. Yoneya, H. Yokoyama, Nature **420**, 159 (2002).
50. A. Rastegar, M. Škarabot, B. Blij, Th. Rasing, J.Appl.Phys. **89**, 960 (2001).
51. M. Škarabot, S. Kralj, A. Rastegar, Th. Rasing, J.Appl.Phys. **94**, 6508 (2003).
52. B. Zhang, F. K. Lee, O. K. C. Tsui, P.Sheng, Phys.Rev.Lett. **91**, 215501 (2003).
53. M.W. Kim, A. Rastegar, I. Drevenšek Olenik, M. W. Kim and Th. Rasing, J.Appl.Phys. **90**, 3332 (2001).

54. M. Boamfa, M.W. Kim, J.C. Maan and Th. Rasing, Nature **421**, 149 (2003).
55. K. Ichimura, Supermolecular Sci. **3**, 67 (1996).
56. Y.Y. Luk, N.L. Abbott, Science **301**, 623 (2003).
57. L. Komitov, Journal of the SID **11**, 437 (2003).

2 Surface-Induced Order Detected by Deuteron Nuclear Magnetic Resonance

Marija Vilfan, Boštjan Zalar, Gregory P. Crawford, Daniele Finotello and Slobodan Žumer

2.1 Introduction

As many as 30 years ago P. G. de Gennes wrote in the first edition of his book *The Physics of Liquid Crystals* [1]: "The dual aspects of a nematic phase (liquid-like but uniaxial) are exhibited most spectacularly in the NMR spectrum; the uniaxial symmetry causes certain line splittings, which are absent in the conventional isotropic liquid phase. On the other hand, the lines are relatively narrow; this implies rapid molecular motions and is a natural consequence of the fluidity".

The previous statement, stressing the fact that nuclear magnetic resonance (NMR), a widely used experimental technique in chemistry, physics, biology, farmacology, and medicine [2], is clearly illustrated in the particular case of deuteron NMR spectra of a selectively deuterated liquid crystal (Fig. 2.1). In the crystalline phase, the spectrum is 'powder-like' and spans a frequency range of about 260 kHz; characteristic of the spectrum in the nematic phase are two sharp lines separated for about 45 kHz; they merge into a single, narrow line of width smaller than 100 Hz in the isotropic phase. The splitting of the two lines in the nematic phase is directly proportional to the scalar nematic order parameter S of the molecule, but it also depends on the orientation and dynamics of the Carbon-Deuterium (C-D) bond and on the orientation of the molecular director with respect to the external magnetic field [3]. The abundance of phases where those characteristic parameters have been determined made liquid crystals a model system to probe with NMR in order to get a detailed understanding of their structures and dynamics.

The sensitivity of deuteron NMR to the molecular orientational order and to director field configurations turned out to be extremely useful in studies of liquid crystals confined into submicrometer pores. Moreover, the large surface-to-volume ratio of these composite systems render the interfacial and surface phenomena, induced by the liquid crystal-surface interactions, accessible even to an essentially integrative technique like NMR. Since the discovery of polymer dispersed liquid crystals (PDLCs) in 1986 [4], NMR of selectively deuterated liquid crystals was used to discriminate unambiguously among various director structures in cavities, resulting from an interplay between elastic forces, morphology and size of the cavity, and surface interactions. These structures include the escaped-radial, planar axial, planar-polar, and

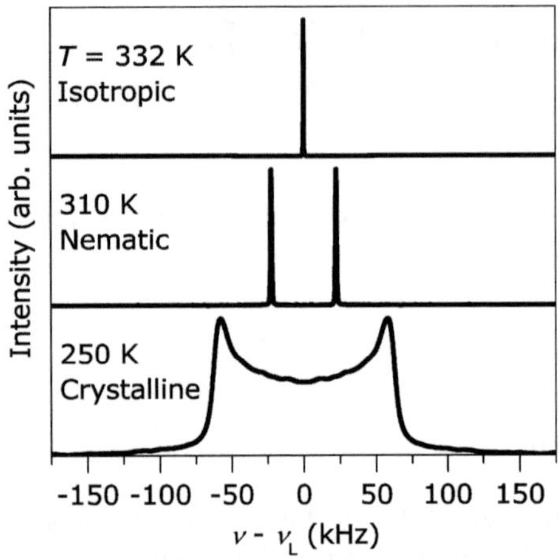

Fig. 2.1. Deuteron NMR spectra of selectively deuterated liquid crystal 8CB-αd_2 at different temperatures.

planar-radial in cylindrical cavities [5–9], bipolar and radial director configurations in spherical cavities of PDLC materials [10], and the complex director structures of liquid crystals in porous glasses, aerosils, and aerogels [11–14]. NMR could also discern different structures occurring in a PDLC droplet upon the application of an external electric field that switches this material from the opaque to the transparent state [15, 16]. This property of PDLC materials leads to their application as switchable, privacy windows [17].

Another part of the NMR research of microconfined liquid crystals focused on the detection and measurements of the weak orientational order induced by solid surfaces above the transition temperature $T_{\rm NI}$ where the bulk liquid crystal turns into the isotropic phase. The onset of orientational order on approaching $T_{\rm NI}$ from above, surface wetting phenomena and the continuous or discontinuous nature of the transition in a cavity have been of particular interest. The contribution of NMR to the knowledge of these phenomena is the topic of the present article. As for the different names that have been used in the literature to denote the high-temperature phase of microconfined liquid crystals, we will use the term 'isotropic' as long as the phase transition into the nematic phase is discontinuous, but it is understood that truly isotropic is only the substance far away from the confining surfaces whereas a certain degree of orientational order persists next to the walls.

The first indication that the surface-induced order in the isotropic phase of confined liquid crystals can be detected by deuteron NMR was the spectrum of selectively deuterated 4'-n-pentyl-4-cyanobiphenyl (5CB) in the spherical cavities of a PDLC material. Far above 35℃, where the bulk turns into the isotropic phase, the spectral line (a single one) of the PDLC was considerably broader than in the bulk at the same temperature. An even more definite and surprisingly clear evidence of surface-induced order was later obtained from the spectra of the same liquid crystal in cylindrical cavities of Nuclepore and Anopore membranes [18,19]. In the isotropic phase, two distinctly separated lines appeared. Though their splitting was much smaller than in the nematic phase, it provided a direct way to measure the surface-induced order and to study pretransitional phenomena. Today we know that cylindrical cavities are – due to their specific symmetry with respect to the magnetic field – the most suitable confining matrix to study the liquid crystal-surface interactions. The required information is obtained directly from the splitting of the spectrum in the isotropic phase. The situation is more complex in all other confining systems where no splitting is observed and the analysis cannot be performed on the basis of NMR spectroscopy data alone. However, applying NMR relaxometry techniques allows a reliable interpretation of line width broadening, giving some quantitative information on molecular order and dynamics even in systems with randomly disordered pores.

In Sect. 2.2 the origin of the splitting in the NMR spectrum of cylindrically confined liquid crystals is explained and the influence of the liquid crystal-surface interactions on the surface order parameter and on the surface orientational wetting is discussed. A particularly interesting topic in this context is the growing of not only orientational (nematic) but also of positional (smectic) order in the isotropic phase. In Sect. 2.3 the study of oligomolecular liquid crystalline layers next to the surface of cylindrical cavities is described. Section 2.4 is devoted to the conventional and to the holographically formed PDLC materials, where the surface order parameter is obtained from the NMR relaxometry data. A simple relation between the orientational order parameter and the measurable relaxometry quantities is given. Short conclusions are found in Sect. 2.5.

2.2 Liquid Crystals in Cylindrical Cavities

In an NMR experiment, the resonant signal appears at the Larmor frequency which is determined by the Zeeman interaction between the nuclear spins and the magnetic field of the spectrometer. The Larmor frequency of deuterons, for example, is 6.53 MHz in a magnetic field of 1 T. Due to other, weaker interactions of nuclear spins, the spectrum usually has a structure that reveals the configuration, the orientation, and partly, also the dynamics of the spin-bearing molecule. The structure of deuteron spectra in liquid crystals originates from the interaction of the electric quadrupole moment of the deuteron (spin $I = 1$) with the electric field gradient (EFG) at the site of this nucleus. The splitting of the two lines in the spectrum of the nematic phase (Fig. 2.1) is thus given by [3,20]

$$\Delta\nu = \frac{3}{2}\,\nu_q\, S\, \frac{1}{2}\,\left(3\cos^2\Theta_{\mathrm{B}} - 1\right)\,\frac{1}{2}\,\left\langle 3\cos^2\beta - 1\right\rangle \tag{2.1}$$

where $\nu_q = e^2qQ/h$ stands for the quadrupole coupling constant of deuterons ($\sim$170 kHz), Θ_{B} is the angle between the director and magnetic field, S is the molecular orientational order parameter and β the angle between the EFG tensor symmetry axis and the long molecular axis. The averaging $\langle \ldots \rangle$ is meant over the conformational changes of the molecule. Equation (2.1) is given in a simple form where the small anisotropy of the EFG tensor and the possible molecular biaxiality have been neglected, as well as the effect of director fluctuations in view of their small amplitude. Occasionally, (2.1) is written in the form

$$\Delta\nu = \Delta\nu_0 \, S \, \frac{1}{2} \left(3\cos^2\Theta_{\mathrm{B}} - 1\right) \qquad (2.2)$$

where $\Delta\nu_0$ denotes the splitting of a perfectly oriented nematic phase along the magnetic field ($\Delta\nu_0 \sim$90 kHz for deuterons in the first position of the alkyl chain).

In spatially constrained liquid crystals the director field is space and time dependent (from the view of a molecule that diffuses between regions with different local director orientations). The orientational order parameter usually varies within the sample. In a local region of the cavity, represented by the position vector $\boldsymbol{r}$, a selectively deuterated compound yields a spectrum of two sharp lines separated by

$$\Delta\nu\left(\boldsymbol{r}\right) = \Delta\nu_0 \, S\left(\boldsymbol{r}\right) \, \frac{1}{2} \left(3\cos^2\Theta_{\mathrm{B}}\left(\boldsymbol{r}\right) - 1\right) . \qquad (2.3)$$

Taking into account molecular translational diffusion, the spectrum of the whole sample is given by the Fourier transform of the free-induction decay function $G\left(t\right)$:

$$G\left(t\right) = \left\langle \exp\left[\mathrm{i}\pi \int_0^t \Delta\nu\left(\boldsymbol{r}\left(t'\right)\right) \mathrm{d}t'\right] \right\rangle , \qquad (2.4)$$

where $\langle \ldots \rangle$ stands for the statistical average. However, as it turns out, the effect of translational diffusion is usually negligible in the nematic phase and the spectrum consists simply of a superposition of contributions from different parts of the sample. It has the form of a 'powder' pattern with two singularities at half of the bulk splitting and two shoulders at twice the distance if the distribution of directors in the cavity is spatially isotropic.

The situation in the isotropic phase is usually quite opposite. In cavities of diameter less than $\sim$200 nm and with a weak surface-induced order, molecular translational diffusion leads to complete motional averaging occurring in the time of recording the NMR spectrum. The measured quadrupole splitting – neglecting a possible biaxiality induced by the surface – incorporates separately the average orientational order parameter of the cavity $\langle S\left(\boldsymbol{r}\right)\rangle$ and the average value $\left\langle 3\cos^2\Theta_{\mathrm{B}}\left(\boldsymbol{r}\right) - 1\right\rangle$. Both quantities are uncorrelated in the averaging procedure. The factor $\left\langle 3\cos^2\Theta_{\mathrm{B}}\left(\boldsymbol{r}\right) - 1\right\rangle$ can be expressed in terms of the angle γ_B between the cylinders' axes and magnetic field, and the angle γ_C between the local molecular director at the surface and cylinder axis [19]. The resulting splitting is then

$$\langle \Delta\nu \rangle = \Delta\nu_0 \, \langle S(\boldsymbol{r}) \rangle \, \frac{1}{2} \left\langle 3\cos^2\gamma_C - 1 \right\rangle \frac{1}{2} \left(3\cos^2\gamma_B - 1 \right) . \tag{2.5}$$

The advantage of cylindrical cavities over other confined systems is that given the uniform orientation of local molecular directors the angle γ_C is the same everywhere at the wall and the frequency splitting is relatively large and easily observed. It measures directly the average order parameter in the cavity $\langle S(\boldsymbol{r}) \rangle$ or, stated in a different way, the adsorption parameter Γ characterizing the effective thickness of the ordered layer (for flat surfaces $\Gamma = \int S(z)\,\mathrm{d}z$ with z denoting the distance from the surface). Measuring $\langle \Delta\nu \rangle$ as a function of temperature reveals the orientational wetting behavior: if Γ diverges as the temperature approaches the nematic-isotropic transition temperature T_{NI}, the wetting is referred to as complete, otherwise it is said to be partial. The nature of the wetting strongly depends on the strength and type of liquid crystal-substrate interactions.

Many experimental studies of liquid crystals in cylindrical cavities have been performed using Anopore membranes as the confining matrix (Fig. 2.2). Anopore membranes are inorganic alumina (Al_2O_3) materials manufactured by an electro-chemical process. Well defined cylindrical channels of 200 nm in diameter are oriented perpendicular to the membrane surface and penetrating straight through its 60 μm thickness. A strong positive attribute of Anopore membranes is a large porosity of nearly 40%. To prepare an NMR sample, the membranes are sliced into approximately 5 mm×20 mm strips, wetted with the selectively deuterated liquid crystal, and heated into the isotropic phase to ensure complete wetting of the channels. Twenty to forty strips are then uniformly stacked and placed in the NMR tube. The Anopore cavity walls are very robust and can be treated with different surfactants or polymers prior to the liquid crystal filling. Such treatment of the walls allows a systematic variation of liquid crystal-surface interactions using different surface coupling agents. The final orientation of liquid crystal molecules at the surface in the isotropic phase cannot be determined from the NMR spectra alone, but it

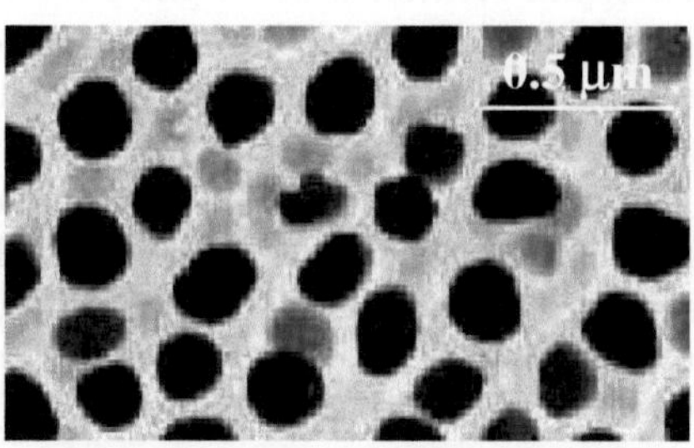

Fig. 2.2. Scanning electron micrograph of an Anopore membrane.

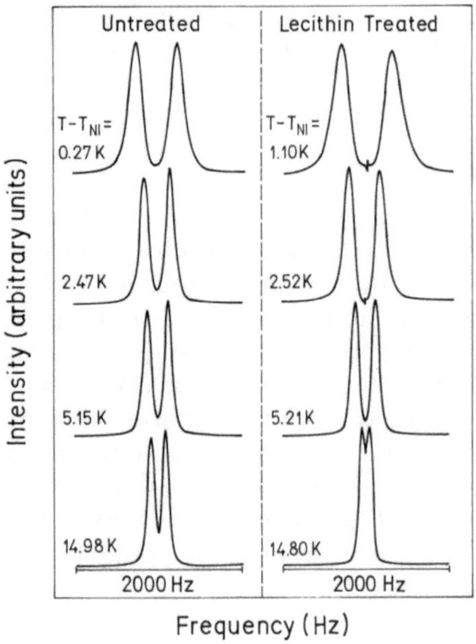

Fig. 2.3. Deuteron NMR spectra of 5CB confined to untreated and lecithin-treated Anopore shown at several temperatures above T_{NI} (from [19]).

can be determined from the structure in the nematic phase. It is assumed that in the isotropic phase the orientation of the surface molecular director is identical to that in the nematic phase.

NMR research of molecular anchoring, of molecular ordering and of the orientational wetting in the isotropic phase has developed in two directions. On the one hand, changes in the surface-induced order caused by varying the surface coupling agents were extensively studied [19,21–24]. On the other hand, the impact of the liquid crystal molecular length on the anchoring behaviour of the alkylcyanobiphenyl family enclosed in the same type of cavities has equally received considerable attention [23–25]. Figure 2.3 shows a few selected spectra of 5CB confined into Anopore with non-treated walls (left panel) and with lecithin treated walls (right panel). The corresponding spectra in the nematic phase show that the orientation of 5CB molecules in the untreated sample is parallel to the walls and parallel to the cylinder axis (parallel-axial structure). A layer of lecithin molecules, attached in advance to the alumina walls, imposes homeotropic conditions, i.e. perpendicular orientation to the adjacent 5CB molecules. The spectra in Fig. 2.3 clearly show that the surface-induced order persists at least 15 K above T_{NI} and that the splitting increases but does not diverge on approaching the transition temperature (Fig. 2.4).

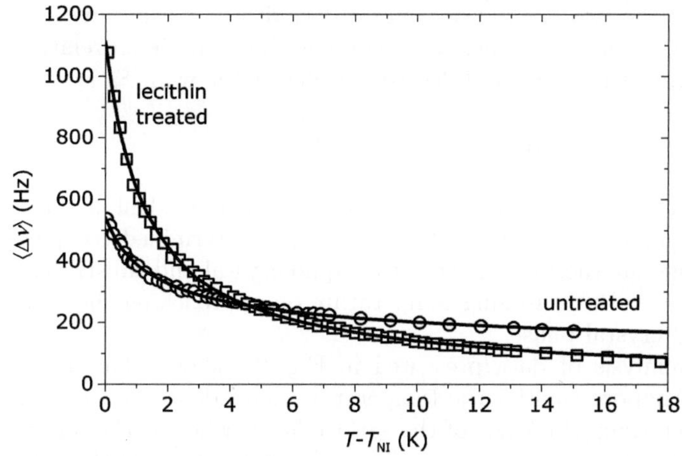

Fig. 2.4. Experimental temperature dependencies of the 5CB doublet splitting $\langle \Delta\nu \rangle$ in untreated (circles) and lecithin-treated (squares) Anopore. Solid lines are theoretical fits using (2.5) and (2.8) (from [19]).

The simplest model used to explain the temperature dependence of $\langle \Delta\nu \rangle$ is based on the Landau–de Gennes theory of the isotropic phase. Sluckin and Poniewierski added two "surface terms" to the free energy density [26]

$$f = f_0 + 1/2\,a\,(T - T^*)\,S^2 - 1/3\,bS^3 + 1/4\,cS^4 + 1/2\,L\,(\mathrm{d}S/\mathrm{d}r)^2 + \left(-w_1 S + 1/2\,w_2 S^2\right)\delta(R - r) \tag{2.6}$$

where f_0 is the order parameter-independent part of the free energy density, R is the radius of the cylinder, r is the distance from the cylinder axis, and a, b, c, L, and T^* are material-dependent Landau parameters determined from calorimetry data. w_1 and w_2 are surface–liquid crystal coupling constants. The first takes into account the orienting effect of the surface and the second the disorienting counterpart due to the rugged or strongly curved interface. In the case of relatively small surface-induced order, the third and the fourth order terms in S in (2.6) can be neglected. Then, a two-step minimization procedure of the free energy yields the spatial profile of the order parameter and its value at the surface, S_0. The spatial profile, $S\,(r)$, is given by

$$S\,(r) = S_0 \cosh\,(r/\xi)\,/\cosh\,(R/\xi) \tag{2.7}$$

where ξ is the nematic correlation length, that exhibits a strong pretransitional increase, $\xi\,(T) = \xi_0\,(T/T^* - 1)^{-1/2}$ with ξ_0 being of the order of the molecular size ($\sim$0.65 nm for 5CB). If the diameter of the cylinder is much larger than the correlation length ($\xi \ll R$), (2.7) can be reduced to a simple form

$$S\,(r) = S_0 \exp\left[-\,(R - r)\,/\xi\right] \tag{2.8}$$

indicating the exponential decay of order with increasing distance from the wall. The characteristic decay constant is the nematic correlation length ξ. In this limit, the calculated degree of order at the wall, S_0 is:

$$S_0 = w_1 \left[w_2 + \sqrt{aL\,(T - T^*)} \right]^{-1} . \tag{2.9}$$

The surface order parameter is temperature independent if the w_2 term in the denominator prevails. In this case, S_0 is determined completely by the short range interactions between the liquid crystal and substrate. Otherwise, it increases with decreasing temperature on approaching the transition into the liquid crystal phase.

The analysis of data presented in Fig. 2.4 shows that for 5CB in untreated Anopore cavities the temperature dependence of $\Delta\nu$ is entirely due to the increasing thickness of the surface layer, whereas the order parameter $S_0 \sim 0.025$ remains constant and is much smaller than in the nematic phase where $S > 0.3$. It should be mentioned that the fit to the experimental data using (2.6) is considerably improved if – in addition to the exponential profile – a thin layer next to the wall with constant order parameter equal to S_0, is added. The thickness of this layer is of the order of the molecular length and accounts for the breakdown of continuum theory at distances comparable to molecular sizes. A layer with constant order parameter ($\sim$1.9 nm thick for the untreated sample and $\sim$1.1 nm thick for the treated) has been taken into account in calculating the solid lines in Fig. 2.4, which represent the fit using (2.5), with $S\,(r)$ given by (2.8), to the experimental data. The magnitude of the surface order parameter in lecithin-treated cavities was found to be larger and temperature dependent, according to the predictions of (2.8) and (2.9). Just above T_{NI}, S_0 is $\sim$0.11. This value, though larger than in untreated cavities, is still considerably smaller than the threshold value 0.27 needed for the divergence of Γ suggesting that, even in the presence of an orienting layer at the walls, the wetting of the surface is not complete. The temperature dependence of S_0 indicates that interactions among liquid crystal molecules affect the surface order parameter. These examples, 5CB in untreated and in lecithin-treated alumina cavities, show the basic features of surface-induced order. They are quite general and valid for all members of the cyanobiphenyl family up to 10CB and for numerous surfactants. These characteristics include a temperature-independent and very small surface order parameter in the case of parallel molecular anchoring, and a larger and temperature-dependent – though not allowing for complete orientational wetting – S_0 in the case of surfactant-induced homeotropic anchoring.

Crawford et al. performed an extensive systematic study of the influence of 5CB–substrate interactions on the ordering of liquid crystal molecules at the interface and on the orientational wetting of the system [21,22]. To vary the liquid crystal-substrate interaction, the chain length of the aliphatic acids (C_nH_{2n+1}-COOH), attached to the walls of Anopore membranes, was varied from $n = 5$ to $n = 20$. Based on the experience with lecithine, one would

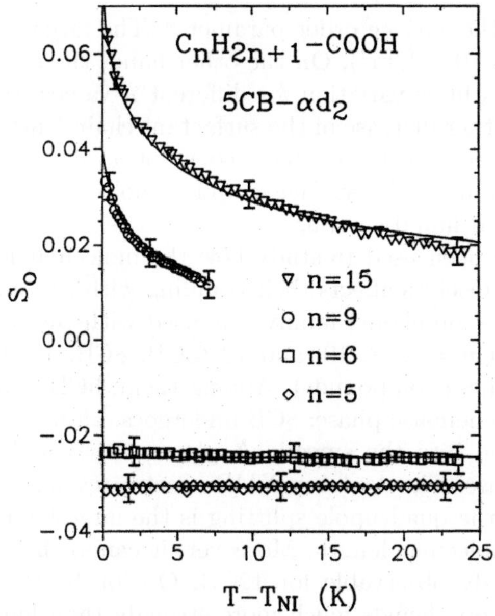

Fig. 2.5. Temperature dependence of the α deuterated 5CB surface order in Anopore, treated with aliphatic acids of various chain lengths (from [22]).

expect a homeotropic anchoring at the treated walls. The spectra in the nematic phase, however, reveal that the homeotropic anchoring is present only for $n > 7$, where a planar-polar structure is observed. Between $n = 6$ and 7, a discrete homeotropic-to-planar anchoring transition occurs. Accordingly, the structure of the nematic phase changes from the planar-polar to the planar-axial, the latter with molecules parallel to the cylinder axis. This phase transition indicates that interfacial coupling is dominated by short range interactions rather than quadrupolar ones [27]. Remarkably, the deuteron NMR spectra in the isotropic phase reproduced the anchoring transition observed in the nematic phase. The surface order parameter of 5CB molecules, oriented parallel to the substrate treated with $n = 5$ and 6 aliphatic acids, is temperature independent (Fig. 2.5) in contrast to acids with $n > 7$, where a much stronger pretransitional increase in the quadrupole splitting is observed leading to a temperature-dependent S_0. This indicates the transition into the homeotropic anchoring regime. With increasing chain length of the surface-coupling molecules, the splitting increases to reach a maximum at $n = 17$. The temperature dependence for this chain length exhibits a logarithmic divergence on approaching T_{NI}, which, according to the Landau–de Gennes theory, may indicate a complete wetting of the surface by the nematic phase. However, the magnitude of S_0 (maximal ~ 0.13) does not reach the critical value 0.27 and the wetting is taken only as quasicomplete. The values of the surface coupling constant w_1 vary with the length of the aliphatic acids in a similar way as the surface order parameter. The largest w_1 was observed for $n = 17$ ($\sim 3 \times 10^{-4}\,\mathrm{J/m^2}$). On the other hand, the disordering coupling constant w_2 shows little variation for different n values. It is interesting to note that with further increase of the surfactant chain length, for $n > 17$, the surface-induced order and the coupling constant w_1 decrease. Obviously, the ordering is the strongest for surface-coupling molecules of length similar to that of the confined liquid crystal.

NMR has also been used to study how the molecular length of confined liquid crystals themselves affects their coupling with the surface [25]. In this study, the alkylcyanobiphenyl family was used with the number of carbons in the alkyl chain $n = 5$, 8, 10, and 12 (5CB, 8CB, 10CB, and 12CB are short names for these compounds). Among them, 5CB is the only one that possesses only the nematic phase; 8CB undergoes, after a 7 K broad nematic phase, a transition into the smectic A phase; 10CB and 12CB only have the smectic A phase. The deuteron NMR spectra revealed that in untreated Anopore cavities the quadrupole splitting is the largest for 5CB, decreasing with increasing molecular length. Moreover, it cannot be resolved at all for 10CB and is hardly observable for 12CB. Obviously, the alumina surface orients short cyanobiphenyls much more strongly than long-chain members of the family. These changes in the anchoring strength can be observed also in the liquid crystalline phase. Whereas the director field of 5CB and 8CB clearly shows the parallel-axial structure, the structure of 12CB can be bistable, either parallel-axial or planar-radial, depending on the thermal history of the sample and on the atmosphere in the tube. It seems that the parallel orienting effect of the alumina surface weakens with increasing chain length and even turns into a homeotropic one for 12CB.

An exactly opposite effect is observed when the same liquid crystals are in contact with an alumina surface covered with a suitable surfactant (for example, aliphatic acid with $n = 15$) which imposes strong homeotropic anchoring to all of them. In this case the surface-induced order just above the N–I or N–SmA transition increases with increasing liquid crystal chain length. The quadrupole splitting, proportional to the surface order parameter, is 1.9 kHz for 5CB, about 2.8 kHz for 8CB, 3.0 kHz for 10CB, and 3.2 kHz for 12CB. The origin of such a large surface order parameter does not lie in the stronger orienting effect of the surface-coupling molecules but in the formation of smectic layers at the interface. The presence of positional order, characteristic of the smectic phases, supports and increases the orientational order in the surface layer. In Fig. 2.6, the formation of positional order and the corresponding increase in the orientational order are clearly seen. The temperature dependence of $\langle \Delta\nu \rangle$, that was rather smooth for 5CB, reveals for 12CB and $n = 15$ aliphatic acid two temperatures at which there is either a discontinuity (near 63℃) or a very sharp increase (close to 60℃) in its growth with decreasing T. The discrete growth of smectic order has been also observed in other systems by x-ray scattering [28], optical methods [29], and atomic force microscopy

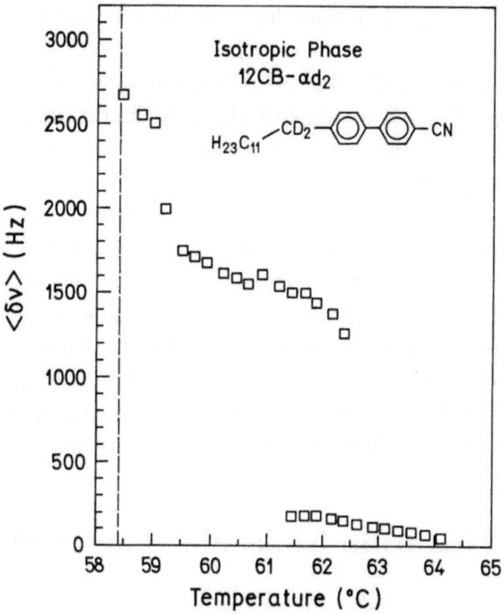

Fig. 2.6. Formation of smectic layers as evidenced by the appearance of discontinuous rises in $\langle \Delta\nu \rangle$ (T) (from [24]). See also ellipsometry experiments in Fig. 4.4.

[30]. The temperature dependence of the discontinuous or sharp rises depends on the length of the chain of the surfactant molecules. The appearance of the rises in Fig. 2.6 can be explained in the following way: high above the SmA–I transition, where only a small splitting is seen, the surface-induced orientational order parameter profile follows that used in describing the N–I transition. A weakly orientationally ordered molecular layer of thickness d_1 and order parameter S_{01} is here followed by an exponential decay of order. Upon cooling towards the SmA–I transition, a roughly ten times larger splitting abruptly appears. We assume that a smectic bilayer of thickness d_2 has formed. The presmectic surface layer of thickness $d_1 + d_2$ has a uniform order parameter S_{02}, followed again by an exponential decay of orientational order. Closer to $T_{\mathrm{SmA-I}}$ another large but continuous increase in the splitting occurs. It indicates the formation of the second smectic bilayer. It was found that the orientational surface order parameter drastically increases upon formation of the first smectic bilayer, S_{02} is several times larger than S_{01}. The change from S_{02} to S_{03}, accompanying the second step, is relatively small. Finally, it should be mentioned that surfactants with $n < 8$ hardly induce any ordering effect, a splitting in the spectrum is not resolved.

In contrast to 12CB where the growth of $\langle \Delta\nu \rangle$ is discrete for all chain lengths, 10CB shows the discontinous change in the splitting only for a few surfactants ($8 < n < 12$), otherwise a steep continuous increase is observed. In 8CB, the effect of smectic ordering is still observable but it is much weaker. The formation of smectic layers above the N–I transition is recognized only through two discontinuities in the slope of an otherwise continuous $\langle \Delta\nu \rangle$ vs. T curve.

Liquid crystals confined into cylindrical cavities are the most suitable systems for deuteron NMR studies of the interfacial liquid crystal-substrate interactions. There are, however, other systems of confined liquid crystals, which are more important for applications (PDLC and H-PDLC materials), or have been the object of intensive theoretical studies. The latter are composite systems with a random network of pores like liquid crystals in nano-pore glasses, aerogels and aerosils. Unfortunately, in all these cases NMR spectroscopy alone cannot yield accurate information on the surface- or constraint-induced order in the high-temperature phase as no quadrupole splitting is observed. The nature of the spectral line-broadening (static or dynamic) has to be established using NMR relaxometry. Current NMR results related to these systems will be briefly discussed in Sect. 2.4.

2.3 Order in Ultrathin Molecular Layers Detected by NMR

In the above cases, the surface-induced order was studied in submicrometer confining geometries where the 3D continuum approach is still valid. In Anopore with $R = 100\,\mathrm{nm}$, the decay of the surface order in the isotropic phase is exponential (except for a thin layer at the wall with constant S_0). By replacing r in (2.8) with $R - r'$ we obtain $S\left(r'\right) = S_0 \exp\left[-r'/\xi\right]$, with r' representing the distance from the surface. Such an $S\left(r'\right)$ spatial profile corresponds to a planar geometry, i.e. to a flat substrate (Anopore surface), covered with an infinitely thick layer of bulk LC. This is the case for completely filled cylindrical cavities. The disadvantage of studying the surface-induced order via deuteron NMR is that the surface layer and bulk contributions are not directly resolvable in the observed line shapes due to translational diffusion of molecules. Recently, systems with partially filled cavities were also studied [31] in order to increase the relative volume of the surface layers and, consequently, to emphasize surface-ordering effects in the NMR line shapes. Assuming that a constant thickness LC film is formed on the inner surface of Anopore cavities, partial filling provides an efficient way to alter the dimensionality of the mesogens from 3D (full cavities) gradually to thick films, multilayer films, monolayer films, and finally to 2D surface molecular depositions with sub-monolayer effective thickness, by simply varying the filling factor. Thin surface depositions ($d_{eff} << R$) in Anopore are not subject to strong elastic deformations. From the point of view of the Landau–de Gennes free energy one therefore deals with a quasiplanar geometry. The advantage of using Anopore instead of planar substrates is its high surface-to-volume ratio. At least 10^{18} selectively deuterated molecules are needed to detect a NMR spectrum. In a planar geometry, this would demand more than $0.5\,\mathrm{m}^2$ of solid substrate, covered with a monolayer of LC, to be cut and fit into the

NMR coil with a typical working volume of about $100\,\mathrm{mm}^3$, a task close to impossible. On the other hand, $0.5\,\mathrm{m}^2$ inner surface area is present in less than $100\,\mathrm{mm}^3$ of $R = 100\,\mathrm{nm}$ Anopore material.

The easiest way to prepare variable thickness films is by soaking the Anopore material with the solution of mesogen in an appropriate solvent. The solvent is subsequently removed by evaporation. In order to form a film with an effective thickness d_{eff}, a solution with concentration, or equivalently, pore filling factor $c = \{1 + \lambda (R - d_{\mathrm{eff}})^2 / (2Rd_{\mathrm{eff}} - d_{\mathrm{eff}}^2)\}^{-1}$ is required. Here $\lambda = \rho_{\mathrm{LC}}/\rho_{\mathrm{solv}}$ stands for the ratio of mesogen vs. solvent mass densities. Specifically, with a 4% solution of 5CB in methanol ($c = 0.04$), a molecular monolayer ($d_{\mathrm{eff}} \sim l_{\mathrm{5CB}} \sim 2\,\mathrm{nm}$) forms at the surface. For surface depositions with an effective thickness below the molecular length l_{LC}, it is appropriate to use the area per molecule, A, as a measure of surface coverage. If 5CB is the mesogen of choice, A is related to c via $A \sim 0.84\,\text{Å}^2/c$ in the $c << 1$ limit [32]. For $c = 0.02 \Rightarrow A = 42\,\text{Å}^2$, the surface coverage falls in the range in which 5CB monolayers that form at the air-water surface resemble a 2D-liquid phase [33].

The orientational order of ultrathin effective thickness layers differs considerably from the order in thick layers, as demonstrated in Fig. 2.7. At low

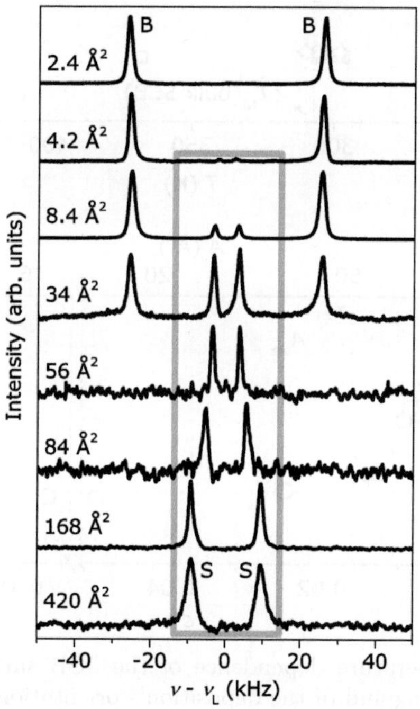

Fig. 2.7. Deuteron NMR spectra ($\theta = 0$, T_{room}) for various 5CB surface coverages A. The 'S' doublets (gray frame), arising from the molecular deposition at the surface, exhibit a considerably smaller frequency splitting than the bulk doublets 'B' (from [32]).

surface coverage ($A > 3\,\text{Å}^2$), a second doublet is observed in the deuteron NMR spectra recorded at $T = T_{\text{room}} = 300\,\text{K}$ and at the orientation where the pore axes are parallel to the direction of the external magnetic field ($\theta = 0$). The bulk nematic phase does not gradually evolve into the surface phase on decreasing the effective layer thickness, i.e. no intermediate phase between bulk and surface phases can be identified. This observation complies with a dewetting scenario [34] according to which thick molecular depositions dewet into a bulk phase 'B', whereas the dewetted areas remain covered with an ultrathin molecular surface layer 'S'. The properties of this component are very different from the bulk. The 'S' doublet is present even at high temperatures where the bulk component is in the isotropic phase. Its frequency splitting $\Delta\nu_{\text{S}}$, and consequently its orientational order (see (2.5)), only slightly decreases with increasing temperature and shows no anomaly at the bulk T_{NI} (Fig. 2.8a). Within this dewetting scenario it is also anticipated

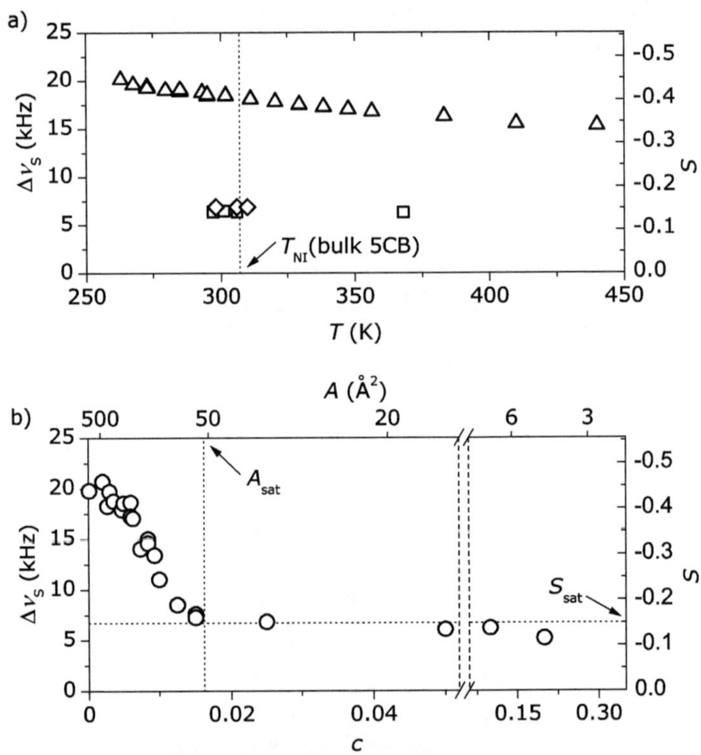

Fig. 2.8. (a) Temperature dependence of the 5CB surface deposition deuteron NMR doublet splitting and of the deposition's orientational order parameter S for the samples with surface coverages $A = 8.4\,\text{Å}^2$ (squares), $33.6\,\text{Å}^2$ (diamonds), and $168\,\text{Å}^2$ (triangles). (b) Deuteron NMR doublet frequency splitting and orientational order parameter S of the 5CB surface depositions as a function of surface coverage A and filling factor c (from [32]).

that whenever bulk coexists with a surface layer, the latter is 'saturated', i.e. its surface molecular density is independent of the amount of coexisting bulk. This is demonstrated in Fig. 2.8b: for A below the 'saturation' threshold $A_{\rm sat} \sim 50\,\text{Å}^2$ where bulk and surface components coexist (Fig. 2.7), $\Delta\nu_{\rm S}$ is saturated at $\Delta\nu_{\rm S}^{\rm sat} \sim 7\,\text{kHz}$, reflecting a saturated orientational order parameter and therefore a saturated density of the surface layer. In this regime, decreasing A is equivalent to increasing the relative volume of the dewetted bulk component while the physical properties of the 'S' phase remain unaffected. It is interesting to note that $A_{\rm sat}$ closely matches the value predicted for monolayer molecular films [31, 33]. It is thus safe to assume that for $A > A_{\rm sat}$ one deals with monolayer and submonolayer effective thickness surface molecular depositions. Let us call these structures "ultrathin depositions". Their $\Delta\nu_{\rm S}(A)$ behavior is drastically different from the behavior of saturated depositions discussed above: the doublet's frequency splitting exhibits an increase inversely proportional to A with a limiting value $\Delta\nu_{\rm S}(A \to \infty) = \Delta\nu_{\rm S}^{\infty} \sim 22.5\,\text{kHz}$. In the $A > A_{\rm sat}$ regime, the increase in A (in contrast to the $A < A_{\rm sat}$ regime discussed above) actually results in the decrease of the molecular density at the solid substrate-air interface. This is somewhat analogous to altering the surface pressure of a molecular layer at the liquid-air interface by changing the area of the film in a Langmuir trough.

The orientational order parameter of molecules in ultrathin depositions can be determined by measuring the angular dependence of the deuteron NMR spectra, shown in Fig. 2.9. These angular patterns are similar to those obtained in the case of planar radial alignment in full pores [22] where the alignment of LC molecules at the Anopore surface is homeotropic. There is, however, an important difference. At the $\theta = 90°$ orientation, the splitting of the inner peaks $\Delta\nu_{\rm S}^{\rm II}$ does not match one half of the splitting of the outer peaks $\Delta\nu_{\rm S}^{\rm I}$ as expected for uniaxial orientational order. This discrepancy is attributed to the presence of biaxiality η that affects the splittings in the following way [32]:

$$\Delta\nu_{\rm S}^{0} = -\frac{1}{4}\,\nu_q\,S\,(1+\eta)\,, \quad \Delta\nu_{\rm S}^{\rm I} = \frac{1}{2}\,\nu_q\,S\,, \quad \Delta\nu_{\rm S}^{\rm II} = \frac{1}{4}\,\nu_q\,S\,(\eta-1)\,. \tag{2.10}$$

Here $\Delta\nu_{\rm S}^{0}$ stands for the splitting of the doublet at $\theta = 0°$. The above measurable quantities determine the orientational order parameter and the biaxiality:

$$S = 2\frac{\Delta\nu_{\rm S}^{\rm I}}{\nu_q}\,, \qquad \eta = 1 + \frac{2\Delta\nu_{\rm S}^{\rm II}}{\Delta\nu_{\rm S}^{\rm I}}\,. \tag{2.11}$$

The observed biaxiality can be of two different origins. Surface interactions can give rise to intrinsic biaxiality, whereas molecular translational diffusion on the surface of the substrate can result in an apparent biaxiality. The strong angular dependence of the spin-spin relaxation time T_2 speaks in favor of the second mechanism [32]. The relaxation is fast at $\theta = 90°$ since the molecular translation along the inner pore surface alters the orientation of

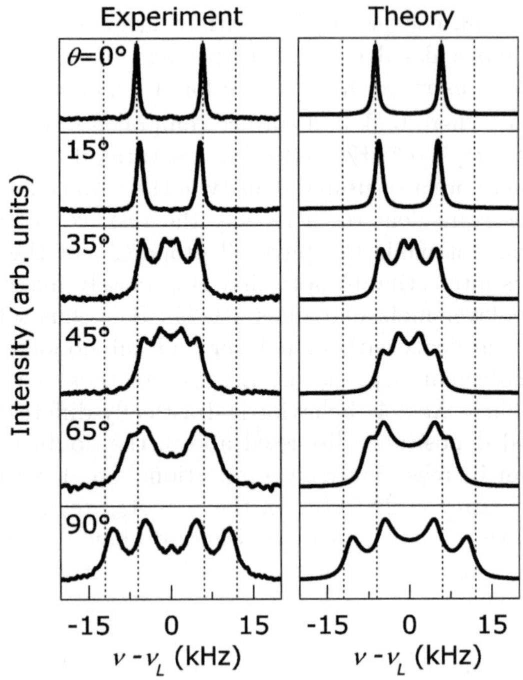

Fig. 2.9. Experimental and theoretical angular dependencies of the deuteron NMR line shapes of the 5CB molecular deposition with $A \sim 200\,\text{Å}^2$. Theoretical lines were calculated with $\eta = 0$ and $D_S = 7 \times 10^{-12}\,\text{m}^2\text{s}^{-1}$ [32]. Vertical dotted lines denote the expected positions of the spectral peaks for the case of uniaxial symmetry ($\eta = 0$ and $D_S = 0$) (from [31]).

the molecular director with the respect to external magnetic field, i.e. it induces fluctuations of the resonance frequency with time. This results in a homogeneous broadening of the spectral peaks (Fig. 2.9). At $\theta = 0°$, the relaxation is slow and no broadening takes place as translation along the surface maintains the angle between director and magnetic field at a constant value. In the slow surface diffusion limit, $D_S < 3/16\, R^2 \nu_q |Q(1 - \eta/3)| \sin^2 \theta$, the diffusion-induced effective biaxiality is related to the diffusion constant via [32]

$$D_S = \frac{1}{9} R^2 \nu_q S \eta \,. \tag{2.12}$$

Using (2.11) and (2.12) one can determine the surface diffusion constant by measuring the effective biaxiality. For saturated ($A < A_{sat}$) 5CB surface depositions, $|S_{sat}| \sim 0.14$ and $\eta_{sat} \sim 0.24$ at T_{room}, yielding $D_S \sim 7 \times 10^{-12}\,\text{m}^2\text{s}^{-1}$. This value is only about an order of magnitude smaller than the bulk 5CB value $D_{5CB} \sim 6 \times 10^{-11}\,\text{m}^2\text{s}^{-1}$, indicating possible collectiveness in the behavior of 2D LC molecular films. A gradual decrease of

D_S with increasing A is also observed, signifying a reduction of cooperative effects in highly dilute depositions. The limiting value $\sim 22.5\,\mathrm{kHz}$ of the 'S' doublet splitting $\Delta\nu_S^\infty$ corresponds to $|S| \sim 0.5$. Although the sign of the order parameter value cannot be determined via deuteron NMR due to the undefined sign of the doublet splitting $\Delta\nu_S$, the above numerical value is suggestive of a model with $S(A \to \infty) \to -0.5$ in which the molecular long axes lie flat on the surface and exhibit quick (on the deuteron NMR timescale $\sim 2\nu_q^{-1}$) uniaxial reorientations about the axis of symmetry perpendicular to the surface. One can further conclude that the orientational order parameter of a saturated surface deposition is negative as well, $S_{\mathrm{sat}} \sim -0.14$, since on increasing the surface coverage from the infinite dilution limit $A \to \infty$ towards A_{sat}, the 'S' doublet splitting and, consequently, the order parameter S never cross zero (Fig.2.8b).

The above results demonstrate that ultrathin depositions of LC molecules similar to a 2D gas can be formed at the inner surfaces of Anopore membranes. Their orientational order parameter is rather large, negative, and changes gradually from $S_{\mathrm{sat}} \sim -0.14$ for the case when the depositions coexist with bulk (saturated depositions) to $S = -0.5$ for extremely diluted depositions. The rather strong dependence of the orientational order parameter on the surface coverage and a surface diffusion with a rate similar to bulk LC indicate a collective, 2D liquid-like behavior.

2.4 Deuteron Spin Relaxation Above T_{NI}

The spin relaxation process of deuteron nuclei is induced by the stochastic, time-dependent part of the quadrupole interaction and is therefore directly related to the intensity and frequency of the thermal motion of spin-bearing molecules. Up to now, the relaxation studies of deuteron spins in confined liquid crystals above T_{NI} included the Zeeman longitudinal or spin-lattice relaxation time T_1, and the transverse spin relaxation time T_2 measurements. The former is the characteristic time for the exponential increase of magnetization towards the direction of the magnetic field if it is turned away from the equilibrium. The process involves the exchange of energy between the spin system and the surrounding 'lattice'. The transverse spin relaxation time T_2 describes the decay of magnetization transverse to the magnetic field and is caused by the interactions within the spin system. When the spin-lattice relaxation time T_1 is measured with conventional NMR spectrometers, it reflects predominantly molecular motions in the range of the Larmor and double Larmor circular frequencies, i.e., in the high MHz regime. T_2, on the other hand, contains also the spectral density function at zero frequency and is sensitive not only to fast molecular motions but also to slower molecular dynamics in the kHz range.

The spin relaxation in the bulk isotropic phase of a liquid crystal is caused by the fast (10^{-11} s to 10^{-9} s) molecular reorientations, including molecular

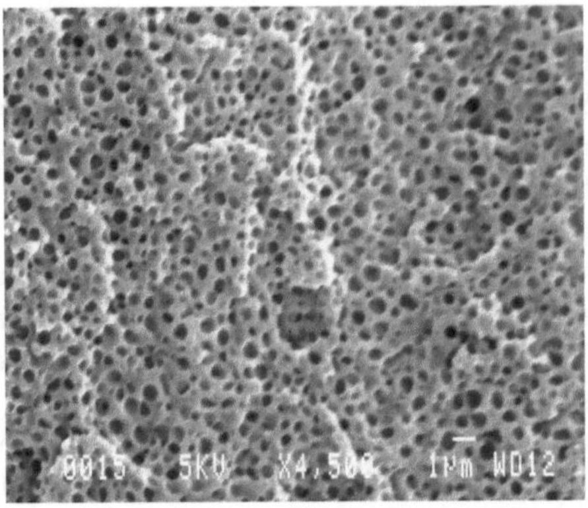

Fig. 2.10. Scanning electron micrograph of a PDLC.

conformational changes and rotations around the long and short axes. First measurements of the deuteron spin-lattice relaxation rate T_1^{-1} in confined liquid crystals were performed on a polymer dispersed liquid crystal (PDLC) (Fig. 2.10). These materials consist of spherical microdroplets of liquid crystals dispersed in a transparent solid polymer (usually of epoxy type) and are used as optical shutters, privacy windows, etc. [17]. They are opaque in the absence of electric field and become transparent as the voltage is switched on. In spite of an extremely large surface–to–volume ratio of PDLC materials, the spin-lattice relaxation rate T_1^{-1} shows hardly any difference that would exceed the experimental error with respect to the bulk (Fig. 2.11). The same lack of the effect of surface-induced order on T_1^{-1} was found later in other liquid crystals embedded in micro- and nano-porous media. This leads to the conclusion that the fast molecular reorientations are not significantly affected by the confinement.

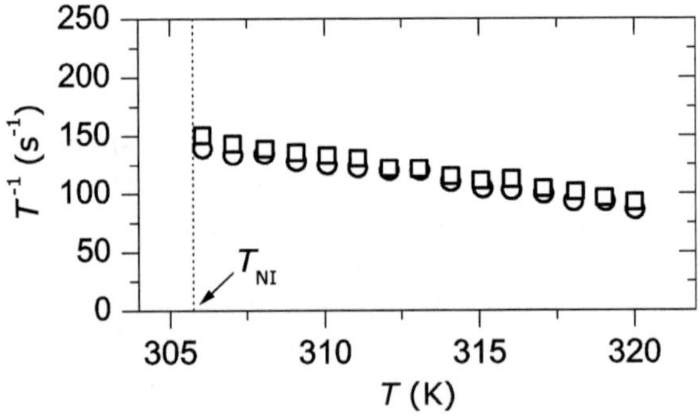

Fig. 2.11. Temperature dependence of the spin-lattice relaxation rate T_1^{-1} in bulk 5CB-αd_2 (circles) and polymer dispersed 5CB-αd_2 (squares).

The situation is different with the transverse spin relaxation rate T_2^{-1}. It is considerably larger in the confined liquid crystals above $T_{\rm NI}$ than in the bulk isotropic phase. The additional relaxation mechanism is obviously related to molecular dynamics in the kHz or low MHz frequency range. This mechanism could be either *order fluctuations*, which produce the well-known low-frequency relaxation mechanism in the bulk nematic phase [3], or *molecular translational diffusion*. Ziherl and Žumer demonstrated that order fluctuations in the boundary layer, which could provide a contribution to T_2^{-1}, are fluctuations in the thickness of the layer and director fluctuations within the layer [36]. However, these modes differ from the fluctuations in the bulk isotropic phase only in a narrow temperature range of about 1 K above $T_{\rm NI}$, and are in general not localized except in the case of complete wetting of the substrate by the nematic phase. As the experimental data show a strong deviation of T_2^{-1} from the bulk values over a broad temperature interval of at least 15 K (Fig. 2.12), the second candidate, i.e. molecular translational diffusion, should be responsible for the faster spin relaxation at low frequencies in the confined state.

Translational diffusion gives rise to a stochastic modulation of quadrupole interaction as in the course of time the molecules travel between non-identical environments adjusting themselves to the local degree and direction of orientational order. In particular, the diffusion of molecules between ordered and

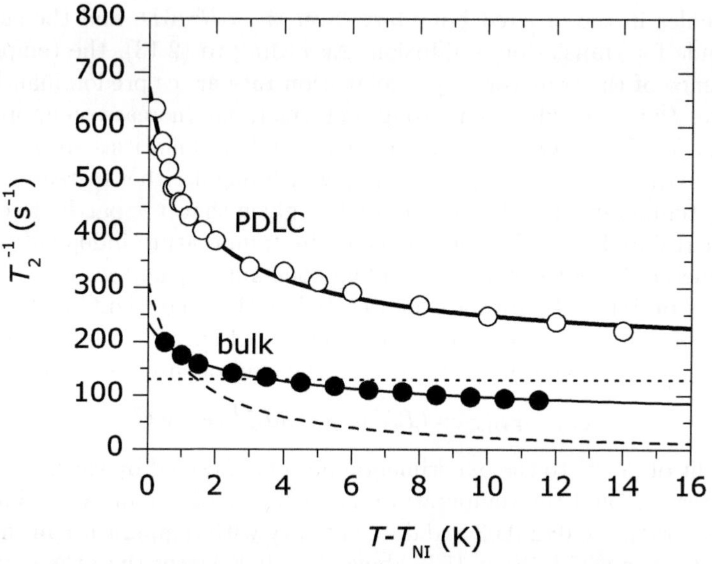

Fig. 2.12. Temperature dependencies of the deuteron spin-spin relaxation time T_2^{-1} for 5CB$-\alpha d_2$ PDLC droplets and the bulk sample. Bold solid line is the fit to (2.14). The three contributions are shown as thin solid line (bulk), dashed line (translational diffusion) and dotted line (T-independent part), respectively.

disordered regions represents the so-called 'exchange' relaxation mechanism, known from the two-state model of bound and free water in porous media [38]. On the other side, the relaxation mechanism produced by molecules that probe different surface orientations in the course of their self-diffusion is known as "reorientations mediated by translational displacements" (RMTD) [39, 40]. Both effects, exchange and RMTD, have been taken into account in the theory where the spin relaxation rates are calculated using the joint probability function for the diffusion of liquid crystal molecules within a sphere with reflecting boundary [37, 41]. Assuming that the sphere is small enough that within the spin relaxation time the molecules probe all different positions in the cavity, the calculated T_2^{-1} is a weighted sum of a discrete set of times $\tau_{ls} = R^2/\beta_{ls}^2 D$, related to the topologically restricted diffusive motion. The characteristic times τ_{ls} are determined by the radius of the cavity, the translational diffusion coefficient D, and β_{ls} denotes the s-th zero of the first derivative of the spherical Bessel function of order l. The rather complex expression for T_2^{-1} given in [37] can be, to a good approximation, written as

$$(T_2^{-1})_{\mathrm{TD}} = \frac{9}{20}\,\pi^2 \left(\frac{\overline{e^2qQ}}{h}\right)^2 \left(\frac{3\,\xi}{R}\right)^2 S_0^2\,\tau\,. \tag{2.13}$$

Here S_0 is the surface order parameter, $\overline{e^2qQ}/h$ is the quadrupole coupling constant averaged over molecular conformational changes and rotations around the long molecular axis, $3\,\xi/R$, represents, approximately, the fraction of molecules in the ordered boundary layer, $\tau = R^2/6D$, and the subscript TD stands for translational diffusion. According to (2.13), the temperature dependence of the transverse spin relaxation rate arises predominantly from the factor ξ^2 which exhibits a strong pretransitional increase on approaching T_{NI}, $\xi^2 \propto (T - T^*)^{-1}$. Other temperature dependent quantities are D with the smooth Arrhenius-type behavior and S_0 which might either increase with decreasing temperature in the vicinity of T_{NI}, given that it is partly determined by mutual liquid crystal interactions, or be temperature independent when determined only by the surface coupling constants w_1 and w_2 (see (2.9)).

The experimental data for α deuterated 5CB in the PDLC material with droplets of diameter ~ 500 nm are presented in Fig. 2.12. The experimental values of T_2^{-1} can be well interpreted as the sum of three contributions:

$$(T_2^{-1})_{\mathrm{PDLC}} = (T_2^{-1})_{\mathrm{bulk}} + (T_2^{-1})_{\mathrm{TD}} + C\,. \tag{2.14}$$

A good fit of (2.14) to the experimental data is obtained by varying only two parameters, S_0 and the temperature independent contribution C. The value obtained for S_0 is 0.08 ± 0.03 and does not vary with temperature in the whole temperature range of about 15 K above T_{NI}. It is larger than the surface order parameter of the same liquid crystal in contact with inorganic Anopore membrane, but still considerably smaller than the bulk nematic value at the transition temperature. This indicates that only a partial orientational wetting takes place in the isotropic phase of the PDLC material and that the orienting effect of the polymer surface is relatively weak. Such a surface has an ordering effect in the isotropic phase but a disordering one in the nematic phase. The temperature independent surface order parameter implies that it is determined by short range substrate-liquid crystal interactions at the interface. Evidently, NMR relaxometry is able to yield some information on the surface-induced order, but it is not capable of distinguishing between the details of the surface alignment, i.e. whether it is uniform planar, random planar, oblique or homeotropic. The origin of the parameter C, which does not show a pretransitional increase, should be a slowing-down of the molecular translational motion in the first molecular monolayer next to the polymer wall, which has not been taken into account in evaluating (2.13). The thickness of this layer does not change with temperature nor does its order parameter S_0. Recent measurements of the proton spin-lattice relaxation time in the rotating frame for the same sample lead to an estimate of the average dwell time of molecules in the immediate vicinity of the surface. It is between 10^{-6} s and 10^{-5} s, exceeding drastically the dwell time associated with the bulk-like translational diffusion elsewhere in the cavity [42].

Recently, a new kind of liquid crystal-polymer dispersion has attracted great attention. These are holographic polymer dispersed liquid crystals (HPDLCs). They are formed by illuminating a suitable monomer–liquid crystal–curing agent–surfactant mixture with two or more coherent laser beams that form an interference fringe pattern in the sample. In the bright regions the polymerization takes place more rapidly than in the dark regions. The resulting sample consists of liquid crystal-rich layers separated by pure polymer layers. Based on HPDLC materials, electrically switchable diffraction gratings [43] and photonic crystals [44] are being developed. To gain an insight into the actual state of liquid crystalline order in a HPDLC material, a comprehensive study using NMR spectroscopy, NMR relaxometry, and dynamic light scattering was performed [45]. The liquid crystal used in this study was a mixture of the commercial BL038, which is frequently chosen for optical applications, and selectively deuterated 5CB. It was found that the transverse spin relaxation rate T_2^{-1} in the isotropic phase of HPDLC is two orders of magnitude larger than in the bulk. Consequently, the surface-induced order parameter is significantly larger than in similar confining systems and the translational diffusion of molecules in the surface layer is at least two orders of magnitude slower than elsewhere in the cavity. T_2^{-1} exhibits an unusual temperature dependence on cooling, suggesting the possibility of a partial separation of the 5CB component in the liquid crystal. Whereas the transition from the isotropic into the nematic phase in standard PDLC materials occurs at almost the same temperature as in the bulk, a large shift towards lower temperature is observed in HPDLC. The transition takes place gradually due to different sizes and presence of other ingredients in the cavities acting as impurities. A part of the droplets is found isotropic even at room temperature what might affect the switching efficiency of HPDLC materials.

Finally, it should be stressed that the inclusion of deuteron relaxation studies in the conventional NMR spectroscopy is important whenever the splitting in the isotropic phase is not clearly seen. Another example are liquid crystals embedded into extremely small, nano-sized cavities of a porous glass. NMR spectroscopy shows in such systems a considerable broadening of the deuteron line width far above the bulk T_{NI}. It might be either static due to different orientation and degree of order in different pores, or dynamic and equal to $1/\pi T_2$. In the first case, the line width linearly depends on the order parameter, whereas in the second it is proportional to S^2. Preliminary data obtained for 5CB in porous glass CPG with 7 nm large pores show that the effect is purely dynamic and that the average order parameter in the cavity attains almost nematic-like values several 10 K above T_{NI} [46].

We expect that in the future relaxation measurements of confined liquid crystals will not be limited only to T_1 in the MHz range and T_2 revealing the zero-frequency limit. A much better understanding of the liquid crystal-surface phenomena could be obtained by a continuous scanning covering 4 decades in frequency, i.e., the low MHz and the full kHz range. Such measurements have been already performed with protons using field-cycling NMR relaxometry. The results are very promising: the T_1 dispersion curve discerns the dynamic processes in the first molecular layer and in the next weakly ordered region [47]. However, the protons are a suitable probe only in systems where the matrix does not contain the same type of nuclei. To probe deuterons in the same way, field-cycling NMR relaxometers with a better sensitivity need to be developed.

2.5 Conclusions

Deuteron NMR sensitivity problems, encountered whenever the orientational order induced by surfaces is probed for planar samples, can be efficiently overcome by using porous hosts with well defined topology of the surface. Anopore is particularly convenient due to an almost ideal cylindrical geometry of the pores' surfaces. In systems completely filled with LC, the order at the surface is probed indirectly through the changes in the bulk NMR line shapes and in the nuclear magnetization relaxation behavior. In the isotropic phase the thin, weakly orientationally ordered molecular layer at the surface results in a small splitting of the deuteron NMR line and in a reduction of the spin-spin relaxation time T_2. By analyzing these changes one can estimate the surface order parameter and the correlation length of its decay with the distance from the surface. Due to diffusion averaging it is however impossible to determine whether the ordering is homeotropic, planar, or tilted. This is not the case in the nematic phase. There the angular dependencies of the deuteron NMR line shapes clearly resolve between different types of arrangement at the surface that also determines the configuration in the whole cavity. In ultrathin surface depositions with effective layer thickness of the order of or lower than LC monolayer thickness the orientational order can be probed directly. The molecular alignment is homeotropic, with a negative order parameter which strongly depends on the surface molecular density and approaches -0.5 in extremely diluted depositions.

In summary, in this Chapter we have shown that the combination of a powerful technique like deuteron NMR and the wide availability of porous media hosts has allowed to gain unparalleled knowledge on liquid crystals-surface interactions.

This work was supported by SILC network and in part by a NSF-INT USA-Slovenia research grant #03-06851.

References

1. P.G. de Gennes, J. Prost: *The Physics of Liquid Crystals* (Clarendon Press, Oxford 1993).
2. E. Fukushima, S.B.W. Roeder: *Experimental Pulse NMR* (Addison-Wesley Publishing Company, Reading, Massachusetts 1981).
3. R.Y. Dong: *Nuclear Magnetic Resonance of Liquid Crystals* (Springer Verlag, New York Berlin 1994).
4. J.W. Doane, N.A. Vaz, B.G. Wu, et al: Appl. Phys. Lett. **48**, 269 (1986).
5. G.P. Crawford, M. Vilfan, J.W. Doane, et al: Phys. Rev. A **43**, 835 (1991).
6. G.P. Crawford, D.W. Allender, J.W. Doane et al: Phys. Rev. A **44** 2570 (1991).
7. G.P. Crawford, D.W. Allender, J.W. Doane: Phys. Rev. A **45** 8693 (1992).
8. R. Ondris-Crawford, G.P. Crawford, S. Žumer et al: Phys. Rev. Lett. **70**, 194 (1993).
9. R. Ondris-Crawford, M. Ambrožič, J.W. Doane et al: Phys. Rev. E **50**, 4773 (1994).
10. A. Golemme, S. Žumer, D.W. Allender et al: Phys. Rev. Lett. **61**, 2937 (1988).
11. G.S. Iannacchione, S.H. Qian, D. Finotello et al: Phys. Rev. E **56**, 554 (1997).
12. G. Iannacchione, G.P. Crawford, S. Žumer et al: Phys. Rev. Lett. **71**, 2595 (1993).
13. T. Jin, D. Finotello: Phys. Rev. Lett. **86**, 818.
14. A. Zidanšek, S. Kralj, G. Lahajnar et al: Phys. Rev. E **51**, 3332 (1995).
15. R. Aloe, G. Chidichimo, A. Golemme: Mol. Cryst. Liq. Cryst. **202**, 9 (1991).
16. M. Ambrožič, P. Formoso, A. Golemme et al: Phys. Rev. E **56**, 1825 (1997).
17. P.S. Drzaic: *Liquid Crystal Dispersions* (World Scientific, Singapore 1995).
18. G.P. Crawford, D.K. Yang, S. Žumer et al: Phys. Rev. Lett. **66**, 723 (1991).
19. G.P. Crawford, R. Stannarius, J.W. Doane: Phys. Rev. A **44**, 2558 (1991).
20. M. Vilfan, N. Vrbančič-Kopač: Nuclear magnetic resonance of liquid crystals with an embedded polymer network. In: *Liquid Crystals in Complex Geometries*, ed by G.P. Crawford, S. Žumer (Taylor&Francis, London 1996) chap 7.
21. G.P. Crawford, R. Ondris.Crawford, S. Žumer et al: Phys. Rev. Lett. **70**, 1838 (1993).
22. G.P. Crawford, R. Ondris-Crawford, S. Žumer et al: Phys. Rev. E **53**, 3647 (1996).
23. P. Ziherl, M. Vilfan, N. Vrbančič-Kopač et al: Phys. Rev. E **61**, 2792 (2000).

24. T. Jin, G.P. Crawford, R.J. Crawford, et al: Phys. Rev. Lett. **90**, 015504 (2003).
25. T. Jin, B. Zalar, A. Lebar et al, submitted.
26. T.J. Sluckin, A. Poniewierski: Phys. Rev. Lett. **55**, 2907 (1985).
27. P.I.C. Teixeira, T.J. Sluckin: J. Chem. Phys. **97**, 1498 (1992).
28. B.M. Ocko: Phys. Rev. Lett. **64**, 2160 (1990).
29. T. Moses: Phys. Rev. E **64**, 010702R (2001).
30. K. Kočevar, I. Muševič: Phys. Rev. E **65**, 021703 (2002).
31. B. Zalar, S. Žumer, D. Finotello: Phys. Rev. Lett. **84**, 4866 (2000).
32. B. Zalar, R. Blinc, S. Žumer et al: Phys. Rev. B **65**, 041703 (2002).
33. J. Xue, C.S. Jung, M.W. Kim: Phys. Rev. Lett. **69**, 474 (1992).
34. F. Vandenbrouck, M.P. Valignat, A.M. Cazabat: Phys. Rev. Lett. **82**, 2693 (1999).
35. N. Vrbančič-Kopač: Nuclear Magnetic Resonance Study of Microconfined Liquid Crystals. Ph.D. Thesis, University of Ljubljana, Ljubljana (1997).
36. P. Ziherl, S. Žumer: Phys. Rev. Lett. **78**, 682 (1992).
37. M. Vilfan, N. Vrbančic-Kopač, B. Zalar et al: Phys. Rev. E **59**, R4754 (1999).
38. E.E. Burnell, M.E. Clark, J.A.M. Hinke et al: Biophys. J. **33**, 1 (1981).
39. R. Kimmich, P. Gneiting, K. Kotitschke et al: Mag. Res. Imag. **19**, 433 (1990).
40. F. Grinberg, R. Kimmich: Mag. Res. Imag. **19**, 401 (2001).
41. S. Žumer, S. Kralj, M. Vilfan: J. Chem. Phys. **91**, 6411 (1989).
42. M. Vilfan, G. Lahajnar, I. Zupančič et al: Mag. Res. Imag. **21**, 169 (2003).
43. A.K. Fontecchio, C.C. Bowley, S.M. Chmura et al: J. Opt. Technol. **68**, 652 (2001).
44. M.J. Escuti, J. Qi, G.P. Crawford: Appl. Phys. Lett. **83**, 1331 (2003).
45. M. Vilfan, B. Zalar, A.K. Fontecchio et al: Phys. Rev. E **66**, 021710 (2002).
46. A. Lebar, G. Lahajnar, M. Vilfan et al, submitted.
47. M. Vilfan: Field-cycling NMR relaxometry of a confined mesogenic fluid. 3rd Conference on Field Cycling NMR Relaxometry, Techniques, Applications, Theories, Torino (2003).

3 Interfacial and Surface Forces in Nematics and Smectics

Igor Muševič

It is well known since the pioneering work of Israelachvilli and Tabor [1], that the structure of an ordinary liquid near a solid wall is quite different from that in the bulk. The breaking of continuous translational symmetry of free space by a solid wall imposes positional ordering of molecules near that wall. As a result, a sequence of several molecular layers is usually formed, that extends from the surface and vanishes in the isotropic bulk.

The surface-induced layering of molecules has some interesting consequences, when a liquid is confined between two flat surfaces, and the separation between the surfaces is reduced to molecular dimensions. When the surface separation is changed, an oscillatory force is detected on both surfaces, which is due to layering of molecules near both walls. This can be intuitively understood as a result of periodic match or mismatch between the number of molecular layers and the total surface separation D. As this force originates from the local structure of a liquid, it is usually called the *structural* (or solvation) force [2].

To put these considerations on a quantitative basis, we recall that the force F is defined as the negative derivative of the total free energy $\mathcal{F}$ of a system with respect to the surface separation D at constant volume and surface area:

$$F = -\frac{\partial \mathcal{F}}{\partial D}. \tag{3.1}$$

The structural force therefore appears whenever the free energy of a confined system depends on surface separation. Following these considerations, it is therefore reasonable to expect a rich variety of structural forces in liquid crystals. Here, the confining wall imposes not only positional order, but also induces orientational order of molecules near the surface. This gives rise to structural forces of both positional (smectic-like) and orientational origin (nematic-like), which are therefore specific to orientationally ordered fluids. Furthermore, the free energy of a confined liquid crystal depends not only on the mean-field molecular distribution, as the collective molecular fluctuations are quite strong in these soft systems. As a result, interfacial forces of fluctuation origin may be quite important in confined liquid crystals.

Finally, we may note that the structural forces are particularly strong, when the free-energy very sensitively depends on surface separation. This situation is met near a second order phase transition, where the system is soft and very susceptible to external perturbations. It is therefore reasonable to expect that structural forces are significantly enhanced near the phase transitions of highly confined complex fluids. In these systems, a small surface field that imposes positional or orientational order can induce large changes in the free-energy, which in turn gives rise to a strong structural force.

By measuring the surface forces one can therefore obtain important information not only on the structure of a liquid crystal, but also on the influence of confining surfaces on orientational and positional ordering on a molecular level. This Chapter describes experimental techniques that are used for the measurements of surface forces and is focused on the results of the experiments that have been recently performed in the isotropic, nematic and smectic-A phases.

3.1 AFM Force Spectroscopy Near Isotropic-Nematic and the Isotropic-Smectic Phase Tansitions

Igor Muševič and Klemen Kočevar

3.1.1 Introduction

In this part we will present the power of atomic force microscopy (AFM) for the nanoscale investigations of the solid-isotropic liquid crystal interfaces. The work described here is focused on direct measurements of forces between solid objects, immersed in an isotropic liquid crystal that is close to the phase transition to an ordered liquid crystalline phase. Solid surfaces induce some liquid crystalline order in the surrounding liquid crystal and this influences the force between the objects. Consequently, it is possible to extract important information about the behaviour of the liquid crystalline order in the few nanometer thick interfacial layer by studying the forces.

The contribution starts with a description of the atomic force microscope setup, in Section 3.1.2. The advantages and drawbacks of the method compared to surface force apparatus (SFA) are also discussed. The results, obtained with this setup are described in the remaining sections. In Sect. 3.1.3 we present experimental evidence for the adsorbed liquid crystal layer on the $N, N-dimethyl-N-octadecyl-3-aminopropyltrimethoxysilyl\, chloride$ $(DMOAP)$ treated glass surface at temperatures where the bulk is in the isotropic phase. This layer is anisotropic and serves as a seed for the partial orientation of the nearby liquid crystal. In Sects. 3.1.4 and 3.1.5 two competing force mechanisms in nematic liquid crystals are discussed. In Sect. 3.1.4 the observation of a typical structural force due to an overlap of nematicaly ordered interfacial regions is presented and the explanation in the framework of a mean-field Landau-de-Gennes theory is given. When the surfaces induce a high degree of liquid crystal order, nematic capillary condensation of the isotropic liquid crystal in the gap between the surfaces is observed, as presented in Sect. 3.1.5.

In smectic liquid crystals it is again possible to observe both types of interactions. Overlap of pre-smectically ordered interfacial regions gives rise to the oscillatory structural force, presented in Sect. 3.1.6 and if the coupling energy between the smectic order and surface is very strong, the liquid crystal in the gap beteen the surfaces condenses into a smectic phase, as discussed in Sect. 3.1.7).

3.1.2 Measuring Interfacial Forces on a Molecular Scale with an Atomic Force Microscope

The surface topography of solids and soft matter can be directly observed using an Atomic Force Microscope (AFM) [3]. The AFM uses a sharp tip and an elastic cantilever to sense the force between the tip and the surface, typically at very close separation of the order of 1 nm or less. Usually the AFM tip is scanned over the surface in a regular manner and at the same time the magnitude of the tip-surface force is kept constant by moving the sample surface up or down, depending of the local topography. In this way, the surface topography is recorded on a lateral scale of hundreds of microns down to atomically resolved images.

As the AFM can detect forces in the range of pN, it is an ideal instrument for measuring the separation dependence of forces between surfaces. In this special, "force spectroscopy mode" of operation, a micron-sized spere is often attached to the force-sensing cantilever of the AFM, in order to increase the force-sensing surface and therefore the total force. When the sphere is brought into close vicinity of a flat surface using the piezo movement of an AFM scanner, forces that act between the two objects induce deflection of the elastic AFM cantilever. The deflection is proportional to the force on the cantilever and by measuring it, one can measure the force once the elastic constant of the cantilever is known. The deflection of the cantilever is measured optically with a resolution of less than 1nm, which for a typical elastic constant 0.01 N/m corresponds to a force of several pN. Due to thermal noise this is also the lower force limit of an AFM.

Whereas the first experiments on structural forces in liquid crystals were performed using a Surface Force Apparatus (SFA, discussed later on in part 3 of this chapter), a temperature controlled AFM was used later [4] to measure the forces between the AFM probe and the surface, mediated by a liquid crystal in between. Compared to other methods, using an AFM for measuring surface forces has some advantages and also some drawbacks. The most important advantage is, that the temperature of the sample can be controlled in a simple way to better than 0.01 K. Furthermore, the sample is not in direct contact with large mechanical parts like in the case of a SFA, and only a small amount of liquid crystal is needed for the experiment. The drawback of the AFM is, that the separation between surfaces is not measured directly

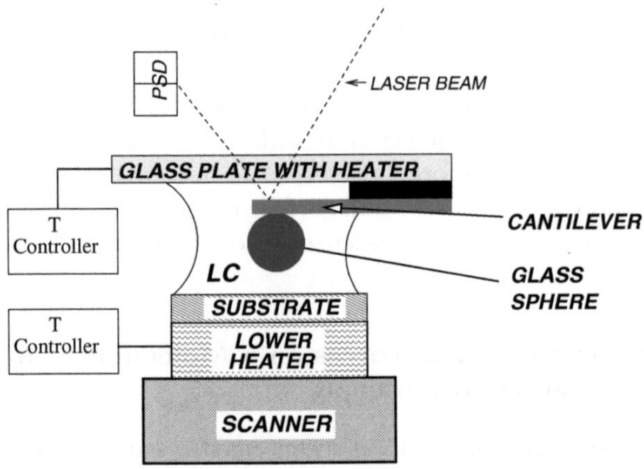

Fig. 3.1. The schematics of the experimental setup. Liquid crystal (LC) is filling the space between the bottom substrate and the upper transparent glass plate, used as a heater. The first temperature controlled heater is placed just below the substrate and is moving together with the scanner. The other is integrated into the upper glass plate. The size of the glass sphere is greatly exaggerated. PSD is the position sensitive detector for the reflected laser beam.

as in a SFA, but is determined indirectly from the point of hard contact and the movement of the piezo scanner.

A commercial AFM (Nanoscope III, Digital instruments) is conveniently equiped with two additional heaters, as shown in Fig. 3.1 [4]. One of the heaters with a precision temperature sensor is mounted just below the substrate and the other is integrated into a glass plate with an Indium-Tin-Oxide transparent layer, that is placed above the AFM cantilever. Using this simple geometry, the temperature of the liquid crystal is controlled with a precision of $\approx 10^{-2}$ K. In this arrangement, the AFM probe and the cantilever are immersed in the liquid crystal and there is no force due to the air-liquid crystal interface. This is another advantage over SFA, where the background forces due to the air-liquid crystal interface have to be subtracted.

Several types of probes can be used in the experiments. The first type are just commercial Si_3N_4 pyramidal tips, usually used for AFM imaging. These imaging tips are extremely sharp with a typical apex radius of ≈ 15 nm, which generates large local pressures under the apex of the tip when it is in contact with the surface. The other type of probe that generates much lower local pressure, are glass spheres of radius $\approx 10^{-5}$ m, which are attached to the cantilever. SEM images of spheres, attached onto the AFM cantilevers are presented in Fig. 3.2.

In an experimental run, the AFM is thermalized for one day, with microheaters turned on. The temperature of the heaters is set above the isotropic phase transition of the liquid crystal under study. After that, liquid crystal material is introduced between the glass substrate and the AFM probe, and

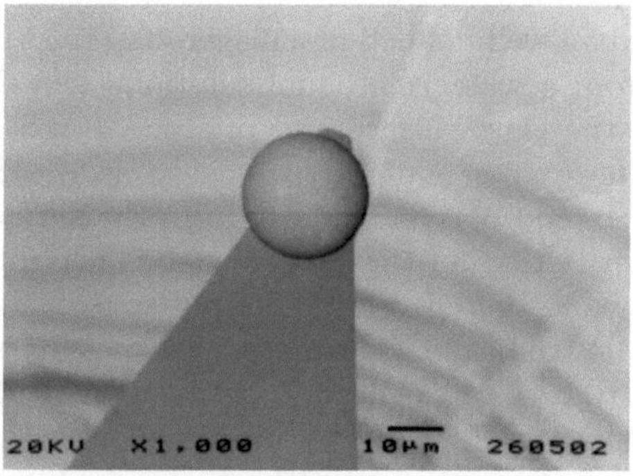

Fig. 3.2. A glass sphere of radius 10^{-5} m, glued onto the cantilever. The microspheres can be attached to the cantilever by using an optical microscope and a precise 3D micropositioning system. A thermoplastic glue is used for attaching.

the complete system is again thermalized for several hours. The experiment is then performed in a single experimental run that typically lasts for 6-8 hours. At each temperature, a large number of force-versus-distance plots are recorded and the results of the fits are averaged for each temperature. The force experiments show that solid-liquid interfaces are subject to ageing that appears over a period of several days. Surface adsorption of impurities from the liquid crystals is observed after one day.

For the Force Spectroscopy, the AFM is used in the "force plot mode" of operation, where the piezo scanner of the AFM and the substrate perform time-periodic linear movements in the direction of the AFM tip. The speed of approach is several nm per second and at the same time, the deflection of the AFM cantilever is monitored. In this way, one obtains the force-versus-separation plots. The zero of the separation is determined from the point of hard contact between the sharp AFM probe and the substrate.

It is important to note that it is only possible to measure forces by optical detection of the cantilever bending, when the liquid crystal is in the isotropic phase. In the semi-ordered liquid crystal phases, the scattering of the laser beam decreases the sensitivity of the experiment. The limited temperature range is the most serious limitation of AFM force spectroscopy and can be overcome using a detection technique, based on piezo-resistive cantilevers, which is not considered here.

3.1.3 First Adsorbed Layer of Cyanobiphenyl Molecules on Silanated Glass

It is well known from the early work on surface alignment of liquid crystals [5] that glass surface covered with $N,N-dimethyl-N-octadecyl-3-aminopropyltrimethoxysilyl\ chloride(DMOAP)$ orients liquid crystal molecules perpendicularly to the glass surface. The quality of alignment is extremely good, as only a well-defined monolayer of DMOAP is deposited onto the glass surface from a water/methanol solution. The DMOAP molecules bind chemically to the glass surface and produce a well defined monolayer coating, whereas the excess material is washed-away. 18 units long alkyl chains of DMOAP extend perpendicularly from the surface and form a two dimensional random lattice on the glass surface. The alkyl chains are not packed very densely and there is a lot of empty space between them [5].

Optical and NMR studies [6, 7] indicate, that molecules of cyanobiphenyl liquid crystals (n-CB, n=5-12) penetrate between the DMOAP alkyl chains and orient parallel to them. The experiments show that a typical trilayer interfacial structure is formed, where the first liquid crystal monolayer is quite tilted, very polar and oriented almost parallel to the surface. As the cyanobiphenyl molecules have a strong tendency to form dimers, the molecules in the second bilayer orient almost perpendicularly to the surface and represent an anisotropic base for the orientation of the rest of the liquid crystal. The trilayer is stabilized by the alkyl chains of DMOAP. The schematic picture of this interface is presented in Fig. 3.3.

Using a sharp AFM tip that generates relatively large local pressures, AFM can be used to observe the adsorbed layer and study some of its properties. A typical force plot with a sharp pyramidal Si_3N_4 tip is shown in Fig. 3.4. At large separations there is no detectable force, whereas at a separation of 10-20 nm, a weak attractive force suddenly appears. Then, at a separation of several nm, the tip touches a relatively soft layer, that is elastically deformed by increasing the force load. At a certain force load, the layer cannot resist any more and ruptures.

Rupturing experiments show, that the thickness of the adsorbed layer tends to be related to the length of a fully extended liquid crystal molecule

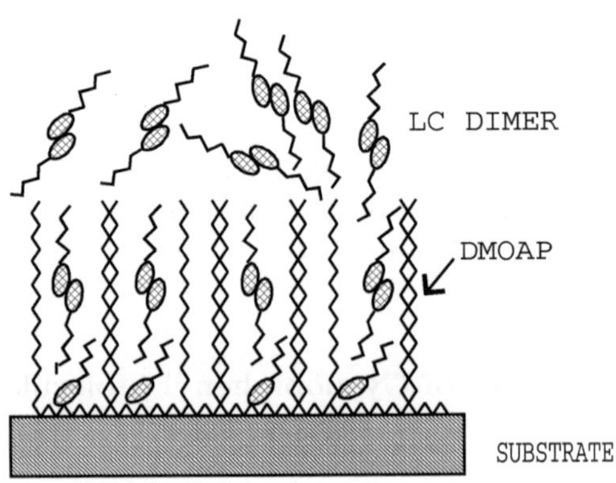

Fig. 3.3. The schematic picture of the DMOAP-LC interface. The alkyl chains of DMOAP stretch perpendicularly from the surface and stabilize the adsorbed trilayer.

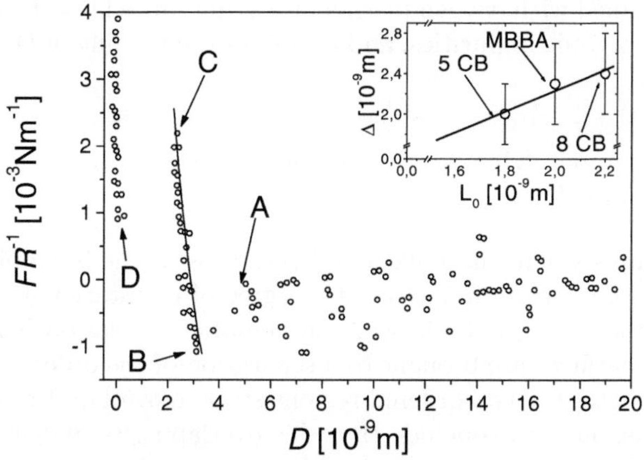

Fig. 3.4. The thickness of the adsorbed trilayer of the liquid crystal 5CB on DMOAP covered BK7 glass substrate. The full line is a fit to the Hertz theory with $E^* = E/(1-\nu^2) = 1.2 \times 10^7\ (1 \pm 0.3) Nm^{-2}$. From point B to the point C, the surface adsorbed molecular trilayer is elastically deformed. At the instability point C, the layer ruptures and the AFM tip is in hard contact with the surface at D. The inset shows the linear relation between the thickness of the adsorbed layer and the length of fully extended liquid crystal molecule.

[8,9], as presented in the inset to Fig. 3.4. Actually, the thickness of this first molecular layer corresponds to a molecular trilayer. Using the AFM tip, the adsorbed molecular layer can be elastically deformed, as shown in Fig. 3.4, and the Young's elastic modulus of the trilayer can be determined, using the Hertz-theory and the elastic foundation model [10]. The moduli are given in Table 3.1 for different liquid crystals and are of the same order as the compressibility modulus of a typical smectic-A phase [11].

The experiments show that this smectic-like adsorbed trilayer does not cover the glass surface completely. Instead, there are voids and islands of submicron dimension, where there is no adsorbed layer [9]. A clear correlation can be found between the cleaning procedure and the degree of coverage. For a good coverage and a strong alignment of the liquid crystal, glass surfaces

Table 3.1. Young's elastic moduli $E^* = E/(1-\nu^2)$, for the adsorbed layer of different liquid crystals used in our study.

liquid crystal	E^* N/m^2
MBBA	$1.7 \times 10^6\ (1 \pm 0.3)$
5CB	$7.8 \times 10^6 (1 \pm 0.3)$
8CB	$1.2 \times 10^7\ (1 \pm 0.3)$

have to be cleaned with oxygen plasma or a sulphuric acid bath. This removes the adsorbed organic impurities, and promotes the binding of DMOAP to the glass surface.

3.1.4 Pre-nematic Mean-field Surface Interaction in the Isotropic Phase

At temperatures where the bulk liquid crystal is in the isotropic phase, the DMOAP treated surface induces some degree of orientational order S, that decays into the isotropic bulk over the nematic correlation length $\xi(T)$. If two identical surfaces are brought to a separation of the order of the nematic correlation length, the two ordered regions start to overlap. Based on general considerations, one can conclude that this overlaping is energetically always favourable [12] and induces an attractive structural force between the two surfaces. As this structural force appears in the isotropic phase and is due to a surface-induced order parameter, it is called the pre-nematic mean field force.

There are basically two possible scenarios for the behaviour of this pre-nematic mean field force when the separation between surfaces is reduced, and both depend on the degree of the surface-induced nematic order. For low surface-induced order, the magnitude of the attractive force just increases when approaching the isotropic-nematic phase transition from above. On the other hand, if a surface induces a high degree of LC orientation, the pre-nematic phase can spontaneously transform into the nematic phase, when the separation is decreased below a certain value. This is the nematic capillary condensation, that is discussed further on in this Chapter.

The pre-nematic surface force is calculated following a simple approach of de Gennes [13], where the free energy density in the lowest order of approximation is

$$f_B = \frac{1}{2}\alpha(T)S^2 + L\left(\frac{\mathrm{d}S}{\mathrm{d}z}\right)^2 . \tag{3.2}$$

Here, the temperature dependent coefficient $\alpha(T) = a(T - T^*)$ drives the weakly first order isotropic-nematic transition at T_{NI}. Together with the coefficient L of the gradient term, these two coefficients determine the nematic correlation length $\xi = \sqrt{L/\alpha}$.

The surface free energy is

$$f_S = \left(-w_1 S_0 + \frac{1}{2}w_2 S_0^2\right)\delta(z = \pm D/2) . \tag{3.3}$$

The first term is negative and therefore favours large surface order, whereas the second term favours disorder, as it is always positive. The combination of both terms therefore favours a certain degree of surface order. The force between two planparallel surfaces is then calculated in several steps.

First, after minimizing the free energy one obtains the order parameter profile across the nematic layer. Then, the free-energy is calculated and the force between the surfaces is just a derivative of the free energy with respect to surface separation D. However, in the AFM force experiments, the forces are conveniently measured between a sphere and a flat surface. In this case, the so-called Derjaguin approximation is used [2], which relates the forces between curved surfaces and the interaction energy per unit surface area $\mathcal{F}(D)$ between two flat surfaces. The pre-nematic force $F(D)$ between a sphere of radius R and a flat surface is $F(D) = 2\pi R(\mathcal{F}(D) - \mathcal{F}(\infty))$:

$$F(D) = Rw_1^2\xi(T)\left(\frac{1}{L + w_2\xi(T)} - \frac{1}{w_2\xi(T) + L\tanh(\frac{D}{2\xi})}\right) . \quad (3.4)$$

In view of the weakness of the pre-nematic force, van der Waals forces between surfaces have to be considered as well. In this case, the calculation of the van der Waals force is quite complicated, because of the surface-adsorbed first molecular layers. In the Derjaguin approximation, the van der Waals force between a flat glass surface (medium 1) and a sphere of radius R (medium 1), acting across two surface adsorbed layers, each of thickness T (medium 2) and a pre-nematic layer (medium 3) in between is [2]:

$$F(D)_{vdW} = \frac{R}{6}\left[\frac{A_{232}}{D^2} - \frac{2A_{123}}{(D+T)^2} + \frac{A_{121}}{(D+2T)^2}\right] . \quad (3.5)$$

Here, A_{232} is the Hamaker constant for the van der Waals interaction between two adsorbed layers (2), acting across a paranematic layer (3). A_{123} is the Hamaker constant for the van der Waals interaction between glass (1) and the paranematic layer (3), acting across one adsorbed layer (2). Similarly, A_{121} is the Hamaker constant for the interaction between two glass surfaces (1), acting across two adsorbed layers (2). For 5CB and BK7 glass the Hamaker constants A_{123} and A_{232} are of the order of $A_{123} \approx A_{232} \approx 10^{-25}$ J and can be neglected with respect to A_{121}. The Hamaker constant A_{121} for the van der Waals interaction between two glass surfaces (1) across a surface adsorbed liquid crystalline layer (2), is quite larger and equals $A_{121} = 8.5 \times 10^{-22}$ J. At a separation of $D = 2$ nm and a thickness of adsorbed liquid crystal layers of $T = 2.4$ nm, the van der Waals force on a 7 μm sphere is attractive and equals 22 pN. It is therefore comparable in magnitude to the attractive mean field nematic force given in (3.4). It is important to note that in the above estimation of the van der Waals interaction, the liquid crystal is considered isotropic. In fact, the van der Waals force is different in anisotropic systems [14], but in our case, the degree of anisotropy is very low.

The pre-nematic force is quite small and difficult to observe. A typical measurement of the pre-nematic mean field force between a DMOAP covered 10 μm glass sphere and DMOAP covered glass substrate, immersed in a $4' - n -$ pentyl $4 -$ cyanobiphenyl, (5CB), is shown in Fig. 3.5.

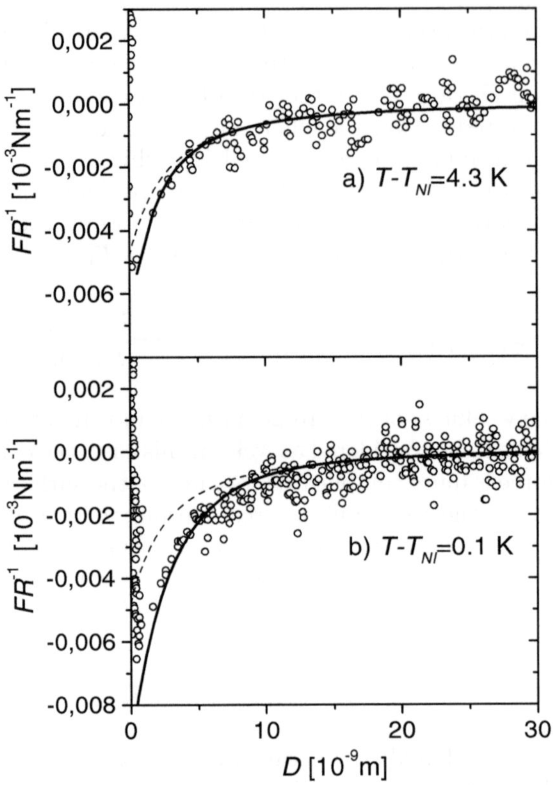

Fig. 3.5. The force between a $10\,\mu$m glass sphere, covered with DMOAP and a DMOAP treated glass substrate in 5CB liquid crystal at two different temperatures close to T_{NI}. The solid line shows the sum of the calculated van der Waals force (dashed line) and the nematic mean field force (3.4). At zero separation the two glass surfaces are not yet in contact, due to adsorbed trilayers of liquid crystal on both surfaces that cannot be removed due to insufficient stress under the apex of the glass sphere [13].

The calculated van der Waals interaction is presented with a dashed line and is nearly temperature independent. On the other hand, it can be clearly seen that the *total force* is temperature dependent, which can only be a consequence of an additional nematic mean-field contribution. The solid line is a sum of the van der Waals and a nematic mean-field force, derived from the Landau-de Gennes theory. The agreement is quantitatively good and gives us the strengths of the two surface coupling coefficients, which are in the case of DMOAP quite large, i.e. $w_1 = 1.4 \times 10^{-4}(1 \pm 0.4)\,\mathrm{J/m^2}$ and $w_2 = 7 \times 10^{-4}(1 \pm 0.3)\,\mathrm{J/m^2}$ [13].

3.1.5 Capillary Condensation of the Nematic Phase in Confinement

A quite different scenario is observed, when the surfaces induce a large degree of nematic order. In this case, a very large spatial gradient of the order is created in the confined liquid, which is energetically unfavourable. At some separation, it is more favourable to phase-transform the isotropic liquid crystal into an ordered state, which is the capillary condensation of the nematic phase.

The capillary condensation is a well known phenomenon in nature and is in most cases associated with the condensation of water in pores and cracks with hydrophilic surfaces. A curved meniscus is formed due to the surface tension of the water-vapour interface. As a result of the change of pressure across this meniscus, a strong attractive force acts between the two surfaces. The phenomenon is known for quite a long time and has been explained by Lord Kelvin back in 1871 [15] with his famous equation [16].

The capillary condensation phenomenon is of course not exclusive to water. It can be found in any confined system, where the surfaces prefer one phase over another and there is a first order phase transition between the phases of the material between the surfaces. A nematic liquid crystal is an example of such a system exhibiting a first order phase transition between the isotropic and the nematic phase. For this system, the nematic capillary condensation has been predicted by P. Sheng in 1976 [17]. Since the isotropic-nematic phase transition is only weakly first order, the phenomenon is not easy to observe. One has to be able to control the distance between the surfaces with a nanometer precision and the temperature within 10^{-2} K, which is unachievable to methods like NMR, SFA, DSC, etc., and very difficult to achieve in dynamic light scattering experiments [18, 19]

Nematic capillary condensation was first reported in an AFM force experiment [20], where it is observed as a sudden appearance of an attractive force during the approach of the two confining surfaces. Signatures of nematic condensation were observed before in a light scattering experiment on thin nematic films performed by Wittebrood [19]. A set of typical AFM measurements of capillary condensation is shown in Fig. 3.6.

The typical pull-in step in the measured force curve is a consequence of a sudden capillary condensation, which occurs at a certain separation for a given temperature. At the moment of capillary condensation, a tiny capillary bridge of the nematic phase is formed between the apex of the sphere and the flat substrate. Since there is a capillary pressure across the curved isotropic-nematic interface, a strong force between both surfaces is observed. The separation dependence of the capillary force can be analyzed within the LdG formalism [21] and fits very well to the experiment.

The inset to Fig. 3.6 shows the experimentally determined (D, T) phase diagram for $8CB$ together with the results of the LdG theory, indicated by the solid line [20]. At a given temperature, the phase boundary is determined

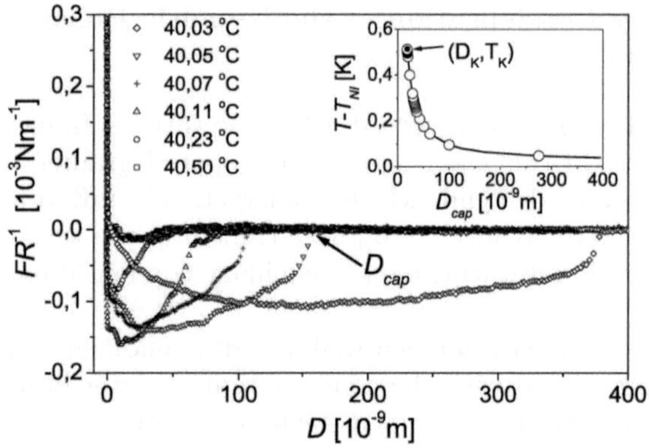

Fig. 3.6. A set of AFM experimental force curves, obtained during the approach of surfaces at different temperatures, just above the isotropic-nematic phase transition. The pronounced pull-in step corresponds to the capillary condensation. The inset shows the experimentally determined (D, T) phase diagram for $8CB$ together with the results of the LdG theory, indicated by the solid line.

by measuring the distance D_{cap}, where the condensation occurs. This separation depends strongly on temperature and cannot be observed above a certain "critical" distance D_K and above a certain temperature T_K, which represent a critical point in the (D, T) phase diagram, predicted by Sheng. [17]. On the other hand, the range where the condensation occurs is of the order of one micron very close to the isotropic-nematic transition temperature. A numerical LdG analysis of the experimental force plots at different temperatures shows the location of the critical point in this system, which is at $D_K = 19(1 \pm 0.15)$ nm and $T_K - T_c = 400(1 \pm 0.1)$ mK. The transition is continuous for higher temperatures, where the discontinuity of the order parameter vanishes.

The experiment of capillary condensation can be as well considered within the framework of the Kelvin equation. For this purpose, the temperature dependence of the inverse distance of condensation, D_{cap}, is presented in Fig. 3.7.

It can be seen that the relation between the inverse of the D_{cap} and $T - T_{NI}$ is almost linear. This has been predicted by Kelvin [15], who described the condensation of water in pores with his famous equation $1/D_{cap} = h_N/2\gamma \cos \Theta T_{NI}$, where γ is the interfacial tension between the condensed and uncondensed phases, Θ is the contact angle of both phases on the third (solid) phase and h_N is the enthalpy density of the phase transition to the condensed phase. The slope of the curve in $(D_{cap}^{-1}, (T - T_{NI})$ is therefore connected to the interfacial tension between the isotropic and nematic phases and the contact angle describes the affinity of a certain surface for either of the phases. The fit to the Kelvin equation is shown in the inset to

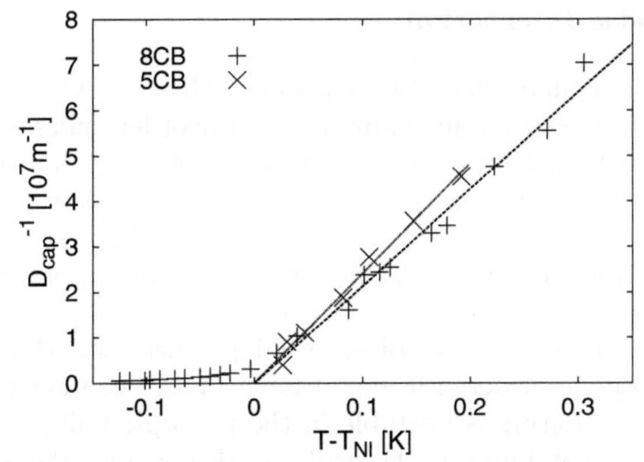

Fig. 3.7. Temperature dependence of the inverse of separation between the surfaces where the capillary condensation occurs D_{cap}^{-1}. Plotted for 5CB and 8CB.

Table 3.2. Interfacial tensions of the nematic-isotropic interface for 5CB and 8CB as obtained by fitting of the observed phase transition temperatures against inverse surface separation, where capillary condensation occurs (inset to Fig. 3.6), to the Kelvin equation.

liquid crystal	γ_{NI} [J/m^2]
5CB	$1.0 \times 10^{-5}(1 \pm 0.1)$
8CB	$1.8 \times 10^{-5}(1 \pm 0.15)$

Fig. 3.6. From this fit, one can extract the interfacial tensions between the isotropic and nematic phases, which are given in Table 3.2

The forces due to capillary condensation of the nematic liquid crystal might become very important for "smart" colloidal systems. The most important advantage of the capillary condensation interaction is that it can be regulated by simply varying the temperature of the liquid crystal. As it is also possible to charge the colloids in the liquid crystal by adding suitable surfactants, the colloidal system can be stabilized at high temperatures, where the liquid crystal is in the isotropic phase. By lowering the temperature, the long-ranged capillary condensation interaction can overcome the range of the electrostatic repulsion, which should lead to destabilization of the colloidal system and binding of particles. With simultaneous use of an external electric field it should be possible to make the capillary condensation interaction unidirectional, which could lead to the formation of linear chains of colloidal particles. The system is also reversible, since raising of the temperature in the bound colloidal system melts the capillary bridges that bind the particles together and therefore the stable colloidal system should restore to the initial state.

3.1.6 Pre-smectic Interaction

Whereas in simple nematic liquid crystals like 5CB only a weak pre-nematic force is observable due to surface-induced nematic order, more complex forces are expected for materials that exhibit a tendency to form smectic layers. An example is 8CB, which has only a narrow (≈ 5 K) nematic phase, which is followed by a smectic-A phase. In the isotropic phase of 8CB there is therefore a kind of natural tendency to form smectic layers under appropriate external perturbations.

A flat solid wall is the kind of external perturbation that favours the creation of positionally ordered layers of molecules in contact with the wall. As this positional ordering is unstable in the isotropic bulk, we expect that positional order is maximum at the wall and decays into the isotropic bulk over a coherence length, which is in this case the *smectic coherence length.*

A very similar reasoning, as in the case of surface induced nematic order, suggests that, if two identical walls are brought in close proximity and there is a smectic-like liquid crystal in between, an oscillatory force is detected when the wall separation is changed. The oscillatory force is due to the overlapping of the two smectic order parameter profiles, emanating from both surfaces. The free energy exhibits local minima, when the number of smectic layers matches the wall separation and there is no need for the distortion of the smectic order. The energetically most unfavourable situation is when the maxima of a density wave from one surface correspond to minima of the density wave from the other surface. The free-energy therefore depends periodically on wall separation, which generates an oscillatory pre-smectic force between the surfaces, first calculated by de Gennes [22]. The calculation is very similar to the one considered for the pre-nematic mean field force and the resulting pre-smectic force is [23].

$$F_{SmA}(D)/A = \frac{L_{Sm}\Psi_0}{\xi_{\|}}\left[\coth\frac{D}{\xi_{\|}} - \frac{\cos(2\pi(D-D_0)/a_0)}{\sinh(D/\xi_{\|})} - 1\right] . \tag{3.6}$$

Here $\xi_{\|} = \sqrt{L_{Sm}/a_{Sm}}$ is the smectic correlation length in the direction along the smectic normal, D_0 is the phase offset, which is due to the uncertainty of the determination of zero surface separation in the AFM force measurements and a_0 is the smectic period. Ψ_0 is the amplitude of the smectic order on the surfaces and L_{Sm} is the coefficient of the smectic gradient term in the LdG free energy density expansion.

The pre-smectic oscillatory interaction was first observed in an AFM force experiment in the isotropic phase of a liquid crystal 8CB [23] and is shown in Fig. 3.8 for different temperatures close to the isotropic-nematic phase transition. The oscillatory force decays gradually with increasing separation, the range of the force is equal to the smectic correlation length inside the isotropic phase.

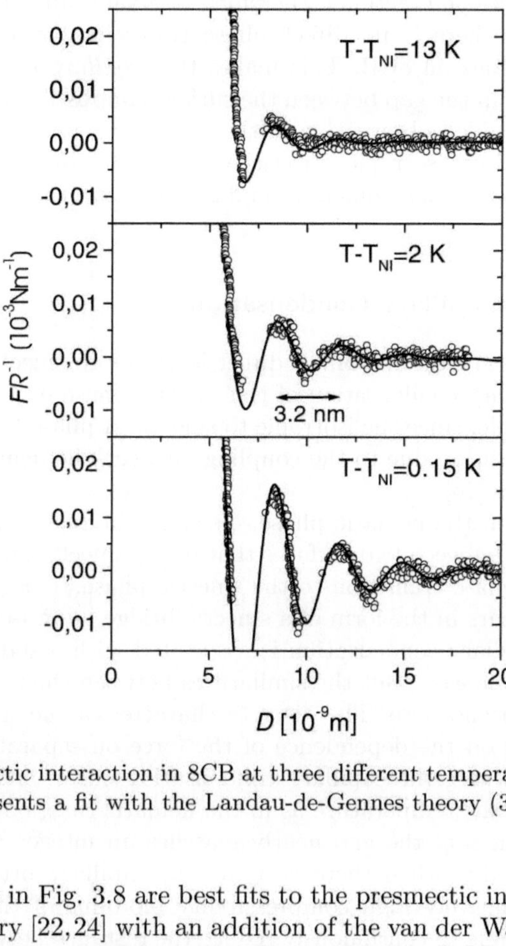

Fig. 3.8. Presmectic interaction in 8CB at three different temperatures above T_{NI}. A solid line represents a fit with the Landau-de-Gennes theory (3.6).

The full lines in Fig. 3.8 are best fits to the presmectic interaction, based on the LdG theory [22,24] with an addition of the van der Waals force [2,14]. Just like in previous nematic cases, the mesoscopic LdG theory superbly describes the structural force between the surfaces. The fitting of the measured presmectic forces to the theory allows for the determination of many important surface parameters, that are difficult to obtain using other methods. For example, the amplitude of the smectic order at the surface and the smectic correlation length can be obtained directly. For 8CB on silanated glass, one obtains a typical surface smectic order parameter Ψ_0 which is of the order of 0.1 and the smectic correlation length $\xi_{\|}$, which is in perfect agreement with X-ray data of Davidov et al. [25]. In addition to that, it has been observed that the smectic order is coupled to the nematic order and this amplifies the presmectic interaction close to the isotropic-nematic phase transition [23].

The pre-smectic interaction is therefore observable already in the isotropic phase, although there is no direct phase transition from the isotropic to the smectic-A phase in 8CB. This makes the *capillary condensation of the smectic-A phase* in the gap between the surfaces impossible and consequently enables more detailed observation of the presmectic interaction. In the case when there is a direct isotropic-smectic-A phase transition, like for example in 12CB, a condensation phenomenon similar to nematic capillary condensation is observable.

3.1.7 Smectic Capillary Condensation

When the liquid crystal exhibits a direct isotropic to smectic-A phase transition, the capillary condensation of the smectic phase in a gap between the surfaces is possible, since the isotropic to smectic-A phase transition is a first order phase transition due to the coupling between the nematic and smectic order parameters [26].

Similarly as in the nematic phase, discussed in Sect. 3.1.5, the isotropic sample confined between two surfaces that prefer smectic ordering, can exibit a spontaneous phase transition to the smectic phase. This smectic capillary condensate appears in the form of a smectic bridge in an isotropic surrounding and the capillary condensation is associated with a sudden pull-in force like in the nematic case. But the similarities between the nematic and smectic condensation end here. The discrete character of the smectic phase has a strong impact on the dependence of the force on separation. First of all, the separation where the capillary condensation occurs, no longer depends continuously on the temperature as in the nematic case, but is only possible when the thickness of the gap nearly matches an integer number of smectic layers. Secondly, when there is a smectic capillary bridge between the surfaces, the relatively high compressibility modulus of the smectic phase makes it impossible to continuously reduce the distance between the surfaces in the experiment. The smectic condensate resists the force, applied on the sphere and deforms only slightly. At a certain threshold force, the smectic bridge melts and the distance bettween the surfaces is abruptly reduced for a thickness of a single smectic layer. Then again, the condition for a smectic capillary condensate between the sphere and the surface is fulfilled, the condensate with one layer less forms and again it resists the compression up to a certain force. As a result, a typical smectic force profile is obtained, as shown in Fig. 3.9.

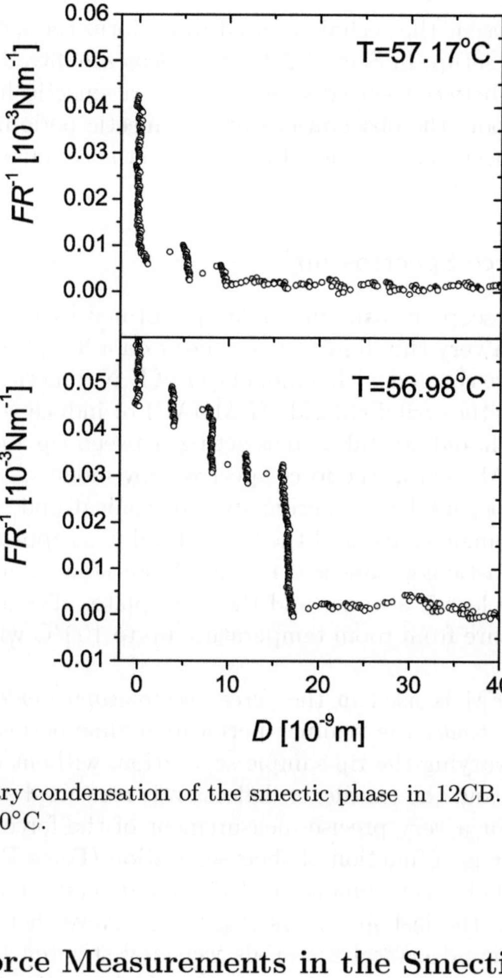

Fig. 3.9. Capillary condensation of the smectic phase in 12CB. The temperature of the T_{Sm-I} is 50°C.

3.2 AFM Force Measurements in the Smectic Phase

Giovanni Carbone, Bruno Zappone and Riccardo Barberi

3.2.1 Introduction

The "Force Spectroscopy Mode" of an AFM is a promising approach to obtain information about the structure of liquid crystal interfaces. It allows not only for a subnanometer control of the thickness of the liquid crystal layer confined between a nanometer size tip and a flat substrate, but can simultaneously measure the structural force, generated by the confined liquid. Such a force, being mediated by the confined liquid crystal, provides new information on the LC interfacial structure with unprecedented spatial resolution.

In this part we discuss the use of the AFM Force Spectroscopy mode to study the smectic periodicity and compressibility of smectic layers. In

Sect. 3.2.2 we present the technique used to perform the measurements and the experimental setup. In Sect. 3.2.3, basic ideas are introduced, describing the force acting between surfaces that confine a smectic liquid crystal. In Sect. 3.2.4 we report the observations of the smectic periodic structure performed on two smectic compounds. In Sect. 3.2.5 the compressibility modulus of 8CB is estimated using a force plot.

3.2.2 AFM Force Spectroscopy

The force spectroscopy measurements are performed by putting an unmodified AFM tip in a very thin liquid crystal layer on a flat glass plate. Both tip and glass plate are covered with a monolayer of N,N-dimethyl-N-octadecyl-3-aminopropyltrimethoxysilyl chloride (DMOAP) to induce good homeotropic alignment of the liquid crystal. Forces acting between tip and glass plate are measured by AFM used in the force spectroscopy mode.

The AFM tips, used in experiments, are conical and 2 μm long, they have a half-cone angle of 10° and the tip-end radius is approximately 10 nm. The cantilever elastic constant is 0.1 N/m. A hot stage is mounted between the AFM's piezoelectric scanner and the glass plate, allowing to control the sample temperature from room temperature up to 100°C with a precision of 0.1°C.

When the AFM is used in the *force spectroscopy mode*, also known as *force-vs-distance mode*, the scanner performs a time-periodic movement in the z direction, varying the tip-sample separation, without changing the x-y sample position, and the cantilever deflection is measured simultaneously.

This allows for a very precise measurement of the force between the tip and the substrate as a function of their separation (Force Plot)[27].

The potential of this technique to study liquid crystal interfaces has been already shown in the last few years [13, 20, 28]. Nevertheless, in the previous works, the liquid crystal materials were investigated in their isotropic phase only, and hence the measurements could be performed with the tip totally immersed in the liquid crystal. When the LC material is in one of its anisotropic phases, it becomes strongly birefringent [26] and hence it is no longer possible to measure the deflection of the cantilever by the usual optical method, because the detection light is strongly scattered by orientational fluctuations.

In order to perform AFM force measurements in the smectic phases using the usual optical method of detection, the cantilever should not be completely immersed in the liquid crystal. Instead, the tip can be partially placed in the liquid, as shown in Fig. 3.10.

Unfortunately, in this experimental geometry, forces due to surface tension (capillary force) play a dominant role, as shown in Fig. 3.11. When the tip reaches the free surface of the film, an attractive force instability appears, suddenly bending the cantilever (jump-in). As the tip-substrate separation further decreases, the attractive capillary force steadily increases, until the

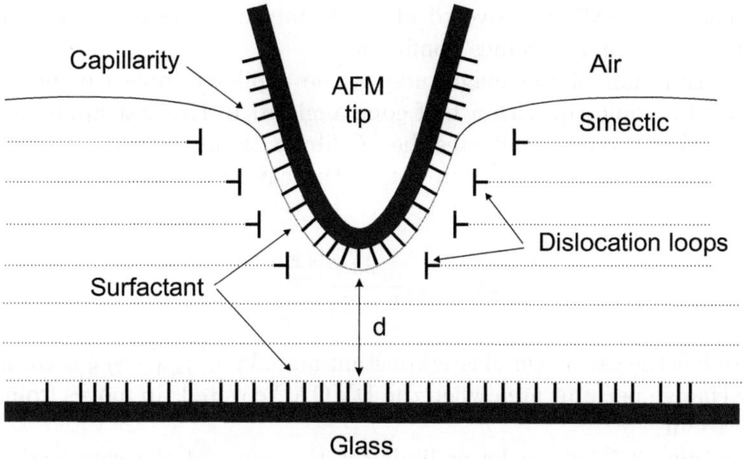

Fig. 3.10. Smectic LC confined between AFM tip and glass plate, both inducing homeotropic alignment.

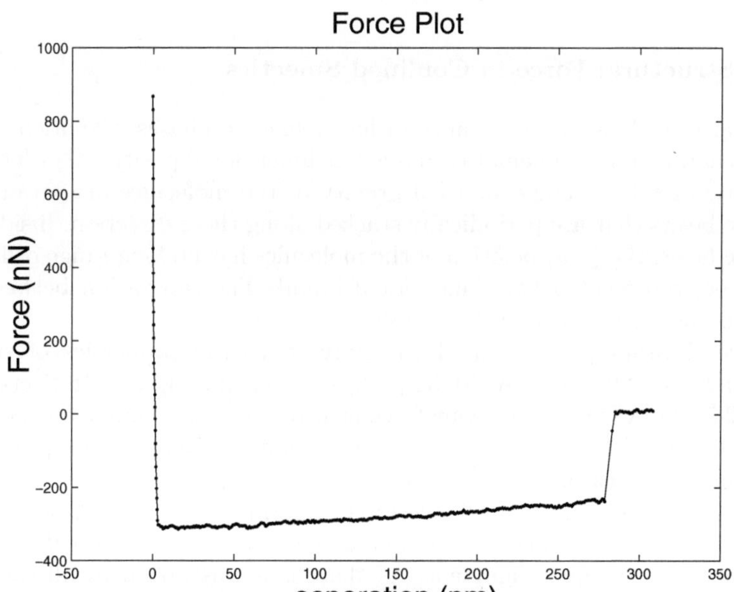

Fig. 3.11. Force plot, performed in air, on a nematic thin film of E18 using a DMOAP coated conical tip. The step on the rigth hand side is the jump-in effect, that is due to a capillary instability. On the left side, the quasi-vertical line is due to the hard contact between the tip and the solid substrate. Between the jump-in and the hard contact, the capillary attraction is the dominant force. It presents a monotonous attractive behaviour. If, before the contact, the tip is clean, i.e. without liquid drops attached, such a plot allows for an accurate measurement of the film thickness, which is in this case about 280 nm.

tip reaches the DMOAP covered glass-LC interface (hard contact). Here a repulsive steric force becomes dominant.

An estimation of the magnitude of the capillary force can be done in the case of a conic tip with a half cone angle α. In the first approximation, assuming that the tip penetrates the LC film without changing its thickness, one can calculate the deflection Δz of the cantilever as a function of the piezoelectric scanner movement z_o:

$$\Delta z = z_o \frac{2\pi\gamma \frac{\mathrm{tg}\,\alpha}{\cos\alpha}}{k - 2\pi\gamma \frac{\mathrm{tg}\,\alpha}{\cos\alpha}} . \tag{3.7}$$

Here k is the cantilever elastic constant and $\Delta\gamma = \gamma_{LA} - \gamma_{LS}$ is the surface energy change per unit area when the DMOAP covered tip moves from liquid crystal to air.

Equation (3.7) sets a lower limit for the value of the cantilever elastic constant that can be used without having the tip instability and jump-into the substrate. On the other hand, selecting a stiffer cantilever with larger k results in reduced force sensitivity, so that a compromise has to be found.

3.2.3 Structural Force in Confined Smectics

There are two kinds of long range order in smectic phases: the orientational order, characteristic of nematics and a one dimensional positional order, characteristic of solids. The centers of gravity of the molecules are arranged in smectic layers that are periodically stacked along the z direction. Inside each smectic layer, the (x,y) position of the molecules has no long range order and the layers represent a two-dimensional liquid. The separation between the layers is the basic period of the smectic structure.

Several techniques are used to study structural properties of smectic phases: X ray [29, 30], SFA [31, 32], ellipsometry [33, 34], etc. In Sects. 3.2.4 and 3.2.5, the AFM spectroscopy force is introduced as a simple and straightforward method to measure the smectic periodicity and the compressibility modulus of a stack of smectic layers.

As described in Sect. 3.2.2, the force experiments are performed by confining the sample between a flat planar surface and a spherical tip, both imposing homeotropic alignment. As there is a frustration to fill the space between the curved and a flat surface with homeotropic alignment, an array of dislocation loops is produced, as shown in Fig. 3.10. When the separation between the confining surfaces changes, the layers are either compressed or streched and the number of smectic layers can vary only via the creation or the annihilation of dislocation loops. The structural force, originating from this mechanism has been analyzed in reference [31] and is also treated in detail in Sect. 3.3 by Zappone et al. It is shown that the force due to the elastic distortion of the central core of the sample is dominating and depends quadratically on the mismatch between the surface separation D and the equilibrium thickness of n_1 smectic layers of total thickness $n_1 d$:

$$F(D) = \pi RB \frac{(D - n_1 d)^2}{n_1 d} . \tag{3.8}$$

Here D is the separation between the surfaces, R is the sphere radius, B is the compressibility modulus, d is the smectic period and n_1 is the number of layers contained in the central cell. By measuring the structural force one can therefore determine simultaneously the thickness of the smectic layers and the smectic compressibility modulus B. The model of reference [31] was originally considered for the crossed cylinders of the SFA geometry, that is equivalent to a plane-sphere geometry with a local radius $R \approx 2$ cm. Some of the approximations used to obtain the (3.8) are however questionable in view of the much smaller size of the AFM tip radius, which is ≈ 10 nm.

3.2.4 Smectic Period from AFM Force Spectroscopy

Force plots in Fig. 3.12 are obtained at room temperature (25°C) in the SmA phase of 8CB (4-n-octyl-4-cyanobiphenyl) that has the following phase sequence: K ← 21.5°C → SmA ← 32.5° → N ← 40.0°C → I. Both run-in

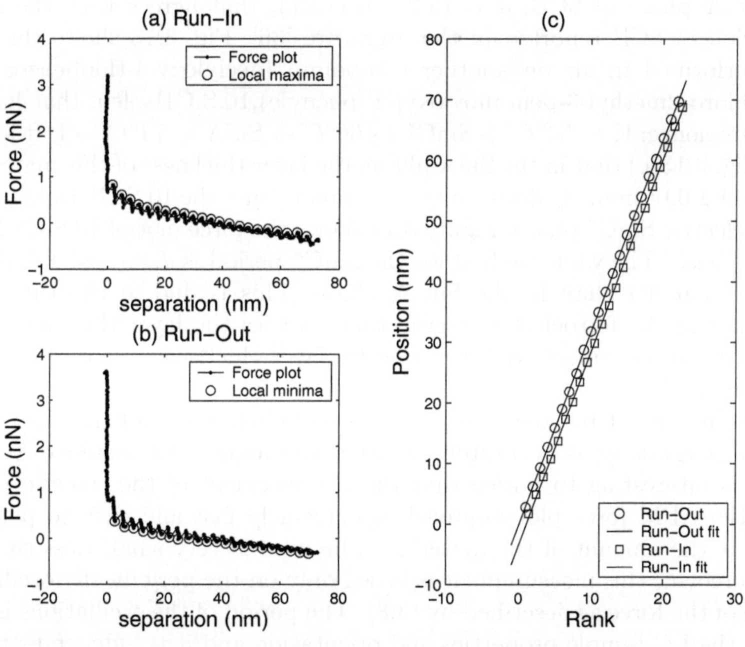

Fig. 3.12. (a) Force plot performed in air using DMOAP covered tip in 8CB on DMOAP covered flat glass. (b) Position of the minima of the force vs. rank. The smectic period is $d = (3.2 \pm 0.1)$ nm.

(decreasing the tip-substrate separation), and run-out (increasing the tip-substrate separation) force plots, show a discontinuous and periodically oscillating force, which is a clear manifestation of smectic periodicity.

The presence of discontinuities (sudden jump-in or jump-out at some value of the separation D) is due to the mechanical instability of the AFM cantilever at the points where the derivative of the force $\mathrm{d}F/\mathrm{d}D$ is larger than the elastic constant k of the cantilever [35]. The periodic oscillations are due to the layered structure of the sample and are described by (3.8). For a given number of layers n, the force has a minimum when the separation D is equal to nd, the thickness of the unstressed n layers. Because of the instability of the tip, this is observable only in the run-out force-plot. Starting from the minimum, the force increases more and more, as the separation between the confining surfaces decreases, and it presents a local maximum at $d = d_c$, when a dislocation loop appears. The tip instability makes this point observable only during the run-in.

For this reason, the value of the smectic period d can be obtained for the run-out force plot, from the slope of the local minima positions versus the rank of the minima . As the position of the maxima is not proportional to the rank i, but to $\sqrt{i(i+1)}$, it is possible to measure the smectic period for the run-in as the slope of the local maxima positions as a function of $\sqrt{i(i+1)}$.

In Fig. 3.12(c) both cases are represented, obtaining a period in the smectic A phase of 8CB, $d = (3.2 \pm 0.1)$ nm, that agrees with the smectic period of 8CB reported in the literature [36]. Fig. 3.13 shows the force plot performed in air on another compound (déclyloxy-4-thiobenzoate de [(2s)-chloro-2méthyl-3-pentanoyloxy]-4'-phényle),$\overline{10}$.S.C1Isoleu, that has the phase sequence: K $\leftarrow 54^oC \rightarrow$ SmC* $\leftarrow 66^oC \rightarrow$ SmA $\leftarrow 74^oC \rightarrow$ I. It follows from Fig. 3.13(c) that in the SmA phase, the layer thickness of this material is $d = (3.08 \pm 0.05)$nm. By decreasing the temperature, the $\overline{10}$.S.C1.Isoleu shows a ferroelectric SmC* phase. Figure 3.14 shows the force plot of $\overline{10}$.S.C1.Isoleu in this phase. The value we find for the SmC* period is $d = (2.85 \pm 0.05)$nm, which is smaller than in the Sm A phase. This is due to the tilt of the molecules in the ferroelectric phase, that reduces the layer thickness. Both values are in agreement with the smectic layer thicknesses, reported in the literature [29].

It is important to point out that this technique does not measure the of smectic periodicity as accurately as other techniques (for instance X-rays), but it is interesting to notice that the measurement of the smectic period using the AFM force plot approach is extremely fast and easy to perform, and that the amount of the material to be used is very small (less than 0.1 μl). Moreover, this measurement is based only on the periodically oscillating nature of the force as described by (3.8). The period of the oscillations is only due to the LC sample properties and orientation and it is independent from parameters like the radius or the shape of the AFM tip. This means that no further fitting operations, depending on unknown parameters, are required.

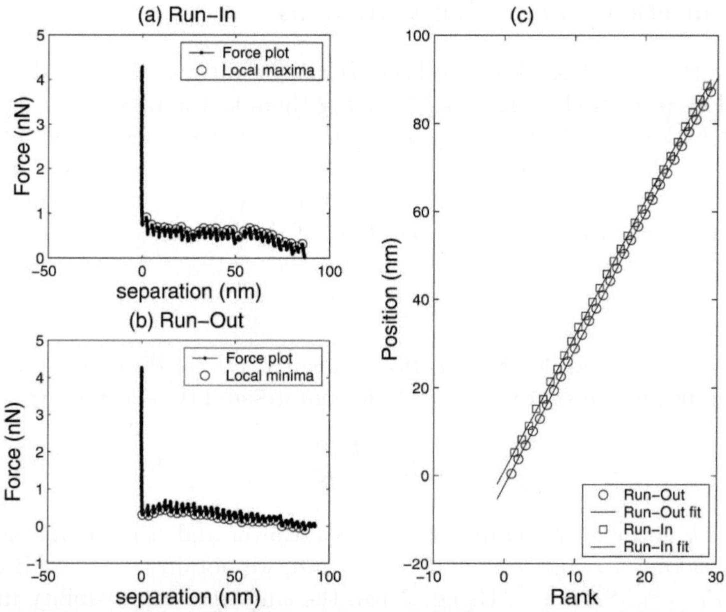

Fig. 3.13. Separation dependence of force on DMOAP covered tip in the Sm A phase of $\overline{10}$.S.C1.Isoleu.(a) shows the force during run-in, (b) Run-out. (c) Position of minima of the force vs. rank. The smectic period is $d = (3.08 \pm 0.05)$ nm.

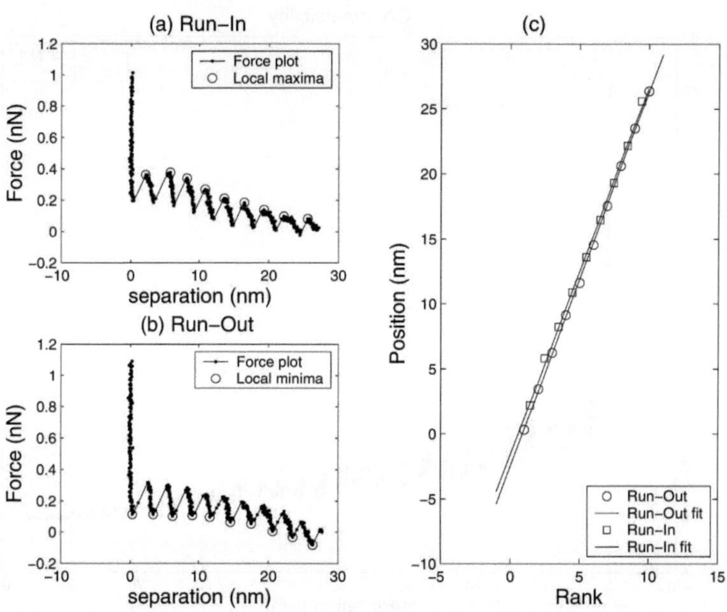

Fig. 3.14. (a) Force plot performed in air using DMOAP covered tip in the SmC* phase of $\overline{10}$.S.C1.Isoleu.(a) run-in (b) run-out (c) analysis of the positions of force minima, which gives the layer thickness $d = (2.85 \pm 0.05)$nm.

3.2.5 Smectic Compressibility Modulus

The smectic compressibility modulus B can be directly estimated from the force plots presented in Sect. 3.2.4, fitting them to the (3.8).

The fitting is performed only on the run-out measurements, because they contain the force minima (Fig. 3.15). To avoid spurious contributions from the slowly varrying surface tension, the i-th oscillation in the force plot is fitted independently by a single parabola of the form:

$$y = y_{oi} + \frac{\alpha}{i}(x - x_{oi})^2 \tag{3.9}$$

where y_{oi} is the cantilever deflection measured at the ith minima, x_{oi} is its position and α is linked to the smectic compressibility modulus B:

$$B = \frac{k\alpha d}{\pi R} . \tag{3.10}$$

Here k is the elastic constant of the cantilever and R is the radius of the tip. By taking $k = 0.1\,\mathrm{N/m}$ and $R = 10\,\mathrm{nm}$, we obtain for the 8CB data in Fig. 3.15, $\alpha = 18.3\,\mathrm{nm}^{-1}$. Using (3.10), the smectic compressibility modulus for 8CB is found to be $B = 1.68 \cdot 10^8\,\mathrm{N/m}^2$. This value is of the correct order of magnitude, but is somewhat larger than the value reported in the literature

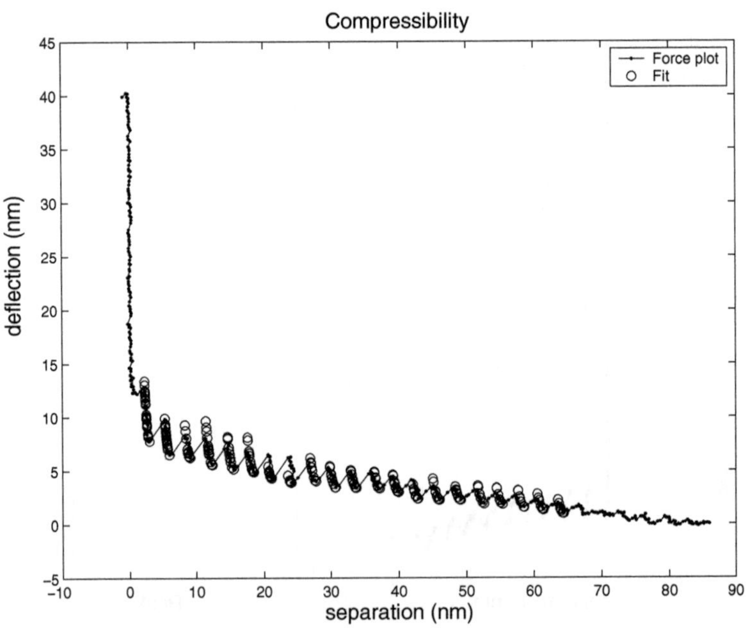

Fig. 3.15. In the force plot each oscillation is fitted using a parabola centered in the oscillation minima.

[11, 37]. Low frequency compressibility modulus in the SmA phase of 8CB is temperature dependent and is $\approx 2 \cdot 10^7\,\mathrm{N/m^2}$ [11]. However, the scattering of the value of B is comparable to its mean value, which is probably due to the not well defined radius of the AFM tip (adsorption and desorption of molecules). Moreover the shape of the tip-end is spherical only in first approximation. To obtain more precise measurements of B, well defined and somewhat larger spherical particles ($R \approx 100\,\mathrm{nm}$) should be used instead of a sharp AFM tip.

3.2.6 Conclusions

It has been shown that the AFM in the force spectroscopy mode is a very simple, accurate and straightforward method to measure the smectic layer thickness with a precision of 0.1 nm using a very small drop of a liquid crystal material. The method is less accurate in measuring the smectic compressibility modulus, which is due to the surface tension on a partially immersed AFM tip and the small, not very well defined radius of the AFM tip.

An improved experimental approach would use a sub-micrometer glass sphere attached to the cantilever [13, 28]. This would increase the force resolution due to larger value of R and a better defined confining geometry. On the other hand, using a larger diameter of the sphere could significantly increase forces due to surface tension and a larger elastic constant of the cantilever should be used to prevent jump-in of the cantilever.

3.2.7 Acknowledgement

The authors want to thank Giuseppe Strangi, from University of Calabria, for the useful discussions. Acknowledgement has to be expressed to H.T. Nguyen, from the recherche center "Paul Pascal", that provided the $\overline{10}$.S.C1.Isoleu and to CalcTec srl, a spin off company of INFM and University of Calabria, that provided the AFM temperature control.

3.3 Surface Forces in Thin Layers of Liquid Crystals as Probed by Surface Force Apparatus – SFA

Bruno Zappone, Philippe Richetti, and Roberto Bartolino

The Surface Force Apparatus (SFA) probes the surface properties of nanometer thin layers of liquids by measuring the forces acting on the confining surfaces. It has been developed in the 70's [1], allowing the first direct confirmation of the Derjaguin-Landau-Verwey-Overbeek (DLVO) theory of surface forces in classical colloidal systems. SFA has provided important contributions in the domain of polymers (depletion forces), colloids and organic solutions (hydrophobic forces, layering structuration of the interface) and, more recently, friction forces on the nanometer scale (see [2] for a complete review).

In liquid crystals specific forces are expected, in view of the particular structure and order of different liquid crystalline phases. First, when there is a long-range positional or orientational order in a confined system, the free energy depends on surface separation, which always produces some long-range *elastic force*, also considered as a *structural force*. An example is the constrained smectic sample, described in Sect. 3.3.2. Second, the surfaces may enhance or lower the order parameter in a thin boundary layer of a few tenths of nanometers. A close proximity of two identical surfaces causes overlapping of these layers, producing short-range *surface-order forces*. An example is the system described in Sect. 3.3.3, where the surfaces induce positional order in a non-layered nematic. Third, the nano-confinement suppresses a number of modes of orientational fluctuations and produces *fluctuation forces*, by a mechanism similar to that producing Casimir forces between non-charged metals [38].

3.3.1 The Surface Force Apparatus

The SFA setup is shown in Fig. 3.16. The core of the apparatus is a curved confinement cell, defined by two cylindrical glass lenses of radius $R \sim 2\,\mathrm{cm}$, facing each other in a *crossed cylinders geometry*. Two flexible sheets of muscovite mica are glued on the lenses and are in direct contact with the liquid to be examined. Mica is used because large sheets ($\sim 1\ \mathrm{cm}^2$) of uniform thickness ($\sim 5\ \mu\mathrm{m}$) can be easily cleaved by hand, exposing the liquid to a very clean and smooth surface (roughness $\sim 0.1\,\mathrm{nm}$).

The cylinders can touch each other at a single *contact point*, around which the mica/mica gap increases as in a sphere-plane geometry of radius R. When the mica sheets are separated, the *minimum separation* D (at contact point) can be varied from several microns down to a few tenths of nanometers. The liquid is inserted in the mica/mica gap either by inserting a small droplet between the cylinders or filling the sealed chamber that contains the cell.

An optical interferometric technique is used to measure the separation D, simultaneously providing also the liquid refractive index n and the curvature R. The mica sheets are made partially reflecting by evaporating a $40 \div 50\,\mathrm{nm}$ silver layer on the inner sides of the mica sheets. The cell is then illuminated by a collimated white light beam, in a direction perpendicular to both axis of the cylinders. The two partially reflective silver layers form a sort of three-layer curved interferometer: selective transmission occurs only for the spectral components resonating with the mica/liquid/mica "cavity". The transmitted wavelengths λ are analyzed by an interferometer with a resolution better than $0.1\,\mathrm{nm}$, to get information about the optical properties of the layers, especially the thickness D and the index n of the liquid.

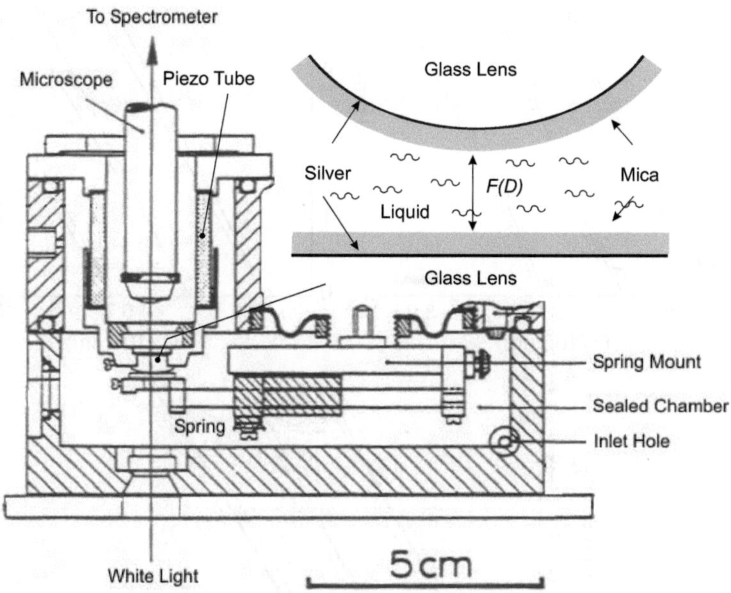

Fig. 3.16. The Surface Force Apparatus. Inset: the crossed-cylinders three-layers interferometer.

Before introducing the liquid, the mica sheets are put into contact and the *contact wavelengths*, (i.e. the wavelengths of maximum transmission in contact) are measured. They depend only on the optical thickness and dispersion properties of the silvered mica sheets, and are used as the reference values to calculate D and n in presence of the liquid. Since the mica sheets are slightly birefringent ($\delta n \sim 0.005$), with the optical axis laying in the plane of the sheet, the contact wavelengths will also slightly depend on the relative orientation of the optical axis of each sheet (Fig. 3.17 at D=0 for examples).

Knowing the optical properties and the orientation of the mica layers, a *conversion chart* can be calculated to determine the separation D from the measured λ, as shown in Fig. 3.17. The conversion chart is calculated from the equations of a three-layer interferometer [39], possibly modified to include birefringence of the mica layers [40]. As a general rule, λ is always shifted towards smaller values with decreasing D. As λ depends also on the index n, measurements for two or more transmission wavelengths are sufficient to calculate D and n at the same time. To do that, one has to invert a set of interferometric equations giving λ as a function of D and n. The resulting typical resolution in separation D is about 0.2 nm and the sensitivity to n is about 0.01. We stress however that the equations are nonlinear and the solutions may become unreliable for some D, due to an instability with respect to the error in λ (see Fig. 3.21 for an example).

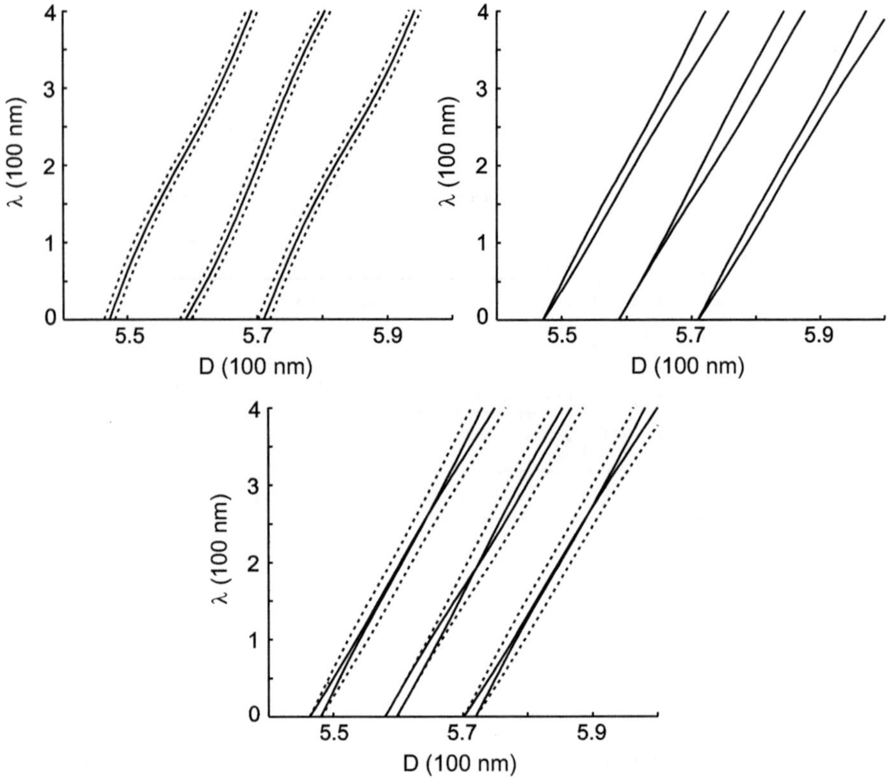

Fig. 3.17. Conversion charts for a positive uniaxial nematic for different alignments with the same pair of mica sheets. (a) Isotropic or homeotropic sample (n=1.3) between mica sheets with parallel (dashed outer lines) and perpendicular (solid inner line) optical axes. The splitting of the line for parallel alignment is due to the sheets birefringence, which is compensated by crossing the sheets over $\pi/2$. (b) Planar sample (n_e=1.7, n_o=1.5) with the optical axis parallel to the optical axis of one sheet and perpendicular to the other. (c) Planar sample with the optical axis parallel (dashed outer lines) and perpendicular (inner solid lines) to both optical axes of the sheets.

For liquid crystals, the conversion chart also depends on the birefringence of the liquid crystal and on the relative orientation of its optical axis with the optical axes of the mica sheets [40]. Figure 3.17 gives some examples of conversion charts for homeotropic or planar nematics in different configurations with respect to the same pair of mica sheets. The sensitivity to the orientation is very useful for rapidly identifying the liquid crystal configuration, but makes the interferometric equations more complicated due to the larger set of unknown parameters (D, mean index n, birefringence δn, mica/director angles at both surfaces, twist distortion between non parallel sheets, etc.). The interferometer resolution or the number of measured wavelengths may become insufficient to solve the equations for all parameters, especially those having a similar effect on the conversion chart.

The separation dependence of forces acting between two SFA cylinders is measured using a detection scheme that is similar to the AFM "Force Spectroscopy" mode Fig. 3.16. The upper glass cylinder is rigidly attached to a piezoelectric tube, that performs vertical translations, while the other cylinder is fixed to a cantilever spring of known stiffness k (typically 10^2 N/m). If, by moving the piezo by an amount ΔD_p, the separation D changes by ΔD, then there is a force acting between the cylinders that is proportional to the difference between the two displacements $F = k(\Delta D - \Delta D_p)$. Since the force is measured as the difference between displacements, it is determined only within a constant $F_0 = k(D_0 - D_{0p})$, usually chosen to give zero force for large surface separations. As in AFM force spectroscopy, the surface-surface separation D is stable only if the condition $\partial F/\partial D \leq k$ is satisfied. If the gradient of the force is larger than the spring constant k, the two surfaces "jump" to reach a stable state. The cantilever instability creates blank regions in the $F(D)$ force curves and jumps connecting stable regions are observed (see Figs. 3.18 and 3.19 as examples.).

In principle, the piezo/cantilever system can reach a force sensitivity of about 10^{-9} N. In practice, thermal dilations induce drift of surfaces with respect to each other, reducing the accuracy of the piezo displacement and the sensitivity. For this reason, the micas and the cantilever are closed in a sealed clean chamber, that is temperature-controlled in the range 15÷45°C with an accuracy of 0.05°C. In this way, thermal drifts are reduced to a few tenths of nm per minute and the contamination of mica and the sample is avoided. The resulting experimental force sensitivity is about 10^{-8} N. The force measurement is performed by progressively approaching or retracting the surfaces. When the equilibrium separation is reached, the piezo displacement and the transmitted wavelengths are recorded. A complete approach/retraction cycle takes typically 30 minutes. A review on the SFA technique can be found in [41].

3.3.2 Structural Forces in the Smectic A Phase

Structural forces due to long-range positional order are quite easily observed in the smectic A liquid crystals. SFA measurements have been performed on lamellar lyotropic smectics [42, 43] and in thermotropic smectics [44–46]. These works extend to a nanometer scale the early studies on elasticity, viscoelastic response and layers instability of smectic A, observed in macroscopic wedge-shaped piezoelectric cells [47, 48].

Fig. 3.18 shows a typical force curve for an aqueous solution of an anionic surfactant and an alcohol (weight composition: 7.2% sodium dodecyl sulfate, 17.5% 1-pentanol, 75.3% water), that forms a lamellar smectic A phase at room temperature and alignes homeotropically on bare mica [43]. The structural force is characterized by periodic damped oscillations, with

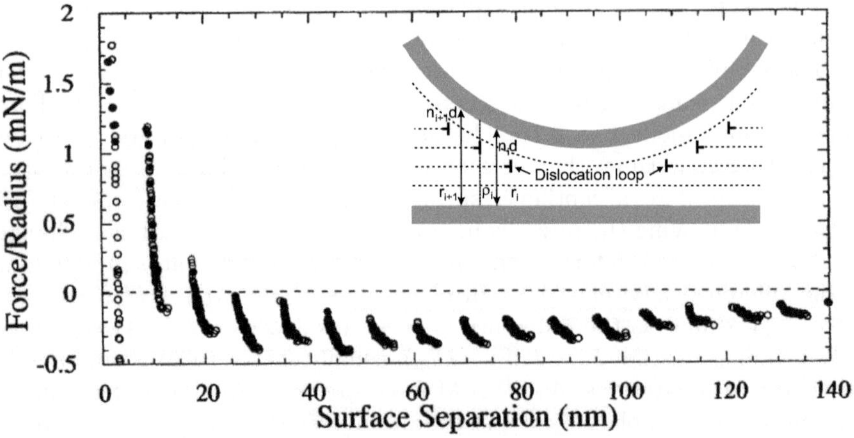

Fig. 3.18. The force F, scaled by the local curvature radius R, as a function of the mica/mica separation D. A complete compression/decompression cycle is shown (filled/open circles). The void regions are caused by the mechanical instability ($\partial F/\partial D > k$). Inset: the array of dislocation loops.

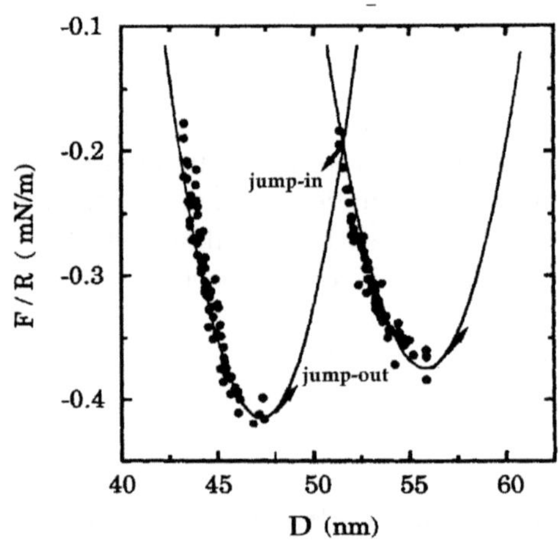

Fig. 3.19. Parabolic fit of the 6th and 7th oscillations of Fig. 3, with $B = 1.8 \times 10^5$ J/m^3 and $d = 8.8 \pm 1$ nm. At the jump-in (compression) the 7th layer is removed from the central stack: the mechanical instability brings the system on the descending part of the first parabola, corresponding to 6 layers. At the jump-out (decompression) the mechanical instability leads to the re-creation of the 7th layer (second parabola).

minima repeating over a distance $d = (8.8 \pm 1)$ nm, that correspond to the interlamellar distance, as measured by X-ray diffraction. The oscillations have a parabolic shape, as shown in Fig. 3.19. They are produced by the elastic response of the central part of the sample (i.e. layers near the contact point), combined with a process of creation and annihilation of layers by the nucleation of dislocations.

Let us first consider a stack of n homeotropic smectic layers between two parallel planes at a distance $D = nd$ apart. By reducing or increasing the separation D from the initial zero-stress configuration, the elastic energy E per unit area of the stack increases as:

$$E(D) = \frac{B}{2}\frac{(D - nd)^2}{nd}\,. \tag{3.11}$$

Here B is the compression modulus of the layers [26]. The formula, translating Hook's law for smectics, is only valid if the gradients of the smectic order across the sample can be neglected. If we continuously change the separation to reach the value $D = (n\pm1)d$, another minimum of energy is reached, if the sample had just $n \pm 1$ layers. The transition to this new configuration can indeed occur by nucleation of dislocation loops. A dislocation line is a defect in the smectic structure, where a layer ends or starts [26]. Under compression, a dislocation loop appears, separating the inside region with $n - 1$ (stretched) layers and the outside region with n (compressed) layers. Under stretching, a dislocation loop divides the inside $n+1$ (compressed) layers from the outside n (stretched) layers. The loop configuration becomes energetically favorable to the initial n-layers configuration, when the stress exceeds a critical value $|\delta D_c| = d/2$ [49]. Under compression (resp. decompression), an unstable loop of finite radius is nucleated, that grows (resp. shrinks) to remove (resp. add) a layer in the stack. The $E(D)$ curve for the geometry of parallel planes is thus a set of intersecting parabolas, each corresponding to a different number n of layers. The minima are located at $D = nd$ and the maxima at $D_c = nd \pm d/2$. The curvature of the parabolas is proportional to B/nd.

For a smectic-A that is confined in the SFA crossed cylinder geometry, there is a competition between the homeotropic alignment on mica and the tendency to form layers of equal thickness d. As a result, dislocations must arise (Fig. 3.18, inset). As the local geometry around the contact point is equivalent to a sphere-plane geometry, the loops are expected to be circular and centered on the contact point. Consider thus an array of torus-like cells, coaxial to the loops (Fig. 3.18, inset). Each cell is defined by an inner radius r_i, corresponding to a thickness $h(r_i) = n_i d$, and an outer radius r_{i+1}, with $h(r_{i+1}) = (n_i + 1)d$, and contains a circular dislocation loop of radius ρ_i. The cells are independent, because the strain patterns produced by the dislocations decays exponentially outside a parabola of equation $r^2 = \lambda z$, where $\lambda \sim$ 10 nm is the penetration length of the smectic [26]. Since $\lambda << r_{i+1} - r_i$ for a SFA cell of radius $R \sim$2 cm, the strain fields do not overlap.

As the inner part of each cell ($r_i < r < \rho_i$) is compressed, while the outer ($\rho_i < r < r_{i+1}$) is stretched, the net force acting on the cell boundaries turns out to be zero. The main contribution to the net elastic force originates from the n_1 central smectic layers ($r < r_1$), that are either all compressed (if $D < n_1 d$) or stretched (if $n_1 d < D$), and produce a net repulsive force [43]:

$$F(D) = 2\pi R E(D) = \pi R B \frac{(D - n_1 d)^2}{n_1 d} . \tag{3.12}$$

This is the main contribution to the total elastic force, which is calculated by integrating the energy density of parallel planes $E(D)$ (3.11) in the range $r < r_1$ of the sphere-plane geometry. The above result is based on the assumption that there is no net force arising from the non-central parts of the cell outside the radius r_1. In fact a line tension must be considered for the loops in the outer cells [26]. As the separation D changes, the length of the loops and thus the tension changes, adding another force contribution to (3.12). However, this force is too small and varies too slowly on the nanometer scale to be measured with the SFA [43].

Compressing the n_1 central layers, a new dislocation loop is nucleated with a finite radius for a critical separation $D_c = \sqrt{n_1(n_1 - 1)}d$. The mechanism is similar to that described before for the parallel-planes geometry. The compression force curve shows a maximum at $D = D_c$ (Fig. 3.19), followed by a jump-in to a new equilibrium separation with $n_1 - 1$ layers. Due to the jump-in, the minima cannot be seen in compression, while the maxima are not observable in decompression. This mechanisms is illustrated in Fig. 3.19, that also includes a parabolic fit to the oscillations giving a compressibility modulus $B = (1.8 \pm 0.5)$ J/m^3.

In Fig. 3.18 we notice that the minima are not located on a zero background force, but rather on an attractive (negative) baseline. The attraction cannot be explained by the elastic model only and is to be attributed to the gradients in the smectic order induced by the surfaces.

3.3.3 Surface Order Forces: Layering in Non-layered Materials

Oscillating forces are not a prerogative of smectics, but have been reported as well for nematics [45,50,51] and even for simple liquids of nearly spherical molecules [52]. In fact, as it was mentioned in the Introduction, the presence of a molecularly flat surface always produces some *layering* (i.e. positional ordering) of the nearby liquid molecules, because it breaks the translational invariance of the liquid. In liquid crystals this tendency is often enhanced by the presence of a fully layered bulk phase at low temperature (smectic, hexatic, columnar phases, etc.).

An example of this phenomenon is given in Fig. 3.20, which shows the force curve obtained for a nematic solution of surfactants in water (molar concentration: water 95%, decylammonium chloride 0.75%, potassium laurate 4.25%). The solution forms biaxial micelles and it is a nematic discotic at 29°C, where the micelles freely rotate about their shortest principal axis, defining the director orientation [53]. In bulk nematic there is a short-range

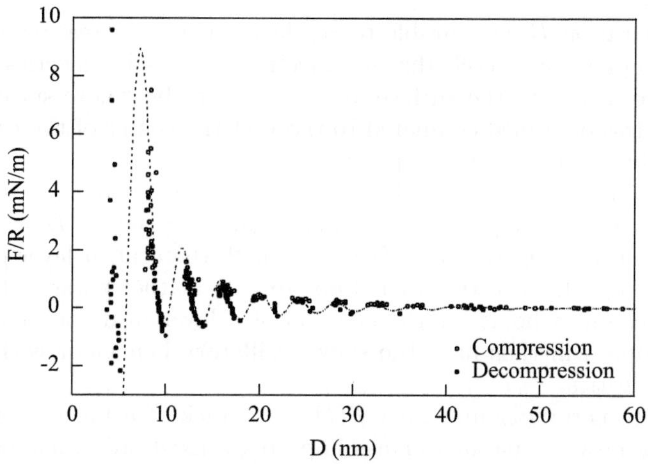

Fig. 3.20. Layering oscillations in a discotic nematic (T=29°). Filled/open circles: compression/decompression. A layer of micelles is adsorbed on each mica surface, resulting in a contact separation of about 9 nm. The fit is done using the (3.13) with $\xi = 11$ nm.

positional correlation of the micelles along their shortest axis, with an inter-micelles distance of $d = 4.2$ nm. The micelles strongly adsorb on bare mica and form a monolayer that causes homeotropic orientation of the nematic director.

The surface induced layering produces an oscillatory periodic force, with period $d = (4.2 \pm 0.1)nm$, that corresponds to the inter-micelles distance in the bulk. Compared to smectics (see Fig. 3.18), the force is spatially damped and only nine oscillations are clearly identified. It is therefore a structural force of pre-smectic origin, induced by the surface. The oscillations are not parabolic and the minima lay on a fully attractive baseline.

Since the smectic order is unstable in the nematic phase (i.e. it is pre-smectic) and consequently not uniformly developed in the sample, we consider the mean-field energy expansion in quadratic powers of the induced smectic order parameter and of its *gradients* across the sample. The expansion will depend on three parameters: the inter-micelles distance d, the *smectic correlation length* $\xi_{\|}$ in a direction parallel to the director, measuring the propagation of smectic order in the sample, and the value Ψ_0 of the order parameter at the surfaces, that is expected to be strong (fixed), since a layer of micelles is strongly adsorbed on mica. There is no force when the separation D is much larger than $\xi_{\|}$, because the order parameter profiles from both surfaces do not overlap. Since $\xi_{\|}$ usually corresponds to a few inter-micelles distance d, while the local curvature is $R = 2$ cm, we just have to consider the energy per unit area between two parallel planes $\mathcal{F}(D)$. The force follows directly from the Derjaguin approximation: $F(D) = 2\pi R\mathcal{F}(\mathcal{D})$.

For separation D comparable to $\xi_{\|}$, decreasing or increasing D will not equally compress or stretch the pre-smectic layers as in smectics. The layers that are nearer to the surface are expected to be "more smectic", thus stiffer and less deformed compared to those at the center of the cell. This is confirmed by the theory, developed in [22, 54, 55].

The mean-field equations also show that there is no need to nucleate dislocations when compressing n layers from $D = nd$ to $D = (n-1)d$. When the surface separation is $D = nd - d/2$, the central layer disappears by melting, i.e. the smectic order drops to zero at the center of the cell. A similar mechanism holds for the creation of a layer in decompression. The resulting force will thus have the same oscillatory behaviour with period d found in smectics.

The free energy per unit area $\mathcal{F}(D)$ of a stack of n homeotropic layers, constrained between parallel planes imposing a fixed order, has been calculated in [54, 55]:

$$\mathcal{F}(D) \propto \Psi_0 \xi_{\|} \left(\tanh \frac{D}{2\xi_{\|}} - 1 - \frac{1 - \cos \frac{2\pi(D-nd)}{d}}{\sinh(D/\xi_{\|})} \right) \tag{3.13}$$

where the separation D does not include the surface adsorbed layer, as it does not behave in the same way as the pre-smectic layers. The range of the pre-smectic force is defined by the correlation length $\xi_{\|}$. The cosine term in (3.13) describes the elastic response and the creation/removal of the layer by compression. The oscillations are not parabolic as in smectics, but sinusoidal. The minima are located at $D = nd$ and the maxima at $D = (n \pm \frac{1}{2})d$. The remaining terms in (3.13) represent a typical *attraction*, that is well known and expected for the symmetric cells with gradient of positional or orientational order [56]. These terms give the the baseline of the minima, whereas the envelope of the maxima, damped over a distance $\xi_{\|}$, is given by:

$$F_{max} \propto 2\pi R \psi_0 \xi \, \| \left(2\pi R \coth \frac{D}{2\xi_{\|}} \right) . \tag{3.14}$$

Figure 3.20 shows a pre-smectic fit based on (3.13). The fit is good for a number of layers with $n > 1$. For the first layer, a short-range micelle-mica interaction should be considered, that cannot be presented by the parameter ψ_0 introduced before.

It is interesting to study how the force profile changes by varying the temperature and approaching other phases. In the system considered in [54, 55], the micelles have a disk-like shape and the sample exhibits a second order smectic-A $\leftrightarrow$ discotic nematic bulk transition. By lowering the temperature, the oscillations become less damped and the attractive background weakens, finally leading to the smectic force profile described in Sect. 3.3.2.

The system of Fig. 3.20 has the following bulk phases sequence: hexatic $\leftrightarrow$ calamitic nematic $\leftrightarrow$ discotic nematic. In the calamitic nematic phase the micelles turn freely around their long axis, which is about 3/2 longer than the smallest one. Whereas the force profile remains essentially the same, the correlation length $\xi_{\|}$ decreases at the discotic $\rightarrow$ calamitic transition: micelles would freely rotate around their long axis, but they have to lay flat to the mica surfaces and maintain their long axis parallel to them. The resulting nematic order is discotic at the surfaces and possibly biaxial across the sample.

3.3.4 Force and Refractive Index in Thermotropic Nematics

Static force [45, 50, 51, 57] and friction properties [45, 46, 51, 57–59] have been extensively studied in thermotropic nematics. The surface anchoring conditions were symmetric, with both mica sheets inducing either planar or homeotropic alignment. Bare mica naturally induces planar anchoring on cyanobiphenyls such as 5CB: the liquid crystal optical axis orients at about $\pi/6$ with respect to the slow optical direction of a (birefringent) freshly cleaved mica sheet [60]. A homeotropic alignment can be obtained by adsorbing a mono-layer of surfactant [45, 50] on mica.

Layering oscillations are observed over a short-range extending up to five or six layers [45, 50]. In homeotropic 5CB and nematic 8CB, the layers are "smectic-A"-like, with molecules perpendicular to the layers. In 5CB the observed layer thickness is 2.5 nm, which roughly corresponds to the length of a molecule, while in 8CB, where the molecules arrange in dimers [61], the thickness corresponds to the length of a dimer 3 nm. The layer thickness in planar samples is 0.5 nm for 5CB and 1 nm for 8CB, corresponding to the diameter of the molecules and of dimers, respectively. This indicates an unusual "parallel" layering, with molecules laying flat in the layer plane.

Elastic forces in thin films of thermotropic nematics have been also investigated, but the results are not as clearly interpreted as for smectics. One can obtain twist deformation between bare mica sheets by twisting their crystallographic axes, or produce a splay/bend distortion by imposing asymmetric planar/homeotropic anchoring conditions on each surface. Even in the symmetric non-twisted planar and homeotropic cells, some director distortion is always expected, because of the curved confining geometry of the SFA [62], leading to some *repulsive* elastic force.

In fact, an elastic contribution to the force has been identified in twisted 5CB samples. The force is repulsive, increases with the imposed twist angle and disappears in the isotropic phase or after aligning the mica sheets, as expected by the Frank-Oseen elastic theory [50]. However, this force is superimposed to a strong monotonic *residual repulsion*, which is twist-independent and vanishes in the isotropic phase. This force is shown in Fig. 3.21a for a 5CB droplet between bare mica sheets. The sheets are aligned with a precision of 5°, as checked from the contact wavelengths before the liquid crystalline droplet was introduced. The uniform planar configuration is also confirmed by the index measurement presented in Fig. 3.21b, giving the correct birefringence of 5CB. The figure clearly shows that an error of the twist angle of

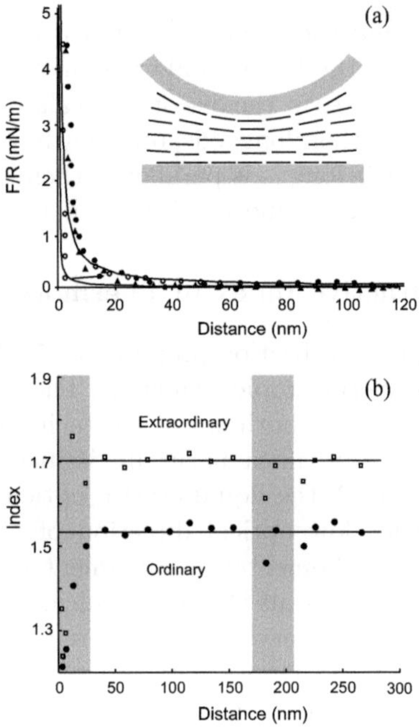

Fig. 3.21. Forces (a) and refractive indices (b) for a 5CB droplet between aligned bare mica sheets (T=27°C). (a) Force runs: approaching the surfaces (filled circles and triangles) a residual repulsion is observed; in retraction (open circles) a jump-out from the minimum of a layering oscillations is shown. The continuous lines represent the elastic twist force $F(D)/R = \pi K_{22}\theta^2/D$ (see [50] for details), calculated assuming an infinite anchoring strength with $K_{22} = 6.5 \times 10^{-12}$ N. The twist θ goes from 10° (lower line) to 60° (upper line). (b) Indexes: $n_e = 1.70 \pm 0.01$ (extraordinary), $n_o = 1.53 \pm 0.01$ (ordinary). Gray regions corresponds to separation ranges where the solutions of the interferometric equations are unstable.

5° cannot explain the residual repulsion, as the twist angle as large as 60° is needed for such a strong repulsion.

A similar repulsive force is also observed in homeotropic samples. For this case a detailed theoretical analysis has been done, considering the residual elastic contribution due to the curved confinement and using a molecular model to describe the layering [62]. The results show that the repulsion is non-elastic. For homeotropic samples it is probably related to a tilting reorientation of the molecules when a layer is removed or created in the cell.

The residual repulsion is strongly reduced in samples with hybrid homeotropic/planar anchoring conditions. Figure 3.22a shows the force curve obtained for a nearly apolar nematic droplet (ME10.5 by Merck) between bare mica (planar anchoring) and a mono-layer of surfactant (hexadecyl-tripropyl-ammonium bromide, homeotropic anchoring). Polar nematics as cyanobiphenyls quickly degrade the hybrid cell, presumably by desorbing

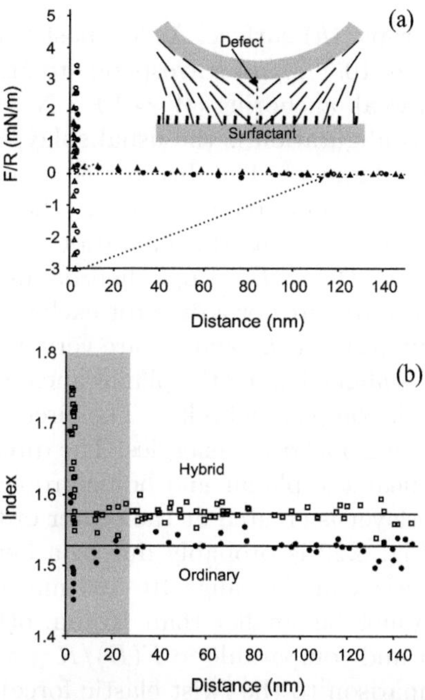

Fig. 3.22. Force (a) and refractive index (b) in a hybrid cell. (a) Forces: complete compression (filled symbols) and decompression (open symbols) cycles for two successive runs (circle and triangles). Notice the weak repulsion, the jump-to-contact and jump-out from the strong adhesion at contact. (b) Index: the birefringence is only a fraction of the full birefringence of the nematic and is constant. Gray regions correspond to instability in the interferometric equations.

the surfactant from the mica surfaces. The repulsion is weaker than in planar samples and for $D < 10$ nm it turns into an attraction, leading to a strong adhesion at contact. A weak planar layering is observed at very small distances (maximum two layers of thickness $0.5 \div 0.7$ nm). The smallest measured thickness ($2.5 \div 3$ nm, corresponding to the maximum compression), is half the value measured for homeotropic samples and is comparable to the thickness of the surfactant monolayer.

Fig. 3.22b shows the refractive index measurement. The birefringence is constant down to $D = 20$ nm and it is only a fraction of the full birefringence ($\delta n = 0.13 \pm 0.02$, as measured for planar samples).

The theory predicts three possible equilibrium configurations for a hybrid nematic between parallel boundaries: the *bent-director* structure, the *uniform-director* configuration and the *director-exchange* configuration [63–65]. These equilibrium structures and the corresponding structural transitions are determined by the cell thickness D and the two *extrapolation lengths* of the anchoring $L_{p,h} = K/W_{h,p}$, defined as the ratio between the mean splay/bend bulk elastic energy K and the anchoring strength $W_{p,h}$ for the planar (p) and homeotropic (h) surface. As for most nematics $K = 10^{-11}$ N, a strong anchoring ($W \sim 10^{-3}$ N/m) corresponds to an extrapolation length $L \sim 10$ nm, whereas a weak anchoring ($W \sim 10^{-5}$ N/m) to $L \sim 1$ µm.

The bent-director configuration is the usual splay/bend structure and is the most stable one for large D. The director is smoothly bent across the sample and is tilted from its preferred direction on the surfaces, due to the finite anchoring strength. By reducing the separation D, the tilt becomes more pronounced and for $D < D_c = |L_p - L_h|$ there is an *anchoring transition* [63], leading to a uniform-director or a director-exchange structure. The uniform configuration is preferred if L_p and L_h are very different. If for example $L_p \gg L_h$, the director is aligned along the planar anchoring direction, and the homeotropic anchoring is completely broken. The director-exchange is created by two equivalently strong anchoring energies. The director is uniformly planar and homeotropic near the planar and homeotropic surface respectively, whereas a biaxial thin layer is created at the center of the cell.

The repulsion in Fig. 3.22 is probably due to a bent director structure. The repulsion is very weak in the range 10÷100 nm, thus both surface extrapolation lengths cannot be smaller than 10 nm, otherwise the repulsion should be much larger and comparable to $F(D)/R = \pi K\theta^2/D$ with $\theta = \pi/2$ (see Fig. 3.21 for comparison to the twist elastic force). On the other hand, if the extrapolation lengths were very different, and the difference $|L_p - L_h|$ would be comparable to the range of the repulsion, we should see an increasing or a decreasing birefringence by approaching the anchoring transition. If the difference $|L_p - L_h|$ was larger than this range, we should be in the uniform configuration everywhere, with a zero or maximum birefringence for homeotropic or planar uniform configurations, respectively. These last two possibilities are incompatible with the constant hybrid birefringence of Fig. 3.22b.

Thus one may conclude that both surface extrapolation lengths are probably large ($L_{p,h}$ >100 nm) and similar ($|L_p - L_h| \sim 10$ nm). In the range $D = 10 \div 100$ nm, the director is highly tilted with respect to the anchoring direction on both surfaces, giving the intermediate birefringence of Fig. 3.22b. The force is weakly repulsive because the stress can be relaxed by both the splay/bend distortion and the surface deviation of the director. At a separation of about $|L_p - L_h| \sim 10$ nm there could be an anchoring transition to a uniform planar director structure. In fact, the anchoring on mica is expected to be stronger than the homeotropic one (crystalline surfaces usually induce strong anchoring [60]) and there is a planar (though poorly defined) layering at separation $D < 10$ nm. In addition, for surfaces enhancing the nematic order, the theory predicts a non-monotonic force in the uniform director configuration, with an attraction for small D [64, 65]. This could explain the strong attraction for $D < 10$ nm, together with a weak van der Waals force and possibly attractive fluctuation force [64,65].

We stress however that this deduction is based on the elastic theory for the nematics, that fails to explain the forces in planar samples. Moreover, the discussion does not take into account the specific curved geometry of the SFA, which is not compatible with the hybrid anchoring conditions. In particular, even by considering a finite anchoring strength, a line defect is expected at the center of the cell (as sketched tentatively in Fig. 3.22), that has not been observed yet.

The elastic theory has been derived for a macroscopic scale, larger than the coherence length ξ_N of the nematic order (typically ∼5 nm, several times larger than the molecule length). One may ask whether this theory is suited to explain SFA experiments, typically dealing with a 10 nm thickness.

Ellipsometry experiments have been performed at a comparable length scale for small nematic droplets on substrates inducing planar anchoring, the air/nematic free interface giving a homeotropic anchoring [67]. The predicted bent-to-uniform director transition is observed at a critical drop thickness of about 30 nm. The microscopic details of the molecule/molecule and molecule/surface interaction become relevant only for thickness < 1 nm.

Optical second-harmonic generation experiments give a more detailed description of the anchoring at a microscopic scale [68,69; see also Chapter 5]. The molecule/surface interaction determines the orientational distribution in a thin surface layer extending up to 1 nm. The bulk uniaxial order develops on top of this layer via a transition layer of thickness $\sim \xi_N$, which is well described by the usual mean-field theory, possibly including non-uniaxial components of the tensor order parameter.

Moreover, the structural transitions in a hybrid geometry predicted by the mean-field theory [64] have been recently confirmed by Monte-Carlo molecular simulations [70] for a confinement thickness of about 20 nm.

In conclusions, the elastic theory is adequate as far as the thickness exceeds the coherence length ξ_N. In future SFA experiments at a smaller length scale (down to about 1 nm) a more complete mean-field description of the spatial variations of the tensor order has to be developed in the framework of the mean-field theory.

References

1. J. N. Israelachvili, D. Tabor, Proc. R. Soc. London, **A331** (1972) 19.
2. J. N. Israelachvili, *Intermolecular and Surface Force*, Academic Press (1985).
3. G. Bining, C. F. Quate, C. Gerber, Phys. Rev. Lett. **56**, 930 (1986).
4. I. Muševič, G. Slak, R. Blinc, Rev. Sci. Instr. **67**, 2554 (1996).
5. F. J. Kahn, G. N. Taylor, H. Schonhorn, Proc. IEEE **61**, 823 (1973).
6. J. Y. Huang, R. Superfine, Y. R. Shen, Phys. Rev. A **42**, 3660 (1990).
7. G. P. Crawford, R. J. Ondris-Crawford, J. W. Doane, S. Žumer, Phys. Rev. E **53**, 3647 (1996).
8. K. Kočevar, I. Muševič, Liq. Cryst. **28**, 599 (2001).
9. K. Kočevar, R. Blinc, I. Muševič, Phys. Rev. E **62**, R3055 (2000).

10. K. L. Johnson, *Contact Mechanics*, Cambridge University Press (1985).
11. J. Yamamoto, K. Okano, Jap. J. Appl. Phys. **30**, 754 (1991).
12. A. Borštnik, H. Stark, S. Žumer, Phys. Rev. E **60**, 4210 (1999).
13. K. Kočevar, I. Muševič, Phys. Rev. E **64**, 051711 (2001).
14. A. Šarlah, S. Žumer, Phys. Rev. E **64**, 051606 (2001).
15. W. T. Thomson (Lord Kelvin), Phil. Mag. **42**, 448 (1871).
16. A. W. Adamson, *Physical Chemistry of Surfaces*, John Wiley and Sons, Inc. (1990).
17. P. Sheng, Phys. Rev. Lett. **37**, 1059 (1976).
18. M. M. Wittebrood, T. Rasing, S. Stallinga, I. Muševič, Phys. Rev. Lett. **80**, 1998 (1998).
19. M. Wittebrood, *Phase Transitions and Dynamics in Confined Nematic Liquid Crystals*, PhD thesis, University of Nijmegen (1997).
20. K. Kočevar, A. Borštnik, I. Muševič, S. Žumer, Phys. Rev. Lett. **86**, 5914 (2001).
21. A. Borštnik Bračič, K. Kočevar, I. Muševič, S. Žumer, Phys. Rev. E **68**, 011708 (2003).
22. P. G. de Gennes, Langmuir **6**, 1448 (1990).
23. K. Kočevar, I. Muševič, Phys. Rev. E **65**, 021703 (2002).
24. P. Ziherl, Phys. Rev. E **61**, 4636 (2000).
25. D. Davidov, C. R. Safinya, M. Kaplan, S. S. Dana, R. Shaetzing, R. J. Birgenau, J. D. Lister, Phys. Rev. B **19**, 165 (1979).
26. P. G. de Gennes, *The physics of liquid crystals*, Clarendon Press, Oxford (1974).
27. B. Cappella and G.Dietler, Sur. Sci. Rep **34**, 1 (1999).
28. K.Kočevar and I.Muševič, Phys. Rev. E **65**, 21703 (2002).
29. M.Zgonik, M.Rey-Lafon, C. Destrade, C. Leon and H.T. Nguyen, J.Phys.France, **51**, 2015 (1990).
30. Seiji Shibahara, Jun Yamamoto, Yoichi Takanishi, Ken Ishikawa, Hideo Takezoe and Hajime Tanaka, Phys. Rev. Lett.,**85**, 1670 (200).
31. P. Richetti, P. Kékicheff and P. Barois, Phys. Rev. Lett., **73**, 3556 (1994).
32. L. Moreau, P. Richetti and P. Barois, J.Phys. II France, **5** 1129 (1995).
33. P. M. Johnson, D. A. Olson, S. Pankratz, Ch. Bahr, J. W. Goodby, and C. C. Huang, Phys. Rev. E, **62**, 8106 (2000).
34. A. Fera, R. Opitz, W. H. de Jeu, B. I. Ostrovskii, D. Schlauf, and Ch. Bahr, Phys. Rev. E, **64**, 021702 (2001).
35. J. N. Israelachvili and G.E. Adams, J.Chem.Soc., Faraday Trans I, **74**, 975 (1978).
36. L. Xu, M. Salmeron and S. Bardon, Phys. Rev. Lett., **84**, 1519 (2000).
37. M. Benzekri, T. Claverie, J. P. Marcerou and J. C. Rouillon, Phys. Rev. Lett., **68**, 2480 (1992).
38. A. Ajdari, L. Peliti, J. Prost, Phys. Rev. Lett. **66** 1481(1991).
39. J. N. Israelachvili, J. Coll. Interf. Sci., **44** 259 (1973).
40. P. Rabinowitz, J. Opt. Soc. Am. A, **12** 1593 (1995).
41. J. N. Israelachvili, P. M. McGuiggan J. Mater. Res., **5**, 2223 (1990).
42. Ph. Richetti, P. Kékicheff, J. L. Parker, B. W. Ninham, Nature, **346** 252 (1990).
43. Ph. Richetti, P. Kékicheff, Ph. Barois, J. Phys. France, **5** 1129 (1995).
44. S. H. Idziak, I. Koltover, J. N. Israelachvili, C. R. Safinya, Phys. Rev. Lett., **76** 18477 (1996).

45. M. Ruths, S. Steinberg, J. Israelachvili, Langmuir, **12** 6637 (1996).
46. A. Artsyukhovich, L. D. Broekman, M. Salmeron, Langmuir, **15** 2217 (1999).
47. R. Bartolino, G. Durand, Mol. Cryst. Liq. Cryst., **40** 117 (1977).
48. R. Bartolino, G. Durand, Phys. Rev. Lett., **39** 1346 (1977).
49. P. S. Pershan, J. Prost, J. Appl. Phys., **46** 2343 (1975).
50. R. G. Horn, J. N. Israelachvili, E. Perez, J. Physique, **42** 39 (1981).
51. M. Ruths, S. Granick, Langmuir, **16** 8368 (2000).
52. R. G. Horn, J. N. Israelachvili, J. Chem. Phys., **75** 1400 (1981).
53. E. A. Oliveira, L. Liebert, A. M. Figueiredo Neto, Liquid Crystals, **5** 1669 (1989).
54. L. Moreau, Ph. Richetti, Ph. Barois, Phys. Rev. Lett., **73** 3556 (1994).
55. L. Moreau, Ph. Richetti, Ph. Barois, P. Kékicheff, Phys. Rev. E., **54** 1749 (1996).
56. S. Marcelja, N. Radic, Chem. Phys. Lett., **42** 129 (1976).
57. V. Kitaev, E. Kumacheva, J. Phys. Chem. B, **104** 8822 (2000).
58. J.Janik, R. Tadmor, J. Klein, **13** 4466 (1997).
59. J. Janik, R. Tadmor, J. Klein, Langmuir, **17** 5476 (2001).
60. B. Jerôme, Rep. Prog. Phys., **54** 391 (1991).
61. A. J. Leadbetter, R. M. Richardson, C. N. Colling, J. Physique Colloques, **36** C1 (1975).
62. A. M. Sonnet, T. Gruhn, J. Phys. Condens. Matt., **11** 8005 (1999).
63. G. Barbero, R. Barberi, J. Phys. (Paris), **44** 609 (1983).
64. A. Šarlah, S. Žumer, Phys. Rev. E., **60** 1821 (1999).
65. A. Šarlah, P. Ziherl, S. Žumer, Mol. Cryst. Liq. Cryst., **364** 443 (2001).
66. F. Vanderbrouck, M. P. Valignat, A. M. Cazabat, Phys. Rev. Lett., **82** 2693 (1999).
67. M. P. Valignat, S. Villette, J. Li, R. Barberi, R. Bartolino, E. Dubois-Violette, and A. M. Cazabat, Phys. Rev. Lett., **77** 1994 (1996).
68. B. Jérôme, P. Shen, Phys. Rev. E, **48** 4556 (1993).
69. B. Jérôme, J. Phys. Condens. Matt., **6** A269 (1994).
70. C. Chiccoli, P. Pasini, A. Šarlah, C. Zannoni, S. Žumer, Phys. Rev. E, **67** 050703 (2003).

4 Linear Optics of Liquid Crystal Interfaces

Igor Muševič

Due to their orientational order and strong optical anisotropy, liquid crystalline phases exhibit many fascinating and attractive linear optical properties. Combining this with the fact that the responsible liquid crystalline structure can be easily manipulated by a combination of orienting surfaces and relatively low electric fields, has led to the successful development of many electrooptical devices. Among them, devices based on a twisted nematic phase have been most widely and successfully used so far. Whereas in the past research was most focused on the the macroscopic quality and "cosmetic" appearance of the surface alignment in liquid crystal displays, we are nowadays more interested in the microscopic mechanisms and surface coupling strengths for the potential future use in more complex systems.

In these electrooptic LC devices, the confining surfaces have to induce a pre-determined and well defined direction of the optical axis at the interface, that subsequently propagates into the bulk due to the long-range nature of the orientational order in liquid crystals. The solid-liquid crystal interface itself can therefore be considered as a discontinuity of the elastic and optical properties of the system. From this point of view, the interface is "visible" to those linear optical techniques that are enough sensitive to detect a discontinuity of the material properties across the interface.

Among them, the most simple method is based on the reflection of an electromagnetic wave at the interface [1, 2]. The solid-liquid crystal interface represents a discontinuity of the dielectric tensor, that gives rise to the reflected wave. The polarization state of the reflected wave depends on the thickness, structure and optical properties of the interfacial region. Being usually much smaller than the wavelength of the light wave used, reflection ellipsometry cannot discriminate details in the ordering profile across the interface, but is only sensitive to an "integral" of the optical parameters across the interface. This integral parameter is called *the surface adsorption parameter* and is also detectable in NMR experiments (see Chap. 2). In the past, several different reflection ellipsometry techniques have been used:

(i) evanescent wave ellipsometry (also Total Internal Reflection Ellipsometry-TIRE) [3–6] where the evanescent wave in a total reflection geometry is used to probe the optical anisotropy of a thin liquid crystal layer at the solid surface;

(ii) Brewster angle reflection ellipsometry (BAE), where a reflection from the interfacial region is analyzed at the Brewster angle [1, 7–9].

(iii) other reflection ellipsometry techniques [10–15]. These reflectivity methods can discriminate optical changes on the scale of a single molecular layer in the direction, perpendicular to the surface. The resolution of both above mentioned methods is particularly enhanced near the isotropic-nematic transition, where a thin pretransitional layer is formed at the surface. The properties of this wetting layer depend strongly on the surface coupling parameters and these can be directly determined using TIRE and BAE.

Thin surface-induced ordered liquid crystalline layers can also be observed by measuring their birefringence, which is a sort of ellipsometry method, used to analyze the polarization properties of directly transmitted waves. The method has been used in the pioneering linear optics experiments at solid-liquid crystal interfaces [16–24] and later in [25, 26].

The interface between the liquid crystal and the solid aligning wall can also be probed either directly or indirectly by Dynamic Light Scattering (DLS). Direct measuring the dynamics of the thin interfacial region was proposed using DLS in the evanescent light wave of a TIRE experiment [27]. This method suffers from a low signal-to-noise ratio because of the dominating static, total internal reflected wave. Another approach is indirect and involves DLS measurements of the *surface extrapolation length* at a particular interface. That method is based on the measurement of the relaxation time of a particular thermally excited wave in a thin layer of a liquid crystal. As the wave is "pinned" at confining surfaces by boundary conditions, only discrete values of the relaxation times are expected. These relaxation times correspond, for example, to waves that have their wavelength equal to one half of the layer thickness, two halfs, etc. Due to the finite surface coupling energy, the wavelength of the excited wave is slightly larger than the thickness of the layer and so is the relaxation rate slightly larger compared to a perfect pinning at the surface. By measuring the thickness dependence of the relaxation rates [28–31], the surface coupling energy can be measured with great accuracy [32], which makes DLS a particularly powerful and important technique. The only drawback of this method is, that the thickness of the cell should be known to a precision, that is equal to or better than the surface extrapolation length in question.

This chapter will presents the principles of DLS and BAE , their advantages and drawbacks, details of their measuring techniques, and results, related to particular interfaces.

4.1 Brewster Angle Ellipsometry of Isotropic Nematic and Smectic Interfaces

Klemen Kočevar, Irena Drevenšek Olenik and Igor Muševič

The state of polarization of light, reflected from an interface, depends strongly on the profile of the dielectric constant across that interface. This simple principle is used in the Brewster angle reflection ellipsometry (BAE), where one measures the ellipticity coefficient of light, reflected from an interface. The method is sensitive enough to detect extremely small changes in the structure of liquid crystalline-solid interfaces. Subnanometer resolution of the adsorption parameter is routinely achieved. The method is therefore very useful for the study of liquid crystal interfaces, where the surface-induced variation of the order can be observed [5, 25, 33–41].

In the first two Sections, the basics of the Brewster angle ellipsometry and the operation principles of the photoelastic modulator (PEM) based ellipsometer will be overviewed. In the following, three different ellipsometric studies of the solid-liquid crystal interface will be presented.

4.1.1 Brewster Angle Ellipsometry of the Isotropic Nematic-glass Interface

Let us first consider an oblique reflection of a light wave at a planar interface between two optically isotropic media. For a given amplitude and polarization of the incident wave, the reflected amplitude and polarization are determined from the continuity relation for the **E** and **H** fields at the interface. This leads to the complex Fresnel's equations [1] for the complex reflectivities of the p-polarized $r_p = E_{r,p}/E_{i,p}$, and s-polarized waves, $r_s = E_{r,s}/E_{i,s}$. The same applies to the transmitted waves. Here, i and r denote incident and reflected light respectively.

After reflection at the interface, the reflected wave is in general elliptically polarized. The shape and orientation of the ellipse depend on the incident polarization, incident angle and reflection properties of the interface, and can be described with the ratio ρ of the complex Fresnel's reflection coefficients for the p and s polarizations:

$$\rho = r_p/r_s \,. \tag{4.1}$$

This complex ratio, also called the *ellipsometric coefficient*, is often conveniently written as

$$\rho = \tan\Psi \cdot e^{i\Delta} \tag{4.2}$$

where $\tan\Psi = |r_p/r_s|$ determines the angle of the rotation of the ellipse. $\Delta = \delta_r - \delta_p$ is the phase difference between p-polarized and s-polarized waves after reflection and therefore describes the ellipticity of the reflected wave.

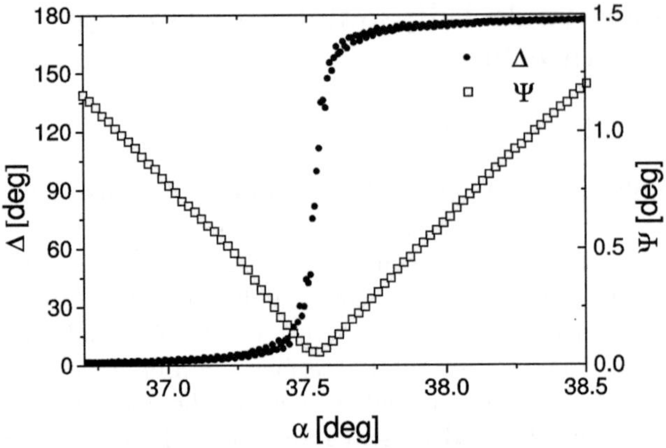

Fig. 4.1. The angular dependence of Ψ (open squares) and Δ (filled circles) around the Brewster angle for water-glass interface. The phase Δ changes for 180° and Ψ reaches minimum at the Brewster angle.

When a p-polarized wave (polarization parallel to the incident plane) is reflected from a perfectly sharp interface, the reflected wave disappears for a particular angle of incidence, which is called the *Brewster angle*, and the wave is *totally refracted* into the second medium. At the same time, the phase of the reflected wave changes over π when the angle of the incident light crosses the Brewster angle. On the other hand, the reflectivity of the s-polarized wave behaves monotonically and the phase does not change with the angle of incidence. Figure 4.1 shows a typical angular dependence of Ψ and Δ around the Brewster angle for an interface between water and glass.

If the interface is not sharp, but is characterized by a certain spatial profile of the dielectric constant of a given medium, $\epsilon(z)$, the ellipticity coefficient ρ is finite at the Brewster angle. In this case, the Brewster angle is defined as the angle, where the phase shift of the p-polarized wave is $\pi/2$. The value of the ellipticity coefficient at the Brewster angle therefore carries information about the sharpness of the interface. As we can measure this value very precisely, we can determine the parameters of the spatial profile of the dielectric constant with great precision.

When we have an interface between a liquid crystal in the isotropic phase and an optically isotropic substrate (glass, air, etc..), the ellipticity coefficient of this interface will be very small, as the surface-induced order is small. By lowering the temperature toward the nematic phase, the magnitude and the range of the surface-induced pre-nematic phase increase, which lowers the sharpness of the interface. As a result, the ellipticity coefficient is expected to increase with decreasing temperature. It can be shown,that one can determine the values of the surface nematic order parameter S_0 and the nematic correlation length with superior accuracy using Brewster-mode ellipsometry [9], measuring the ellipticity coefficient as a function of temperature.

In this case, the optics of the interface is slightly complicated in view of the birefringence of the surface induced pre-nematic phase. The interface is now consisting of (i)the isotropic bulk liquid crystal, (ii) the surface induced, weakly optically uniaxial layer with a thickness of the nematic correlation length and with the optical axis perpendicular to the surface, and, (iii) an isotropic substrate (such as glass). If the optical anisotropy and the thickness of the anisotropic layer are small, the ellipticity coefficient at the Brewster angle, ρ_B, can be calculated in the Drude approximation [1]

$$\rho_B = \frac{\pi}{\lambda}\frac{\sqrt{n_{glass}^2 + n_{iso}^2}}{n_{glass}^2 - n_{iso}^2}\int_{-\infty}^{\infty} dz \left[n_{glass}^2 + n_{iso}^2 - \frac{n_{glass}^2 n_{iso}^2}{\epsilon_3(z)} - \epsilon_1(z)\right] . \quad (4.3)$$

Here, λ is the vacuum-wavelength of the reflected light, n_{iso} is the index of refraction of the liquid crystal in the isotropic phase, n_{glass} is the refractive index of the substrate, $\epsilon_3(z)$ is the component of the dielectric tensor of the liquid crystal along the normal to the interface and $\epsilon_1(z)$ is the corresponding component in a transverse direction. The value of the ρ_B is therefore connected to the thickness, optical anisotropy and z direction variation of refractive index of the interfacial layer.

The components of the dielectric tensor of the liquid crystal depend linearly on the order parameter [42]. This can be written as

$$\begin{aligned} \epsilon_1(z) &= n_{iso}^2 + \frac{1}{3}\Delta\epsilon S(z) \\ \epsilon_3(z) &= n_{iso}^2 - \frac{2}{3}\Delta\epsilon S(z) \end{aligned} \quad (4.4)$$

where $\Delta\epsilon$ is the maximum anisotropy of the dielectric constant for optical frequencies at a saturated nematic order parameter, $S = 1$. In the case of 5CB $\Delta\epsilon = 0.6$.

By considering a simple spatial dependence of the order parameter, i.e. $S(z) = S_0 \exp(-z/\xi)$, the integral in (4.3) can be calculated and the ellipticity coefficient for the surface-induced layer of a pre-nematic liquid crystal on an isotropic substrate is

$$\rho_B = \frac{\pi}{\lambda}\frac{\sqrt{n_{glass}^2 + n_{iso}^2}}{n_{glass}^2 - n_{iso}^2}\xi_N \left[n_{glass}^2 \ln\left(1 - \frac{2\Delta\epsilon S_0}{3n_{iso}^2}\right) - \frac{1}{3}\Delta\epsilon S_0\right] . \quad (4.5)$$

In general, the ellipticity coefficient is temperature dependent because of the temperature dependence of the correlation length $\xi(T)$ and the surface order parameter $S_0(T)$. It increases by approaching the isotropic-nematic phase transition from above. By measuring $\rho_B(T)$ one can therefore directly determine the product $\xi(T)S_0(T)$, and, if we assume a power law dependence of $\xi(T)$, the temperature dependence of the nematic order parameter at the surface $S_0(T)$ can be extracted.

In order to obtain the values of the surface coupling energies from the temperature dependence of S_0, we have to calculate the value of S_0 for a semi-infinite sample, where the interface is located in the xy plane at $z = 0$. The total free energy per unit surface area is $F = F_B + F_S$, where $F_B = \int_0^\infty f_B(z)\mathrm{d}z$ and $F_S = w_1 S_0 + \frac{1}{2} w_2 S_0^2$. After minimization with respect to S_0, the surface value of the order parameter is

$$S_0 = \frac{w_1}{L/\xi + w_2} . \tag{4.6}$$

4.1.2 Photoelastic Modulator Based Ellipsometer

A classical ellipsometer, where the analyzer has to be rotated to record the whole ellipse of the reflected light wave, can be improved by using a photoelastic modulator (PEM) [43]. In this case, the polarizer and analyzer are crossed in a fixed position and a photoelastic modulator is inserted behind the polarizer as shown in Fig. 4.2. The measurements with a photoelastic modulator are faster whereas the sensitivity is slightly worse compared to rotating analyser method.

The photoelastic modulator is an optical element, which periodically modulates the polarization of the light. It has to be aligned carefully, so that one of its optical axis is parallel to the direction of s polarization. To demonstrate

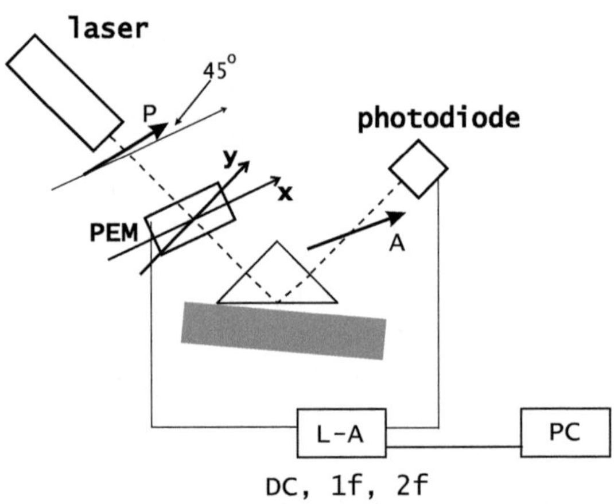

Fig. 4.2. The schematic experimental setup for reflection ellipsometry. The light from the laser source is guided through the polarizer (P) and PEM with its optical axis at an angle of 45° with respect to the direction of the polarization. The light beam is reflected from the lower side of the prism and is detected after passing through the analyzer, which is crossed with respect to the analyzer. DC, 1st and 2nd harmonic of the modulated light intensity are simultaneously measured, using the lock-in amplifier and computer for data acquisition. The prism has been used to eliminate parasite reflection from the air-glass interface. The x axis is parallel to the s polarization of light.

the action of a PEM, we will follow the polarization of light in the ellipsometer using the Jones formalism. In the coordinate system, where the x axis is parallel to the s polarized wave, the polarization state after the light passes the polarizer can be written as

$$\begin{pmatrix} E_x \\ E_y \end{pmatrix} = \frac{E_0}{\sqrt{2}} \begin{pmatrix} 1 \\ 1 \end{pmatrix} . \tag{4.7}$$

A photoelastic modulator delays the phase of one polarization with respect to the other with a time-dependent phase factor $\phi(t)$. Here $\phi(t) = A_0 \cos \Omega t$, where Ω is the frequency of the photoelastic modulator and A_0 is the amplitude of the phase modulation. After passing the PEM, the polarization of light is modulated in time,

$$\begin{pmatrix} E_x \\ E_y \end{pmatrix} = \frac{E_0}{\sqrt{2}} \begin{pmatrix} e^{\mathrm{i}\phi(t)/2} \\ e^{-\mathrm{i}\phi(t)/2} \end{pmatrix} . \tag{4.8}$$

The surface reflects both polarizations differently and typically delays the phase of one polarization with respect to the other, so that the reflectivities r_s and r_p are generally complex numbers. The state of polarization becomes

$$\begin{pmatrix} E_x \\ E_y \end{pmatrix} = \frac{E_0}{\sqrt{2}} \begin{pmatrix} r_s e^{\mathrm{i}\phi(t)/2} \\ r_p e^{-\mathrm{i}\phi(t)/2} \end{pmatrix} . \tag{4.9}$$

The analyzer projects the polarization to a $(1, -1)$ direction and finally the state of polarization in front of a detector is

$$\begin{pmatrix} E_x \\ E_y \end{pmatrix} = \frac{E_0}{2} (r_s e^{\mathrm{i}\phi(t)/2} - r_p e^{-\mathrm{i}\phi(t)/2}) \begin{pmatrix} 1 \\ -1 \end{pmatrix} . \tag{4.10}$$

The intensity detected by the detector is

$$I = \frac{1}{2} \boldsymbol{E}^* \cdot \boldsymbol{E} = \frac{E_0^2}{8} [r_s^* r_s + r_p^* r_p - (r_s^* r_p e^{-\mathrm{i}\phi(t)} + r_s r_p^* e^{\mathrm{i}\phi(t)})] . \tag{4.11}$$

Here the factor 1/2 is due to the temporal averaging by the detector over the oscillations of the electric field of the light. If the reflectivity ratio is $r_p/r_s = \chi e^{\mathrm{i}\delta}$ and r_s is assumed to be real (equivalent phase factors can be extracted from both polarizations without having an effect on the final intensity), the intensity is

$$I = \frac{E_0^2 r_s^2}{8} [1 + \chi^2 - 2\chi \cos(\phi(t) - \delta)] . \tag{4.12}$$

The term $\cos(A_0 \cos \Omega t)$ can be expanded as a series in $\cos n\Omega t$, where n is an integer and the factors are Bessel functions $J_n(A_0)$. Finally the intensity from (4.12) can be separated into DC and harmonic terms

$$
\begin{aligned}
I_{DC} &= \frac{E_0^2 r_s^2}{8}(\chi^2 + 1 + 2\chi J_0(A_0)\cos\delta) \\
I_{\Omega} &= \frac{E_0^2 r_s^2}{2}\chi J_1(A_0)\sin\delta \\
I_{2\Omega} &= \frac{E_0^2 r_s^2}{2}\chi J_2(A_0)\cos\delta\,.
\end{aligned}
\tag{4.13}
$$

The amplitude of the modulation is chosen so that $J_0(A_0) = 0$, which is at $A_0 \approx 0.383\lambda$, where λ is the vacuum wavelength of the light. In this case, the reflectivity coefficient r_p/r_s, often referred to as *ellipticity coefficient*, can be extracted from the measured data. Writing $\chi = \tan\Psi$ the following relations can be analytically expressed

$$
\begin{aligned}
\tan\delta &= \frac{I_{\Omega} J_2(A_0)}{I_{2\Omega} J_1(A_0)} \\
\sin 2\Psi &= \frac{I_{\Omega}}{2 I_{DC} J_1(A_0)\sin\delta}
\end{aligned}
\tag{4.14}
$$

from which the magnitude $\tan\Psi$, and the phase Δ of the ellipticity coefficient can be determined, if DC, 1st and 2nd harmonic terms are simultaneously measured.

4.1.3 Ellipsometry of the Glass-Isotropic Nematic Liquid Crystal Interface

An isotropic liquid crystal is partially ordered close to the orienting solid surface. The surface induced perturbation of the order in an isotropic phase is localized in a thin layer with the thickness of the nematic correlation length ξ_N. Since the values of the nematic correlation length in the isotropic phase are typically only a few nanometers, ellipsometry represents a perfect tool for the study of interfacial order.

For the ellipsometric study of the solid-isotropic liquid crystal interface, we have prepared solid surfaces for homeotropic alignment. This has been achieved by coating a pre-cleaned glass prism with $N,N-dimethyl-N-octadecyl-3-aminopropyltrimethoxysilyl\,chloride(DMOAP)$ silane. The prism was positioned on the glass slide to form a wedge cell as shown in Fig. 4.2. In an experimental run, the liquid crystal was cooled down from the isotropic to the nematic phase. Simultaneously, the ellipsometer was continuously scanning very close to the Brewster angle and DC, 1st and 2nd harmonic signals were measured. Following (4.14), $\rho_B = \tan\Psi(\Theta_B)$, the value of Ψ measured at the Brewster angle Θ_B.

A typical measurement with the nematic liquid crystal $4'-n-octyl-4-cyanobiphenyl$ (8CB) is presented in Fig. 4.3. It is obvious that the thickness of the surface induced ordered region and the amount of order increase close to the phase transition. This increase is due to the growing nematic correlation

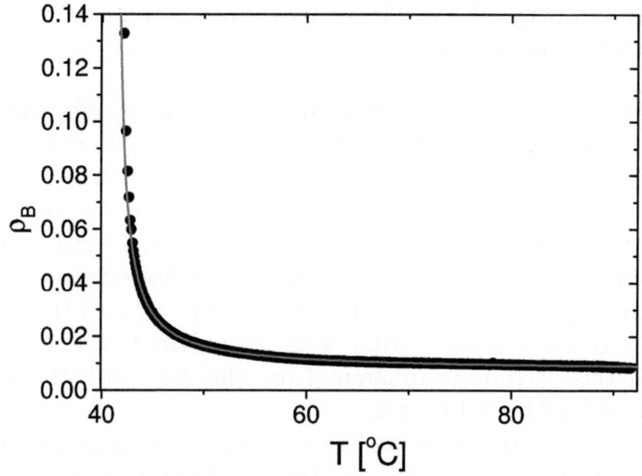

Fig. 4.3. The temperature dependence of the ellipsometric coefficient at the Brewster angle ρ_B for $4' - n -$ *octyl* $- 4 -$ *cyanobiphenyl* (8CB). Phase transition temperature T_{NI} is at 315.3 K. The solid line is a fit to (4.5) and (4.6) with $w_1 = 1.4 \times 10^{-4}\,\mathrm{Jm^{-2}}$ and $w_2 = 5.6 \times 10^{-5}\,\mathrm{Jm^{-2}}$.

length ξ_N and the increased amount of order in the surface perturbed region. The fit to the model of (4.5) and (4.6) describes the measured increase in ρ_B very well and it is possible to determine the phenomenological model parameters of the anchoring energy w_1 and w_2, since the value of the nematic correlation length is known from various measurements. The values for 8CB on various silanated substrates are collected in Table 4.1. From these values it can be seen how the cleaning of the glass surface with oxygen radio-frequency plasma prior to silanating increases the strength of the surface ordering.

Table 4.1. The surface coupling ordering (w_1) and disordering (w_2) energies for 8CB on DMOAP coated LaSF and BK7 glass, determined from the Brewster angle ellipsometry.

substrate	w_1	w_2
BK7	$1.4 \times 10^{-4}\,\mathrm{J/m^2}$	$5.6 \times 10^{-5}\,\mathrm{J/m^2}$
LaSF	$1.5 \times 10^{-4}\,\mathrm{J/m^2}$	$6.0 \times 10^{-5}\,\mathrm{J/m^2}$
plasma treated BK7	$3.9 \times 10^{-4}\,\mathrm{J/m^2}$	$5.6 \times 10^{-4}\,\mathrm{J/m^2}$

4.1.4 Ellipsometry of the Glass-Isotropic Smectic Liquid Crystal Interface

As discussed above, ellipsometry is directly sensitive only to the interfacial variations of the nematic order parameter, which is connected to the optical refraction indices. The interfacial smectic order, which has no direct influence on the optical properties, can only be observed due to its coupling to the nematic (orientational) order. The same experimental setup as described in Sect. 4.1.3 has been used to study the interface between smectic liquid crystal dodecylcyanobiphenyl (12CB) in the isotropic phase and the silanated glass. Although only orientational order is observed, the temperature dependence of ρ_B is in this case quite different from the case with the nematic liquid crystal, as evident from Fig. 4.4.

By lowering the temperature, ρ_B increases in steps, which is a clear sign of a layering transition, i.e. the spontaneous formation of a discrete number of smectic layers. The layering transition has already been observed on solid-liquid crystal interfaces in X-ray experiments by Ocko [44] and in X-ray as well as ellipsometry experiments on free surfaces (i.e. the interface with air) [5, 37].

The smectic layering transition [45] is a phenomenon, where smectic layers grow one by one, parallel to the surface, when the temperature approaches the isotropic-smectic phase transition from above. Ellipsometric observation of these steps is possible due to the coupling between the smectic order and the nematic (orientational) ordering. Consequently, the orientational order in

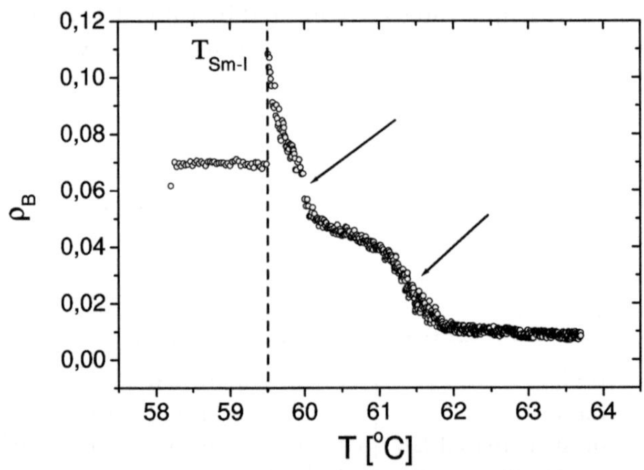

Fig. 4.4. Temperature dependence of ρ_B for 12CB on silanated BK7 glass surface. Two layering transitions of equal height can be identified above the T_{Sm-I}. The first is gradual and occurs between 2.2 and 1.6 K above the phase transition and the second is more discontinuous and occurs ≈ 0.22 K above the phase transition temperature. For comparison see NMR results in Fig. 2.6.

a developed smectic phase is considerably higher than in the rest of the liquid crystal and therefore the value of the ellipsometric coefficient ρ increases for a certain value each time a new smectic layer is formed.

In our measurement, two pronounced steps in ρ_B are observed. The steps are of equal heights, which suggests that layers of similar thicknesses and optical properties are deposited successively at the solid-isotropic liquid crystal interface. This resembles the layer-by-layer growth of smectic layers at the interface. The measurement suggests that the first layer grows onto the silanated surface gradually, since the increase of the ellipsometric ratio with decreasing temperature is relatively smooth. The ellipsometric ratio starts to grow 2.2 K above the T_{Sm-I} and first saturates at 1.6 K above the phase transition temperature. Since locally the layer growth is expected to occur instantaneously, as predicted theoretically [46, 47], this slow growth seems to be a consequence of the non-uniform surface. Once the surface is completely covered with a single smectic layer, the surface becomes uniform and one would expect that the successive layer would nucleate at once. This is in fact observed in our measurement and the nucleation of the next layer causes a step-like increase of the ellipsometric ratio of equal height $\Delta\rho_B \approx 0.03$ at $T = T_{Sm-I} + 0.22\,\mathrm{K}$. Next to this, the ellipsometric ratio increases toward the phase transition and then settles at a constant value in the smectic phase.

Using the ellipsometric measurement, the value of the nematic order inside the surface nucleated smectic layer can be estimated. Knowing the thickness of the 12CB smectic layer and assuming constant dielectric properties within a smectic layer, the jump in the ellipsometric ratio ρ_B can be simply expressed using (4.5) and (4.4). Since both steps are of equivalent height of $\Delta\rho_B \approx 0.03$, both layers have the same dielectric anisotropy of $\delta\epsilon = 0.043(1 \pm 0.01)$. The corresponding nematic order parameter S can be calculated, using the relation $\delta\epsilon = \Delta\epsilon S$, where $\Delta\epsilon$ is the dielectric anisotropy in the case of a perfect orientational order, i.e. $S = 1$. This value is not known, but can be approximated from the values for 5CB and 8CB, which are 0.587 and 0.485 respectively. Using the linear extrapolation in the molecular length, the maximum dielectric anisotropy of 12CB is estimated to be $\Delta\epsilon \approx 0.35$. From this, we conclude that the nematic order parameter within the layer is $S \approx 0.12$, which seems to be high enough to drive the layering transition according to Somoza [47].

4.1.5 Evaporation of 8CB on PVCN Coated Glass Substrate

Brewster angle ellipsometry (BAE) and surface optical second harmonic generation (SHG, see Chap. 5) were used to study the growth of 8CB films, evaporated in air onto glass (BK7) substrates, covered with a 15 nm thick film of poly(vinyl cinnamate) (PVCN) [48]. As the thickness of 8CB on PVCN layers was far below the optical wavelengths, the Drude formula for the ellipsometric coefficient at the Brewster angle, ρ_B, (4.3), was used. The ellipticity coefficient of the 8CB adsorbate was calculated as

$$\rho_m = \rho_a - \rho_0 \tag{4.15}$$

where ρ_a is the measured signal from the substrate covered with PVCN and 8CB adsorbate and ρ_0 is the initial signal from the substrate with PVCN only.

The results of BAE measurements (Fig. 4.5) suggest that the growth of the first layer takes place in the time regions A and B and the growth of the second layer (bilayer) in the time region C.

The observed stabilization of the BAE signal (see time region D of Fig. 4.5) indicates that a continuous layer-by-layer growth of the 8CB film terminates after some initial stages of adsorption. Detection of droplets by optical microscopy suggests that any further evaporation leads to dewetting of the bulk liquid crystalline phase. This is in agreement with the results of drop spreading and molecular compression experiments which show that trilayer nCB films are always very stable [49,50], while for larger thicknesses the material forms thick patches or bulk droplets sitting on top of an interfacial film. Combining the existing knowledge on thin 8CB films on substrates and a typical sequential growth of films observed in our experiments, we assume that an analogous "trilayer" structure (molecular monolayer covered with an approximate bilayer) is also present in our case. The problems that we want to address are what is the tilt angle of the 8CB molecules in these layers and what is their polarity.

The ellipticity coefficient of the interfacial 8CB film can be formally separated into two contributions

$$\rho_m = \rho_{m1} + \rho_{m2} \tag{4.16}$$

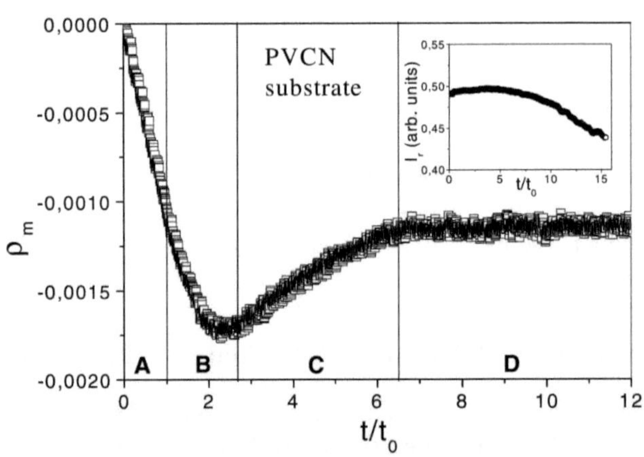

Fig. 4.5. The ellipticity coefficient of 8CB film evaporated onto a PVCN substrate as a function of deposition time. The time scale is given according to the SHG data (See Chapter 5.). Vertical lines denote borders between various growth stages as resolved from the SHG measurements. The inset shows the time dependence of the average intensity of the reflected light.

ρ_{m1} and ρ_{m2} correspond to the contribution of the first and the second layer respectively. Taking into account that the layers are optically uniaxial and that their optical axes are perpendicular to the surface normal [51, 52], the two contributions are given as

$$\rho_{mi} = \frac{\pi d_i}{\lambda} \frac{\sqrt{1+\epsilon_s}}{(1-\epsilon_s)} \left[(\epsilon_{xi} - \epsilon_{zi}) + \frac{(\epsilon_{zi}-1)(\epsilon_{zi}-\epsilon_s)}{\epsilon_{zi}} \right] \tag{4.17}$$

where λ is vacuum laser wavelength, ϵ_s is the optical dielectric constant of glass, d_i is the thickness of the i-th layer, and ϵ_{xi} and ϵ_{zi} are optical dielectric constants of the i-th layer for polarizations parallel and perpendicular to the surface, respectively. Due to large optical anisotropy of liquid crystals, the first term in the parenthesis of (4.17) is dominant for all the molecular tilt angles Θ except those, close to the "magic" angle Θ_{mag}, which corresponds to an apparent optically isotropic structure, i.e. to $\epsilon_{xi} = \epsilon_{zi}$. From this it follows that a decrease of ρ_m, observed in the time regions A and B on Fig. 4.5 indicates an average molecular tilt of $\Theta_1 > \Theta_{mag}$ for the first layer, and an increase of ρ_m in the region C signifies that $\Theta_2 < \Theta_{mag}$ for the second layer.

To obtain the values of Θ_{mag}, Θ_1, and Θ_2, several further assumptions are necessary. We take an approximation that the density and local angular distribution of the 8CB molecules in the adsorbed layers are the same as in the bulk smectic *A* phase, i.e. bulk values of optical dielectric constants are used in the calculation of ϵ_{xi} and ϵ_{zi}. To account for the molecular tilting, the films are regarded to form a multi-domain structure with the molecular directors being azimuthally isotropically distributed on a cone with the tilt angle Θ_i. These assumptions result in

$$\begin{aligned} \epsilon_{xi} &= \tfrac{1}{2}\left(\epsilon_\perp \cos^2\Theta_i + \epsilon_\parallel \sin^2\Theta_i\right) + \tfrac{1}{2}\epsilon_\perp \\ \epsilon_{zi} &= \epsilon_\perp \sin^2\Theta_i + \epsilon_\parallel \cos^2\Theta_i \end{aligned} \tag{4.18}$$

where $\epsilon_\perp$ and $\epsilon_\parallel$ are principal dielectric constants of the bulk smectic phase in the direction perpendicular and parallel to the director axis, respectively. From this it follows that the first term of (4.17) vanishes at $\Theta_{mag} = 54.7°$. We used the value of ρ_m measured at the end of region B to calculate Θ_1 and the difference between the values of ρ_m, detected at the end of regions C and B (Fig. 4.5) to calculate Θ_2. Taking $\epsilon_\perp = 2.28$, $\epsilon_\parallel = 2.82$, $\epsilon_s = 2.31$ [42] and the thicknesses of the layers as found by x-ray measurements on precursor 8CB films ($d_1 = 0.9\,\text{nm}$, $d_2 = 3.3\,\text{nm}$ [53, 54]), we then obtain $\Theta_1 = 78° \pm 2°$ and $\Theta_2 = 41.1° \pm 0.3°$.

Our BAE measurements indicate that the molecules in the first layer lie nearly flat on the surface, while the molecules in the second layer are pointing more toward the surface normal. The structure of the second layer is, however, very different from a homeotropically oriented bilayer of the bulk smectic phase which would give $\Theta_{2,bulk} = 0°$. Assuming that the molecular angular distribution $f(\Theta)$ is constant during the entire deposition of the first monolayer, the ratio between the values of ρ_m, detected at the end of regions A and B (Fig. 4.5) gives an estimate of the corresponding surface coverages which is $N_{1A}/N_{1B} \approx 0.6$. A discrepancy between this value and the corresponding ratio of the evaporating times $t_{1A}/t_{1B} \approx 0.4$ indicates a temporal slowing down of the adsorption kinetics. However, it can be partly attributed also to the mismatch of the time scale with respect to the SSHG measurements (Chapter 5) or to the smearing of the details of the signal due to inhomogeneity of the growth process within the illuminated region ($A \approx 2\,\mathrm{mm}^2$).

Conclusions

Due to a pronounced optical anisotropy of ordered liquid crystals, reflection ellipsometry is a powerful experimental tool for the study of surface-modified liquid crystalline order. The sensitivity of the technique enables quantitative measurements of the nematic order parameter profile at the interfaces. From observation of the nematic wetting of the solid-liquid crystal interface, we have determined the values of the coupling energies between the surface and the nematic order. Even the pretransitional effects at solid-smectic liquid crystals can be studied with ellipsometry due to the strong coupling between the nematic and the smectic order. The time evolution of the complex structure of the liquid crystal adsorbate on a solid substrate has also been successfully monitored using BAE.

4.2 Surface Anchoring Coefficients Measured by Dynamic Light Scattering

Mojca Vilfan and Martin Čopič

The equilibrium configuration in a liquid crystal sample is strongly influenced by the sample boundaries. The confining surfaces can induce order, disorder, or can align liquid crystal molecules in a given direction. Surface interactions not only have influence on the static properties of a confined liquid crystal, but can also have a strong effect on the director dynamics. By studying temporal fluctuations of the director field in confined samples, information about the surface-liquid crystal interaction can be obtained.

4.2.1 Thermal Fluctuations in Nematic Liquid Crystals

In a uniaxial nematic liquid crystal, the spatial orientation of the optical axis is determined by the orientation of the director. Due to thermally excited orientational director fluctuations, the spatial direction of the optical axis is not constant in time. As a result, any light illuminating the sample is strongly scattered by the fluctuations of the nematic director field. Optical experiments, especially light scattering experiments, provide thus an excellent tool for studying the thermally excited orientational fluctuations in liquid crystalline samples.

First detailed dynamic light scattering (DLS) experiments using bulk liquid crystal samples have confirmed the theoretically predicted existence of two dissipative fluctuation eigenmodes in the nematic liquid crystalline phase: the first mode being a combination of splay and bend distortion and the second one a combination of twist and bend fluctuations [55, 56]. Both modes are overdamped and the relaxation rate $1/\tau$ of each mode depends on the fluctuation wave vector $\boldsymbol{q}$ and viscoelastic properties of the sample [57]:

$$\frac{1}{\tau_\alpha(\boldsymbol{q})} = \frac{K_\alpha\, q_\perp^2 + K_3\, q_\parallel^2}{\eta_\alpha(\boldsymbol{q})} \,. \tag{4.19}$$

Here, $\alpha = 1, 2$ denotes the splay–bend and twist–bend mode, respectively, $K_{1,2,3}$ are the Frank elastic constants, $\eta_{1,2}$ are the rotational viscosities, $q_\parallel$ is the component of the fluctuation wave vector parallel to the director and $q_\perp$ the component perpendicular to it.

Standard applications of DLS experiments in bulk samples include measurements of viscoelastic coefficients and dispersion relations of orientational fluctuation modes. As the experimental details are given further in the text, let us first discuss the nature of orientational fluctuation modes in nematics.

4.2.2 Fluctuation Modes in Confined Liquid Crystals

In a bulk liquid crystal sample, the wave vector $\boldsymbol{q}$ of a given fluctuation is in principle arbitrary in both magnitude and direction. There are however two limitations: in the limit of small wave-vectors, the dimensions of the sample determine the cut-off, whereas in the limit of large wave-vectors (small wavelengths), the wave vector magnitude is limited by the requirements of the continuum theory. The relaxation times can be calculated using (4.19) and the obtained spectrum of the fluctuation relaxation times is continuous.

In confined geometries, however, this is not the case. In a planar sample, for example, the magnitude of the wave vector component parallel to the boundaries is arbitrary, whereas the fluctuation wave vector component perpendicular to the boundaries can only have certain values. In this case, the allowed wave vector components are determined by the sample thickness, viscoelastic properties of the liquid crystal, and also by the interaction of a liquid crystal with the aligning surface. If the director is strongly bound to the aligning substrate, it cannot deviate from the induced direction (easy axis) and the fluctuation amplitude at the boundary is zero. On the other hand, if the surface only weakly anchors the director orientation, the director can fluctuate to a certain degree around the easy axis.

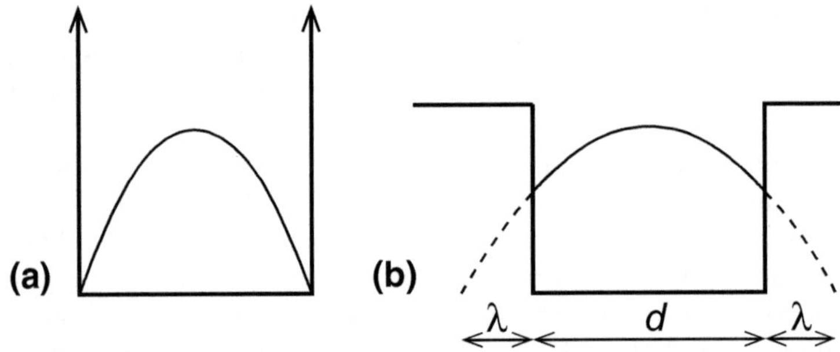

Fig. 4.6. Amplitude of the fundamental fluctuation eigenmode in a planar nematic sample resembles the eigenfunction of a particle in a one dimensional potential well. Infinite anchoring corresponds to infinitely deep well (a) and finite anchoring to finitely deep well (b). The extrapolation length λ is also indicated.

The described phenomenon is analogous to the wave function eigenstates of a particle in a one dimensional quantum potential well. Infinitely strong orientational interaction with the surface (infinite anchoring) corresponds to an infinitely deep potential well, and a finite anchoring to a finite one (Fig. 4.6), where the width of the well corresponds to the thickness d of the planar sample. The fluctuation modes are of sinusoidal shape and the perpendicular wave vector components are determined by secular equations:

$$q \tan\left(\frac{qd}{2}\right) = \frac{W}{K} \quad \text{and} \quad \frac{1}{q} \tan\left(\frac{qd}{2}\right) = -\frac{K}{W} \tag{4.20}$$

for symmetric and asymmetric modes, respectively [29,58]. If the splay–bend mode is considered, the constant K_1 should be taken into account and the zenithal anchoring coefficient W_ϑ, whereas for the twist–bend mode the elastic constant K_2 and azimuthal anchoring coefficient W_φ should be inserted into (4.20).

The quantity K/W appearing in the upper equation is often introduced as the extrapolation length λ. It can be interpreted as the extension of the sample thickness at which the amplitude of the fluctuation mode would drop to zero (see Fig. 4.6). Infinite anchoring means no extension since the fluctuations are already zero at the boundaries, whereas very small anchoring means a nearly negligible effect of the surface and therefore a very large extrapolation length.

Each branch of (4.20) with given parameters K and W gives an allowed value for the fluctuation wave vector component q at a given sample thickness d. The set of allowed wave vectors is discrete and the index m will be used to distinguish between different branches, i.e. different fluctuation *modes*. It will be shown later that in most cases the fundamental mode ($m = 1$) is sufficient for determining the anchoring energy coefficients, since the contribution of the higher modes in thin samples can be neglected.

Secular equations can be solved analytically only if the anchoring is infinitely strong. After assuming symmetric boundary conditions and setting the origin of the coordinate system in the centre between the boundaries, the allowed vector components are in this case equal to $q_m = m\pi/d$; odd values of m for symmetric and even values for asymmetric modes. In general, the secular equations cannot be solved analytically and for further analysis approximations are necessary. If the anisotropic interaction with the surface is strong but finite, the extrapolation length is small compared to the sample thickness and to the inverse fluctuation wave vector component. In this case, the approximate fluctuation wave vector component for $m = 1$ is

$$q \simeq \frac{\pi}{d + 2\lambda} . \tag{4.21}$$

The relaxation time can be calculated using (4.19) and if $q_\parallel$ is reduced by choosing a proper scattering geometry, the relaxation time has an additional term proportional to the sample thickness d:

$$\tau = \frac{1}{\pi^2}\frac{\eta}{K}\left(d^2 + 4\lambda d\right) . \tag{4.22}$$

From these measurements, the ratio K/η can also be determined independently. Fig. 4.7 shows an example of measured relaxation time as a function of sample thickness in the case of strong anchoring of 5CB on Nylon. The circles are the measured data, the solid line is the best fit to (4.22) and the dashed line is the best, purely quadratic, fit without the linear correction due to finite anchoring. As can be seen from Fig. 4.7, the method is sensitive enough to

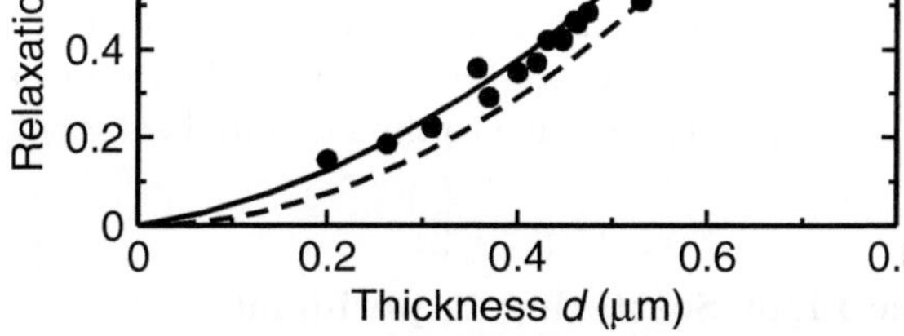

Fig. 4.7. Relaxation time τ for the fundamental twist fluctuation mode (*dots*) as a function of sample thickness d. The aligning layer was rubbed Nylon, the liquid crystal was 4–n–pentyl–4'–cyanobiphenyl (5CB) in the nematic phase ($T = 32°$C). Comparison between the best fit of the theoretically derived equation (*solid line*) and the best fit assuming infinite anchoring strength (*dashed line*) is made [32].

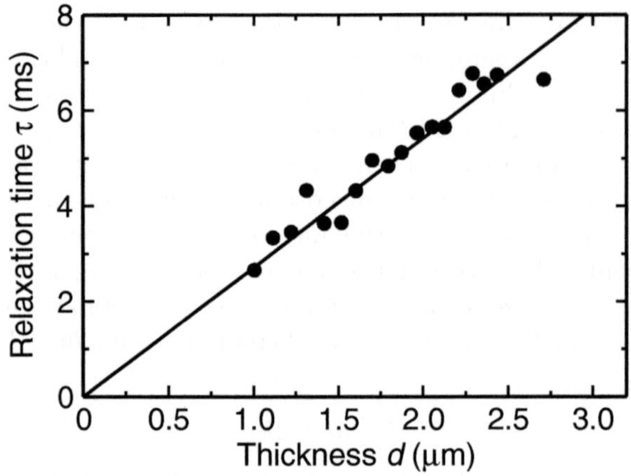

Fig. 4.8. Relaxation time τ for the fundamental splay fluctuation mode as a function of sample thickness (*circles*) and best fit of the theoretically predicted relation for weak anchoring (*solid line*). The aligning layer was UV illuminated photoactive poly–(vinyl–cinnamate), the liquid crystal was 5CB in the nematic phase ($T = 32°\mathrm{C}$) [59].

measure anchoring coefficients when the surface anchoring is strong and the extrapolation length small – in this case the obtained extrapolation length is $\lambda_\varphi = 50 \pm 10\,\mathrm{nm}$, which corresponds to $W_\varphi = (5.6 \pm 1.2) \times 10^{-5}\,\mathrm{J/m^2}$. The only requirement is a small thickness of the sample, as it needs to be of the same order of magnitude as the measured extrapolation length. More details on sample requirements and preparations follow later in this chapter.

In the case of weak anchoring, when $W \to 0$ and $\lambda \to \infty$, series expansion of the relaxation time of the fundamental mode for $qd \ll 1$, to first order yields

$$\tau = \frac{\eta}{2W} d\,. \tag{4.23}$$

Such linear dependence in thin samples has also been experimentally observed (Fig. 4.8) and the obtained anchoring for the shown example is $W_\vartheta = (9.1 \pm 0.9) \times 10^{-6}\,\mathrm{J/m^2}$ with $\lambda_\vartheta = 480 \pm 45\,\mathrm{nm}$. Note that when measuring weak anchoring, the sample can be slightly thicker, up to a few microns, as again only the sample size relative to the measured extrapolation length is relevant.

4.2.3 Dynamic Light Scattering Experiment

In order to perform the measurements of the anchoring energy coefficients described in the previous subsection, the orientational fluctuation relaxation time needs to be determined as a function of sample thickness. This can be best done using dynamic light scattering.

The experiment is based on the fact that the propagation of light through a medium is determined by the refractive index of the material and thus its dielectric tensor. In a nematic liquid crystal, which is an optically uniaxial medium with large birefringence, there are two propagating modes: ordinary and extraordinary wave with orthogonal polarisations. Due to thermally excited director fluctuations, the dielectric tensor field also fluctuates and the incident light is scattered by these optical inhomogeneities.

The electric field of the scattered light depends strongly on the polarisations of the incoming $\boldsymbol{i}$ and outgoing $\boldsymbol{f}$ light beams with respect to the temporally varying dielectric tensor $\varepsilon(\boldsymbol{r},t)$ [60]:

$$E_{\mathrm{s}}(\boldsymbol{q},t) \propto \int_{V_{\mathrm{s}}} \mathrm{e}^{\mathrm{i}\boldsymbol{q}\cdot\boldsymbol{r}}\,(\boldsymbol{f}\cdot\varepsilon(\boldsymbol{r},t)\cdot\boldsymbol{i})\,\mathrm{d}\boldsymbol{r}\ , \tag{4.24}$$

with scattering vector $\boldsymbol{q}$ being the difference between the wave vectors of the incoming $\boldsymbol{k}_{\mathrm{i}}$ and outgoing $\boldsymbol{k}_{\mathrm{f}}$ light beams, and the integral is performed over the scattering volume V_{s}. If the time-dependent dielectric tensor is written in the wave-vector space of the fluctuation eigenmodes, and a cuboid is assumed as the scattering volume with side lengths $a_{x,y,z}$, the amplitude of the scattered light is [32]:

$$E_{\mathrm{s}}(\boldsymbol{q},t) \propto \sum_m (\boldsymbol{f}\cdot\varepsilon(\boldsymbol{q}_m,t)\cdot\boldsymbol{i})\,\Pi_{j=x,y,z}\frac{\sin\left(\frac{q_{mj}-q_j}{2}a_j\right)}{\frac{q_{mj}-q_j}{2}a_j}\ . \tag{4.25}$$

If the dimensions of the sample are much larger than the wavelength of the scattered light, the amplitude becomes a Dirac delta function $\delta(\boldsymbol{q}_m - \boldsymbol{q})$ and only the component of the fluctuations with wave vector equal to the scattering wave vector contributes to the scattering. If the size of the sample is (at least in one direction) comparable to the wavelength of light, several modes contribute to the scattered light, each mode with a corresponding amplitude, given by (4.25). In very thin samples, only the fundamental mode is observed.

Further important information for the experiment can be obtained from the differential scattering cross-section [57]:

$$\frac{\mathrm{d}\sigma}{\mathrm{d}\Omega} \propto \sum_{\alpha=1,2}\frac{(i_\alpha f_\parallel + i_\parallel f_\alpha)^2}{(K_\alpha\, q_\perp^2 + K_3\, q_\parallel^2)}\ , \tag{4.26}$$

where $\parallel$ stands for the projection on the director, $\perp$ for the component perpendicular to it, and i_α and f_α denote the projections of the polarisations of the incoming and outgoing beams on the eigenvectors of the two fluctuation modes $\boldsymbol{e}_\alpha$. (4.26) shows that the maximum intensity of the scattered light by the thermal fluctuations in a nematic liquid crystal is achieved when the incoming and the outgoing polarisations are orthogonal.

Selection rules for the observation of a given fluctuation mode can also be obtained from (4.26): the twist–bend fluctuation mode can be observed and

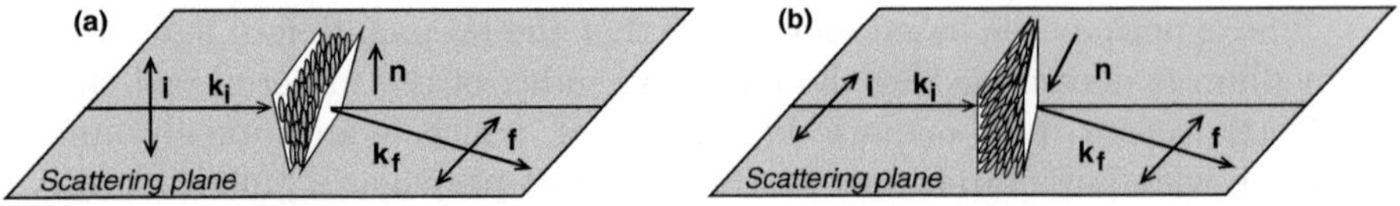

Fig. 4.9. Scattering geometry for measurements of (a) azimuthal and (b) zenithal anchoring coefficient. $\boldsymbol{n}$ is the director, $\boldsymbol{i}$ and $\boldsymbol{f}$ the polarisations of the incoming and outgoing beams, and $\boldsymbol{k}_{\rm i}$ and $\boldsymbol{k}_{\rm f}$ the corresponding wave vectors.

the azimuthal anchoring coefficient determined, if the director $\boldsymbol{n}$ is perpendicular to the scattering plane and parallel to the polarisation of the incoming light $\boldsymbol{i}$; the outgoing beam polarisation $\boldsymbol{f}$ should lie in the scattering plane (see Fig. 4.9a). The only way to detect the splay–bend fluctuation mode and to measure the zenithal anchoring coefficient is when the director and both polarisations of the incoming and outgoing light beams lie in the scattering plane (Fig. 4.9b).

In order to observe the light scattering by thermal fluctuations and obtain the relaxation times of these fluctuations, the scattered light is studied via the autocorrelation spectroscopy technique [60]. The normalised intensity autocorrelation function $g^{(2)}(\boldsymbol{q}, \tau)$ is defined as

$$g^{(2)}(\boldsymbol{q}, t) = \frac{\langle I(\boldsymbol{q}, t')I(\boldsymbol{q}, t' + t)\rangle}{\langle I(\boldsymbol{q}, t')\rangle^2} , \tag{4.27}$$

where $I(\boldsymbol{q}, t')$ is the intensity of the scattered light, proportional to the square of the scattered electric field $E_{\rm s}$ at a given time t'. In general, two different measurement regimes can be considered: *homodyne* and *heterodyne*. In the former case, the intensity of the scattered light by the fluctuations is much larger than the intensity of the statically scattered light in the system, and the dynamically scattered light interferes at the detector. In the latter case, the contribution of the dynamically scattered light is small compared to the statically scattered light (by imperfections in a sample), and as the two contributions are statistically independent, the signals are summed up at the detector. This is observed when the amplitudes of the fluctuations are small, or if the experiments are performed using thin cells with only a small amount of the scattering medium.

Calculating the autocorrelation function of the scattered light for the heterodyne regime, the following relation can be obtained:

$$g^{(2)}(\boldsymbol{q}, t) - 1 \propto \sum_{\alpha=1,2} \frac{(i_\alpha f_\parallel + i_\parallel f_\alpha)^2}{(K_\alpha\, q_\perp^2 + K_3\, q_\parallel^2)} \quad \mathrm{e}^{-t/\tau_\alpha} . \tag{4.28}$$

With proper selection of the scattering geometry, the two dissipative modes can be observed separately and the decay time of the autocorrelation function $g^{(2)}$ equals the relaxation time of the chosen fluctuation mode. However, the amplitude of the above expression relative to the static scattering should not exceed 0.2 or 0.3 for the heterodyne regime to be valid.

Sometimes more than just one fluctuation mode is observed. In this case the decay is no longer a pure exponential, but is composed of several exponentially decaying functions. If they differ by more than an order of magnitude, they can be treated separately, but in the case of similar relaxation rates, one observes one slightly stretched exponential function $g^{(2)}(\boldsymbol{q},t) - 1 \propto \exp(-(t/\tau_r)^s)$ with τ_r being an average relaxation time and s the stretching exponent [61]. In well aligned nematic samples with thickness of up to a few microns, or in bulk samples, only one mode can be observed and $s = 1$. The parameter s is reduced to values around 0.8 in disoriented samples with a distribution in the director orientation [62] or for sample thicknesses between a few microns and bulk. Due to the reduced s, also the average relaxation time has to be corrected according to [61]: $\langle\tau\rangle = (\Gamma(1/s)\tau_r)/s$, where Γ denotes the Gamma function. The correction is for most measurements, however, fairly small and can often be neglected.

All the discussion so far was concerning only the fundamental fluctuation mode. This is dominant mode for small sample thickness, typically up to a micron or two. In thicker samples, the influence of higher fluctuation modes becomes apparent: the increase in the relaxation time is no longer purely parabolic (or linear, depending on the anchoring strength), but reaches a maximum and decreases towards the bulk value with increasing thickness (Fig. 4.10).

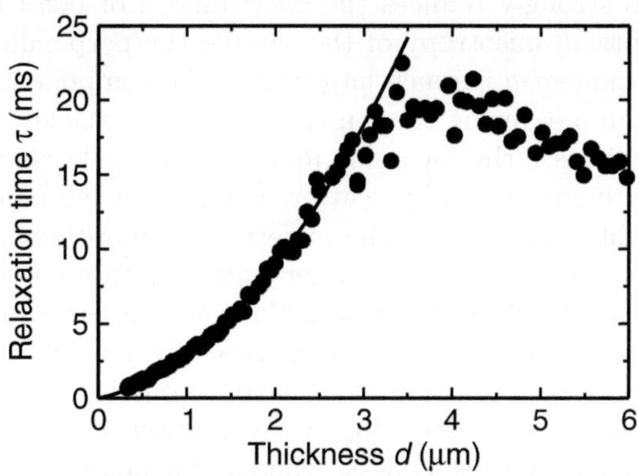

Fig. 4.10. Relaxation time τ of the twist fluctuation mode (*circles*) as a function of sample thickness d. For thicknesses below $\approx 3\,\mu$m a parabolic behaviour is observed (*solid line*) and in this range the anchoring coefficient can be determined. For larger thicknesses the influence of higher fluctuation modes becomes apparent and the relaxation time decreases and approaches its bulk value [32].

4.2.4 Measuring the Anchoring Coefficients

Dynamic light scattering experiments use a laser with good temporal stability as a light source. The beam with a chosen polarisation is focused on the sample and the glass plates of the sample are perpendicular to it. The scattered light is then detected at a certain angle in the scattering plane with a chosen polarisation using an analyser and directed to a single-photon detector. To reduce the noise from the detector, the scattered light can be split into two beams and detected separately using two detectors. This means that rather then autocorrelating one signal, the signal is split into two and the cross-correlation function from the two beams is calculated using a correlator.

For thermal stability or temperature dependence measurements, the sample can be mounted in a heating stage with the possibility of exact positioning and micrometer movements. The sample used in the experiment needs to have a wedge-like shape, which enables measurements as a function of sample thickness by only moving it in the direction perpendicular to the slope, which should coincide with the director orientation. Additional requirement is an overall small thickness of the sample and moreover, the thickness needs to be known. It can best be determined by interferometric methods using a spectrophotometer.

The scattering angle in the experiment has to be very small, typically $2-3°$. Small scattering angle ensures that the inverse scattering vector is large compared to the sample thickness and that only the fundamental mode is observed. In the case of azimuthal anchoring measurements, small scattering angle also strongly reduces the contribution of bend fluctuations, as due to strong optical anisotropy of the sample the perpendicular scattering wave vector component q is much larger than the component parallel to the glass plates. When measuring the zenithal anchoring, the situation is slightly different, as in this case the incoming and outgoing polarisations are almost parallel. Therefore the statically scattered light is a major contribution to the scattered light which makes the experiments more demanding and the signal needs to be measured over a longer time to obtain good signal-to-noise ratio. Due to parallel polarisations, also the wave vector component parallel to the glass plates is not as small as in the case of azimuthal anchoring. However, the contribution to the relaxation rate is constant and independent of the sample thickness. When studying the relaxation time, the constant term can be neglected for small thickness, whereas for larger thickness or strong anchoring, the relaxation time is limited by this term.

A comment should be added on the rotational viscosity of the sample in the experiment. It can easily be shown that in the case of azimuthal anchoring measurements, the viscosity entering the analysis equals the rotational viscosity γ_1. When measuring the zenithal anchoring coefficient, the effective viscosity has to be corrected according to the scattering geometry. However, in first approximation the pure splay mode viscosity is sufficient.

As shown, the dynamic light scattering experiment provides an excellent tool for measuring the anchoring energy coefficients. Its major advantage is the fact that the same sample can be used for measurements of both azimuthal and zenithal anchoring coefficients. This makes a comparison of the two coefficients more reliable. Also no external torque acts on the liquid crystal, which remains in the undistorted configuration during the measurement.

4.2.5 Aging of Photoaligning Layers

Liquid crystal alignment achieved by using rubbed polymers is usually very stable. This is not the case with photoaligning substrates, where often instabilities have been observed during a period of several weeks after preparing the sample. This aging phenomenon can be successfully studied by dynamic light scattering as well.

The first observed difference is a change in the typical $\tau(d)$ curve. In samples older than one day, an offset appears in the relaxation time, and the linear relation (4.23) is no longer valid. Such an offset can be explained by introducing surface viscosity [63]. This gives an additional offset parameter h, and by using the following relation $\tau = (2h + d)\eta/2W$, both W and h can be determined independently (Fig. 4.11). An increase in the anchoring

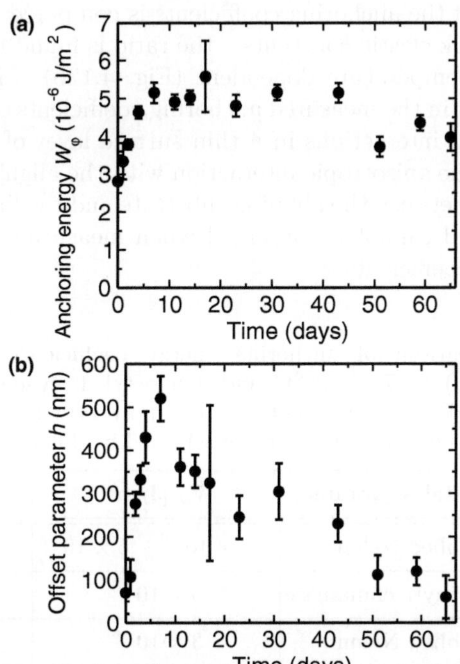

Fig. 4.11. (a) Aging, as observed through the time dependence of the azimuthal anchoring coefficient W_φ for 5CB on UV aligned poly–(vinyl–cinnamate); (b) Aging, as observed through the offset parameter h [59].

during the first days is observed and interpreted as an induced additional ordering of the 'soft' polymer by the strong order of the liquid crystal; after the process stabilises, the anchoring remains almost constant. The parameter h shows much more drastic behaviour. As the values are far too large for it to be the actual surface viscosity, the interpretation is that the offset behaviour can appear if the aligning layer first swells (initial increase) and then slowly diffuses into the liquid crystal. This process changes viscoelastic properties of the liquid crystal and therefore also the relaxation dynamics.

4.2.6 Temperature Dependence of the Anchoring Coefficients

Dynamic light scattering enables not only direct measurements of the anchoring energy coefficients in a sample, but also allows to perform the measurements under varying conditions, e.g., changing temperature. The temperature dependence of anchoring energy coefficients can be monitored in the whole range of the nematic phase and even in the vicinity of the phase transitions using the same liquid crystalline sample. An example of the temperature dependence of both anchoring coefficients is shown in Fig. 4.12a.

A steady decrease on approaching the transition into the isotropic phase is observed, with finite values at $T_{\rm NI}$. Another interesting observation can be made if the ratio of the anchoring coefficients is compared to the ratio of the corresponding Frank elastic constants – the ratio is found to be almost equal and only weakly temperature dependent (Fig. 4.12b). This could indicate that the anisotropy in the measured anchoring coefficients originates from the anisotropic internal interactions in a thin surface layer of the liquid crystal rather than from the anisotropic interaction with the aligning substrate. The actual interaction between the aligning substrate and the first molecular layer of the liquid crystal cannot be observed when measuring the 'macroscopic' anchoring energy coefficients.

Table 4.2. Comparison of anchoring energy coefficients of 4–n–pentyl–4'–cyanobiphenyl (5CB) at $T = 32°$C and 4–n–octyl–4'–cyanobiphenyl (8CB) at $T = 37°$C on rubbed Nylon and photosensitive poly–(vinyl)–cinnamate. Different anchoring coefficients on rubbed Nylon are achieved by different rubbing strengths.

Liquid crystal/substrate	W_φ [J/m^2]	W_ϑ [J/m^2]
5CB/ rubbed Nylon	$5 \times 10^{-6} - 5 \times 10^{-5}$	$10^{-5} - 10^{-4}$
5CB/ poly–(vinyl)–cinnamate	5×10^{-6}	10^{-5}
8CB/ rubbed Nylon	3×10^{-6}	–

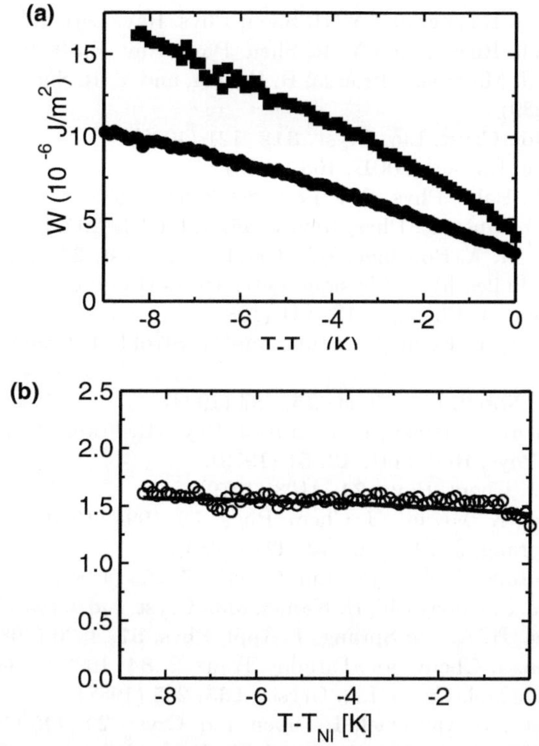

Fig. 4.12. (a) Temperature dependence of azimuthal W_φ (*dots*) and zenithal W_ϑ (*squares*) anchoring coefficients for nematic 5CB on rubbed Nylon with T_{NI} being the transition temperature into the isotropic phase; (b) A comparison of the ratios of the two anchoring coefficients W_ϑ/W_φ (*circles*) with the ratio of the corresponding Frank elastic constants K_1/K_2 [65](*solid line*) [64].

4.2.7 Conclusions

Anchoring energy coefficients of liquid crystals on different substrates can be determined using dynamic light scattering experiments. With no additional torque acting on the director, aging effects can be observed and the behaviour of the system under changing conditions, e.g., temperature can be studied, and a reliable comparison between the obtained anchoring coefficients can be made.

References

1. R. M. A. Azzam, and N. M. Bashara, *Ellipsometry and Polarized Light*, North Holland, Amsterdam, 1977.
2. J. Lekner, *Theory of Reflection*, Martinus Nijhoff Publishers, (1987).

3. H. Hsiung, Th. Rasing, and Y. R. Shen, Phys. Rev. Lett. **57**, 3065 (1986).
4. H. Hsiung, Th. Rasing, and Y. R. Shen, Phys. Rev. E **59**, 1983 (1987).
5. W. Chen, L. J. Martinez-Miranda, H. Hsiung, and Y. R. Shen, Phys. Rev. Lett. **62**, 1860 (1989).
6. T. Moses, Mol. Cryst. Liq. Cryst. **319**, 121 (1998).
7. D. Beaglehole, Physica **100B**, 163 (1980).
8. R. Lucht, Ch. Bahr, Phys. Rev. Lett. **80**, 3783 (1998).
9. K. Kočevar, I. Muševič, Phys. Rev. E **65**, 021703-1 (2002).
10. D. Langevin, M. A. Bouchiat, Mol. Cryst. Liq. Cryst. **22**, 317 (1973).
11. S. Faetti, V. Palleschi, J. Physique Lett. **45**, L-313 (1984).
12. J. P. Nicholson, J. Physique **48**, 131 (1987).
13. S. Immerschitt, T. Koch, W. Stille, and G. Strobl, J. Chem. Phys. **96**, 6249 (1992).
14. S. Faetti, M. Nobili, Liq. Cryst. **25**, 487 (1998).
15. M. Warenghem, M. Ismail, D. Hector, J. Phys.III, France **2**, 765 (1992).
16. K. Miyano, Phys. Rev. Lett. **43**, 51 (1979).
17. K. Miyano, J. Chem. Phys. **71**, 4108 (1979).
18. J. C. Tarczon, K. Miyano, J. Chem. Phys. **73**, 1994 (1980).
19. H. A. van Sprang, J. Physique **44**, 421 (1983).
20. H. A. van Sprang, Mol. Cryst. Liq. Cryst. **97**, 255 (1983).
21. H. Yokoyama, S. Kobayashi, H. Kamei, Mol. Cryst. Liq. Cryst. **107**, 311 (1984).
22. H. Yokoyama, H. A. van Sprang, J. Appl. Phys. **57**, 4520 (1985).
23. H. Yokoyama, J. Chem. Soc. Faraday Trans. 2, **84**, 1023 (1988).
24. H. Yokoyama, Mol. Cryst. Liq. Cryst., **165**, 265 (1988).
25. P. de Schrijver, W. van Dael, J. Thoen, Liq. Cryst. **21**, 745 (1996).
26. Jong-Hyun Kim, R. G. Petschek, and Ch. Rosenblatt, Phys. Rev. A, **60**, 5600 (1999).
27. M. Čopič, C. S. Park, N. A. Clark, Mol. Cryst. Liq. Cryst. **222**, 111 (1992).
28. M. M. Wittebrood, D. H. Luijendijk, S. Stallinga, Th. Rasing, I. Muševič, Phys. Rev. E **54**, 5232 (1996).
29. M. M. Wittebrood, Th. Rasing, S. Stallinga, I. Muševič, Phys. Rev. Lett. **80**, 1232 (1998).
30. M. Škarabot, S. Kralj, R. Blinc, I. Muševič, Liq. Cryst. **26**, 723 (1999).
31. A. Rastegar, I. Muševič, M. Čopič, Ferroelectrics **181**, 219 (1996).
32. M. Vilfan, A. Mertelj and M. Čopič, Phys. Rev. E **65**, 041712-1 (2002).
33. C. S. Mullin, P. Guyot-Sionnest, Y. R. Shen, Phys. Rev. A **38**, 3745 (1989).
34. T. Moses, Y. R. Shen, Phys. Rev. Lett. **67**, 2033 (1991).
35. Ch. Bahr, D. Fliegner, Phys. Rev. A **46**, 7657 (1992).
36. P. De Schrijver, C. Glorieux, W. Van Dael, J. Thoen, Liquid Crystals **23**, 709 (1997).
37. R. Lucht, Ch. Bahr, J. Chem. Phys. **108**, 3716 (1998).
38. D. Schlauf, Ch. Bahr, Phys. Rev. E **60**, 6816 (1999).
39. P. M. Johnson, D. A. Olson, S. Pankratz, T. Nguyen, J. Goodby, M. Hird, C. C. Huang, Phys. Rev. Lett. **84**, 4870 (2000).
40. P. M. Johnson, D. A. Olson, S. Pankratz, Ch. Bahr, J. Goodby, C. C. Huang, Phys. Rev. E **62**, 8106 (2000).
41. J. Ortega, J. Etxebarria, C. B. Folcia, Eur. Phys. J. E **3**, 21 (2000).
42. R. G. Horn, J. Phys. **38**, 105 (1978).
43. S. N. Jasperson, S. E. Schnatterly, Rev. Sci. Instr. **40**, 761 (1969).

44. B. M. Ocko, Phys. Rev. Lett. **64**, 2160 (1990).
45. Z. Pawlowska, T. J. Sluckin, G. F. Kvenstel, Phys. Rev. A **38**, 5342 (1988).
46. A. M. Somoza, L. Menderos, D. E. Sullivan, Phys. Rev. Lett. **72**, 3674 (1994).
47. A. M. Somoza, L. Menderos, D. E. Sullivan, Phys. Rev. E **52**, 5017 (1995).
48. I. Drevenšek Olenik, K. Kočevar, I. Muševič, Th. Rasing, Eur. Phys. J. E **11**, 169 (2002).
49. S. Herminghaus, K. Jacobs, K. Mecke, J. Bischof, A. Fery, M. Ibn-Elhaj, S. Schlagowski, Science **282**, 916 (1998).
50. F. Vandenbrouck, S. Bardon, M. P. Valignat, A. M. Cazabat, Phys. Rev. Lett. **81**, 610 (1998).
51. I. Drevenšek Olenik, M. W. Kim, A. Rastegar, Th. Rasing, Phys. Rev. E **60**, 3120 (1999).
52. I. Drevenšek Olenik, M. W. Kim, A. Rastegar, Th. Rasing, Phys. Rev. E **61**, R3310 (2000).
53. S. Bardon, R. Ober, M. P. Valignat, F. Vanderbrouck, A. M. Cazabat, J. Daillant, Phys. Rev. E **59**, 6808 (1999).
54. L. Xu, M. Salmeron, S. Bardon, Phys. Rev. Lett. **84**, 1519 (2000).
55. Orsay Liquid Crystal Group, J. Chem. Phys. **51**, 816 (1969).
56. Orsay Liquid Crystal Group, Phys. Rev. Lett. **22**, 1361 (1969).
57. P. G. de Gennes and J. Prost, *The Physics of Liquid Crystals* (Clarendon, Oxford, 1995).
58. S. Stallinga, M. M. Wittebrood, D. H. Luijendijk, and Th. Rasing, Phys. Rev. E **53**, 6085 (1996).
59. M. Vilfan, I. Drevenšek Olenik, A. Mertelj, and M. Čopič, Phys. Rev. E **63**, 061709 (2001).
60. B. J. Berne and R. Pecora, *Dynamic Light Scattering* (Dover, 2000).
61. C. P. Lindsey and G. D. Patterson, J. Chem. Phys. **73**, 3348 (1980).
62. A. Mertelj and M. Čopič, Mol. Cryst. Liq. Cryst. **282**, 35 (1996).
63. A. Mertelj and M. Čopič, Phys. Rev. Lett. **81**, 5844 (1998).
64. M. Vilfan and M. Čopič, Phys. Rev. E **68**, 031704 (2003).
65. G.-P. Chen, H. Takezoe, and A. Fukuda, Liq. Cryst. **5**, 341 (1989).

5 Solid-Liquid Crystal Interfaces Probed by Optical Second-Harmonic Generation

Irena Drevenšek-Olenik, Silvia Soria, Martin Čopič, Gerd Marowsky and Theo Rasing

5.1 Introduction

Optical techniques play an important role in surface and interface analysis as they are practically non-destructive, can be used to study any interface accessable to light and offer a unique combination of high spectral, temporal and spatial resolution that allows studying ultrafast molecular dynamics and other transient phenomena [1–3]. Their drawback is, however, that most of the conventional linear optical techniques, like ellipsometry or UV-VIS, IR and Raman spectroscopy, are not surface specific. Accordingly, it is usually very difficult to resolve a weak surface signal from the predominant signal generated in the bulk of the material. This situation can substantially be improved by using nonlinear optical techniques. Among them, optical second-harmonic generation (SHG) and sum frequency generation (SFG) are particularly being noticed because of the many demonstrations of their versatility [4–11].

SHG arises from the nonlinear polarization $\mathbf{P}(2\omega)$ induced in a medium by an incident laser field $\mathbf{E}(\omega)$. In the electric dipole approximation, $\mathbf{P}$ is given by:

$$\mathbf{P}(2\omega) = \varepsilon_0 \chi^{(2)} : \mathbf{E}^2(\omega) \tag{5.1}$$

where the second order nonlinear optical susceptibility $\chi^{(2)}$ is a material parameter which is described by a proper 3rd rank tensor. For centrosymmetric media it follows directly from (5.1) that $\chi^{(2)} = 0$. However, at an interface the surface nonlinear susceptibility $\chi_S^{(2)}$ is non-vanishing because there the inversion symmetry is necessarily broken.

For that reason a second-harmonic field of electric dipole origin can be generated within the interface zone of centrosymmetric media. SHG is therefore an inherent surface and interface specific probe for materials with a centrosymmetric bulk structure. This is for example the case for the media of interest in this book, like polymeric films, glass substrates and liquid crystals both in their isotropic and nematic phase, as the latter is characterized by a directional non-polar order parameter. This implies that for the molecules of interest (mostly the cyano-biphenyl liquid crystals) that have a polar head group, the molecules are arranged in a kind of anti-parallel ordering, that is only broken at the substrate interface where the molecules preferentially point their head groups either up or down. It is exactly this ordering (both in-plane and out-of-plane) of the first molecular layer that can be probed by SHG.

The same description as leads to (5.1) applies also to SFG, if $\mathbf{E}^2(\omega)$ is replaced by $\mathbf{E}(\omega_1)\mathbf{E}(\omega_2)$ and 2ω by $\omega_1 + \omega_2$. By using e.g. a strong visible frequency for ω_1 and a tunable infrared laser for ω_2, the latter technique allows to perform interface specific optical spectroscopy. This for example can provide important information about specific bond orientation [6,36] but will not be discussed further in this book.

In this Chapter we will show how SHG can be used as an in-situ probe to study the organisation and the structure of liquid crystal molecules on polymer surfaces. The interface sensitivity allows to probe the surface order parameter of both the polymer substrates as well as of the adsorbed liquid crystal (LC) molecular layer. After introducing the basic theoretical principles and experimental approaches in Sect. 5.2, the rest of this Chapter will present several examples of SHG studies of the ordering of LC molecules on various substrates. For the latter, both rubbed (Sect. 5.3) and photo-aligned substrates (Sect. 5.4) were used, providing also an understanding of why these substrates have such different anchoring energies. The ordering of LC molecules on photo-aligned polymer samples will be discussed in Sect. 5.5. Finally, we will show in Sect. 5.6 how the SHG techniques can be used to follow the growth and ordering of LC molecules on substrates in-situ.

5.2 Second Harmonic Generation from Surfaces and Interfaces

For a thin film deposited on the surface of a centrosymmetric medium, the net surface second-harmonic response can be written as

$$\chi_S^{(2)} = \chi_0^{(2)} + \chi_a^{(2)} + \chi_{int}^{(2)} \tag{5.2}$$

where $\chi_0^{(2)}$ and $\chi_a^{(2)}$ are intrinsic susceptibilities of the substrate and the adsorbate, respectively, and $\chi_{int}^{(2)}$ denotes a contribution related to the substrate-adsorbate interaction. In the case of organic molecules deposited on inorganic dielectric media, the second term in (5.2), i.e. the contribution $\chi_a^{(2)}$, is usually dominating. For example, for liquid crystal molecules containing several phenyl rings (like the cyanobiphenyls), the electronic states are strongly delocalised and hence the nonlinear optical response can be relatively large: $\chi_S^{(2)} \sim 10^{-19}$ m^2/V depending on ω [12,13]. On the contrary, in inorganic dielectric materials, like for instance fused silica, the values of $\chi_S^{(2)}$ are typically below 10^{-21} m^2/V and the associated SHG signal is difficult to be detected even by the most sensitive photon-counting techniques. Measurements of the modification of the SHG signal with respect to the inorganic dielectric substrate hence provide a convenient possibility to study chemisorption, surface adhesion, molecular aggregation and other phenomena related to the formation of monolayer or even sub-monolayer thin films of organic materials on various surfaces.

The value of $\chi_a^{(2)}$ observed in the selected experimental configuration depends on the molecular hyperpolarizability β of the adsorbed molecules and on their surface density N_a and angular distribution $f(\Omega)$:

$$\chi^{(2)}_{a,lmn} = N_a \int T^{l'm'n'}_{lmn}(\Omega)\beta_{l'm'n'} f(\Omega) d\Omega. \tag{5.3}$$

Here $T(\Omega)$ is the transformation matrix between the molecular $(l', m', n' = 1, 2, 3)$ and the laboratory frame $(l, m, n = x, y, z)$ expressed in terms of Euler angles $\Omega = (\theta, \Phi)$ (see Fig. 5.1). The statistical angular distribution function is given by:

$$f(\Omega) = \frac{e^{-W(\Omega)/kT}}{\int e^{-W(\Omega)/kT} d\Omega} \tag{5.4}$$

where $W(\Omega)$ represents the net intermolecular and interfacial interaction energy. From (5.3) it follows that the SHG in general probes the in-plane as well as the out-of-plane ordering (tilt angle) of the adsorbed molecules with respect to the substrate. Compared to scanning probe techniques like STM and AFM, that mostly provide information about in-plane organization, this makes SHG a very powerful tool to study the interactions of LC molecules at interfaces.

As mentioned above, the value of the molecular hyperpolarisability β is large for the well known alkyl-cyanobiphenyl (nCB) molecules, that possess strongly delocalized molecular electronic orbitals with oppositely attached electron-donor and electron-acceptor groups. Due to the rod-like shape of these molecules, the component of β associated with the perturbation of the electronic density along the long molecular axis $\hat{e}_3$ is usually much larger than the other components. In first approximation therefore only β_{333} has to be

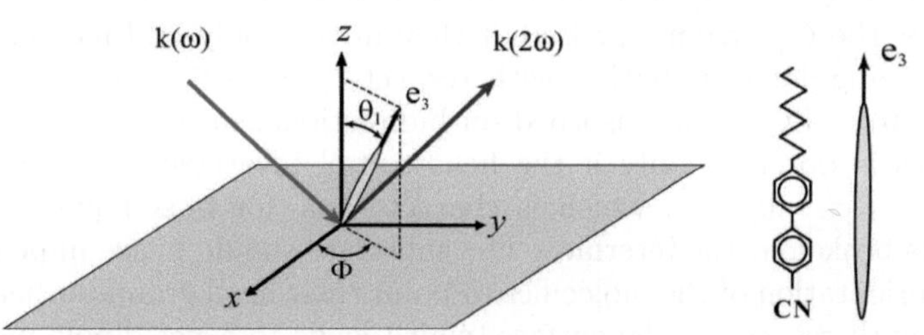

Fig. 5.1. Schematic representation of the SHG experiment denoting the orientation of a rod-shape liquid crystal molecule at the substrate surface. $k(\omega)$ and $k(2\omega)$ denote the wavevector of the incoming fundamental and reflected second harmonic waves. On the right hand side a 5CB molecule and its orientation vector e_3 are sketched. Directions x, y, z indicate the laboratory frame.

taken into account when analysing the SHG signal from such adsorbed LC films. This greatly simplifies (5.3) which is reduced to:

$$\chi^{(2)}_{a,lmn} = N_a < (\hat{e}_l\hat{e}_3)(\hat{e}_m\hat{e}_3)(\hat{e}_n\hat{e}_3) > \beta_{333} \tag{5.5}$$

where <> denotes averaging over molecular angular distribution and $\hat{e}_l$, $\hat{e}_m$, $\hat{e}_n$ are the unit vectors of the laboratory frame.

The number of non-vanishing components of $\chi^{(2)}_a$ depends on the intrinsic symmetry properties of the tensor and on the spatial symmetry of the adsorbed film. The latter can be strongly influenced by the symmetry of the underlying substrate structure. For example unidirectionally rubbed substrates are characterized by the point group symmetry C_{1v} (one vertical mirror, defined by the surface normal and by the rubbing direction), whereas for instance grooved films or photo alignment layers posses two vertical mirror planes, described by the point group C_{2v}. The latter case will lead to more restrictions to $\chi^{(2)}$ and consequently to less non-zero elements (in general, the higher the symmetry, the fewer non vanishing tensor elements [14]).

For azimuthally isotropic structures, as for instance glass substrates covered by a surfactant, the symmetry properties of the adsorbed LC film are described by the point group $C_{\infty v}$. The angular distribution of the adsorbed LC molecules in that situation is independent of the azimuthal angle φ and the non-vanishing components of $\chi^{(2)}_a$ are:

$$\chi^{(2)}_{a},_{zzz} = N_a < \cos^3\theta > \beta_{333}, \tag{5.6}$$

$$\chi^{(2)}_{a},_{zii} = \chi^{(2)}_{a},_{izi} = \chi^{(2)}_{a},_{iiz} = \frac{1}{2} N_a < \cos\theta\sin^2\theta > \beta_{333}; \mathrm{i{=}x,y}\ . \tag{5.7}$$

A detailed overview of all relevant $\chi^{(2)}_a$ components for nCB liquid crystals deposited onto unidirectional substrates can be found in [15], while bidirectional substrates are considered for instance in [16].

A common property of (5.6)–(5.7) as well as of the nonzero components of $\chi^{(2)}_a$ for the C_{2v} symmetry is that they involve only odd moments of the molecular angular distribution with respect to $\cos\theta$ [16]. This means that the SHG from LC layers adsorbed on bidirectional or azimuthally isotropic substrates is observed only if the head-to-tail invariance (i.e. the relation $f(\theta,\varphi) = f(\pi-\theta,\varphi)$), which is characteristic for bulk liquid crystalline phases, is broken at the interface. The substrate should hence impose a preferential orientation of the molecules to point their head groups either "down" or "up" with respect to the surface, which leads to a net dipole moment of the adsorbed LC layer along the surface normal $\hat{e}_z$. For the point group C_{1v} the SHG can appear also if the head-to-tail symmetry is present [15], but then the angular distribution of the LC molecules should have non-vanishing odd moments of $\cos\varphi$, which results in a net dipole moment of the adsorbed LC film along the $\hat{e}_x$ axis.

A standard way to resolve various components of $\chi_a^{(2)}$ is to measure the SHG intensity $I_{2\omega}$ for various polarization combinations of the fundamental and the second-harmonic optical field. The incident angle θ_1 of the fundamental beam with respect to the surface normal is conventionally set to 45°, to be sensitive to both in- and out- of-plane components, and $I_{2\omega}$ is analysed in the direction of the specular reflection. By rotating the sample around the surface normal $\hat{e}_z$ the azimuthal dependence of $I_{2\omega} \propto |\chi_s|^2$on the rotation angle Φ is detected.

A typical experimental set-up for SHG is shown in Fig. 5.2. For excitation nowadays one usually uses either a mode locked Ti: Sapphire laser, that delivers 100 fs pulses at a rate of about 80 MHz, or a pico or nanosecond Nd: YAG laser, often in combination with an optical parametric oscillator (OPO). Whereas the first type of system gives a superior mode quality and stability but limited tunability, the second system is more convenient for SHG spectroscopy.

The fundamental wavelength is filtered out from the SHG signal by using appropriate colour filters and a grating monochromator, whereas colour filters in front of the sample block any second harmonic signal generated by the optics. After proper spatial and spectral filtering, the outcoming specularly reflected SHG light is detected by a photomultiplier in connection with a photon counter. To facilitate the anisotropy experiments, the sample is mounted on a rotation stage, while an evaporation source or UV light source

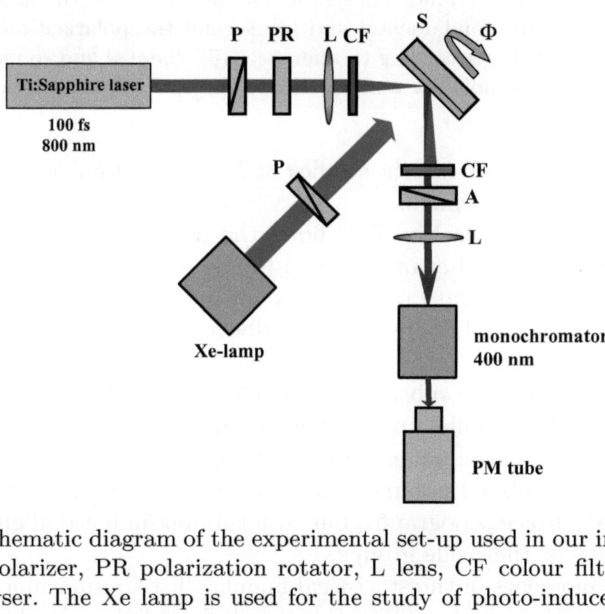

Fig. 5.2. Schematic diagram of the experimental set-up used in our investigations: P denotes polarizer, PR polarization rotator, L lens, CF colour filter, S sample, and A analyser. The Xe lamp is used for the study of photo-induced changes in photo-polymers. It can be replaced by an evaporation source to study in-situ the adsorption of liquid crystal molecules on substrates.

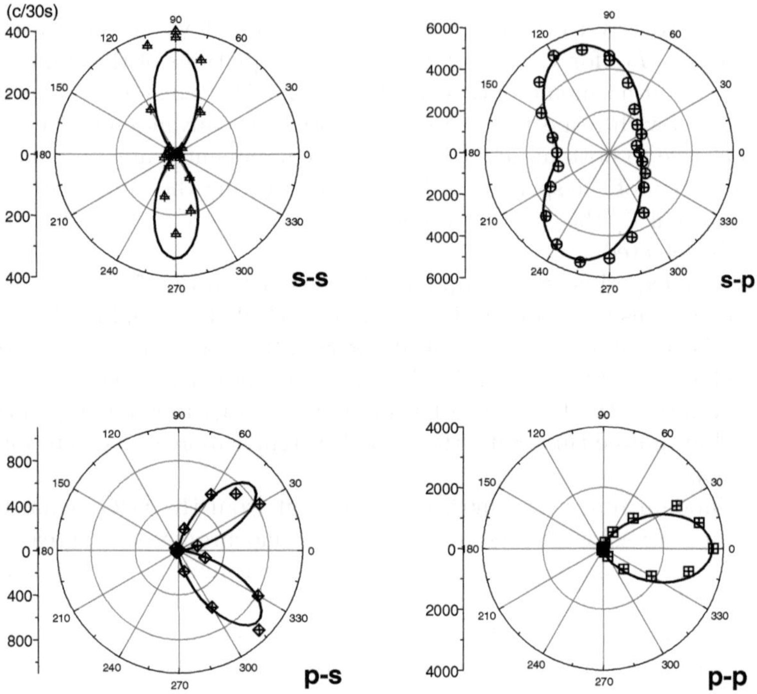

Fig. 5.3. Azimuthal (Φ) dependencies of the SHG intensity for an 8CB monolayer adsorbed on a rubbed polyimide substrate. The first index in the label designates the polarization of the fundamental and the second the polarization of the SHG beam. Solid lines are fits according to reference [15] where s- and -p indicate polarisation directions perpendicular and parallel to the optical plane respectively.

can be implemented for *in-situ* studies of LC adsorbtion or photo-induced alterations of photo-polymers.

As a typical example Fig. 5.3 shows the dependencies of $I_{2\omega}(\Phi)$ for a monolayer of 4'-n-octyl-4-cyanobiphenyl (8CB) molecules evaporated onto a rubbed polyimide substrate prepared by an industrial-level rubbing procedure (Philips Research Laboratories, Eindhoven). A profound anisotropy of the signal with respect to the rubbing direction ($\Phi = 0$) is evident. An interesting feature of these substrates was that despite their strong alignment effect on the LC molecules, practically no grooves were observed on their surface by the AFM technique. This result supports the hypothesis that the LC alignment on rubbed polymer surfaces is not necessarily caused by their unidirectional surface topography, but by a rubbing-induced alignment of the polymer chains or their side groups.

Different moments of the statistical angular distribution function $f(\theta, \varphi)$ contain information on the orientational ordering of the adsorbed LC molecules and can be deduced via particular components of $\chi^{(2)}_{a,lmn}$. For nCB mesogens it is generally found that the rigid cyanobiphenyl core makes an interfacial angle of $\sim 70°$ with respect to the surface normal. This appears bo be quite independent of the substrate treatment, and thus is presumably determined by the conformational restrictions of the molecule [17]. For that reason f is assumed to have the form

$$f(\theta,\varphi) = \left(\frac{\delta(\theta-\theta_0)}{2\pi\sin\theta_0}\right)\left(\sum_{n=0}^{3} a_n \cos n\varphi\right) \tag{5.8}$$

involving five fitting parameters. The first of them, i.e. the tilt angle θ_0 of the adsorbed LC molecules with respect to the surface normal, can then be calculated from

$$\theta_0 = \tan^{-1}\left(\pm\sqrt{\frac{\chi_{a,zxx}+\chi_{a,zyy}}{\chi_{a,zzz}}}\right). \tag{5.9}$$

The surface in-plane order parameter $Q_a =< \cos 2\varphi >$, which is a measure of the substrate alignment strength for the adjacent LC molecules, is given by

$$Q_a = a_2 = \frac{\chi_{a,zxx}-\chi_{a,zyy}}{\chi_{a,zxx}+\chi_{a,zyy}}. \tag{5.10}$$

Although being very simplified, these relations provide a convenient way to analyze the effects of various parameters such as rubbing strength or rubbing speed on the liquid crystal ordering. Numerous SHG studies of the liquid crystal-solid interface using the above described formalism are reported in the literature and have substantially contributed to our general understanding of the surface anchoring and surface alignment phenomena [4, 15, 18–29]. To improve the analysis and achieve a more realistic form of $f(\theta,\varphi)$, one may apply additional statistical principles, like for instance the maximum entropy method [30, 31]. The balance between the benefits and possible shortcomings of using more advanced fitting methods is for the moment, however, still a matter of debate and there is also no consensus yet about which of the various possible forms of $f(\theta,\varphi)$ is most relevant.

The investigations of solid-liquid crystal interaction by SHG are not only restricted to LC monolayers and thin films. As most of the bulk liquid crystalline phases are centrosymmetric, the surface SHG response can be detected also from thick liquid crystal "cells", in which the material is sandwiched between two substrates. By this kind of measurements it is possible to study a relation between the bulk and the surface liquid crystalline ordering and consequently probe various pretransitional phenomena and other surface related effects.

It is, however, necessary to mention that, in contrast to thin LC films, in thick LC cells the quadrupolar and other bulk contributions to the net SHG signal are not negligible and have to be properly taken into account in the data analysis. For example, SHG measurements at the nematic-isotropic phase transition of 8CB have shown that the bulk contribution to the SHG signal has an opposite phase with respect to the surface contribution [32].

They as well revealed that the LC monolayer adjacent to the rubbed polymer substrate remains aligned also in the isotropic phase [20], consistent with NMR and AFM experiments (see Chaps. 2 and 3). The opposite is also true: on several rubbed polymer films that induce a strong bulk alignment, SHG experiments show the first LC layer has an isotropic azimuthal distribution. An example of that will be shown in Sect. 5.3 below.

5.3 SHG Study of Surface Nematic Order Induced by Rubbed Silane Derivatives

In this part we show how SHG can be used to probe the LC ordering, induced by different silane treated surfaces, which are well known substrates for LC alignment.

Three different silanes were used to generate chemically stable alignment films on indium tin oxide (ITO) coated glasses: methyl-trimethoxysilane (MMS), methyl-propoxysilane (MPP) and methyl-triethoxysilane (MES) which present different lengths of their saturated hydrocarbon chains [33,34]. By investigating the SHG spectra from the ITO coated glasses, both clean or coated by silane, it was shown that none of these silanes generated a significant SHG signal before deposition of 4'-n-pentyl-4-cyanobiphenyl (5CB).

As expected, the signal from a 5CB monolayer on unrubbed substrates was azimuthally isotropic and the tilt angle of the 5CB molecules in the monolayer was determined to be $\theta_0 \approx 65^o$ for all the unrubbed silanes. By rubbing these substrates before 5CB deposition one expects to induce a C_{1v} symmetry and consequently an anisotropic SHG signal.

Figure 5.4 shows the azimuthal dependencies of SHG signals for the $p-p$ polarization combination of 5CB on rubbed MMS (a), and MES (b). For

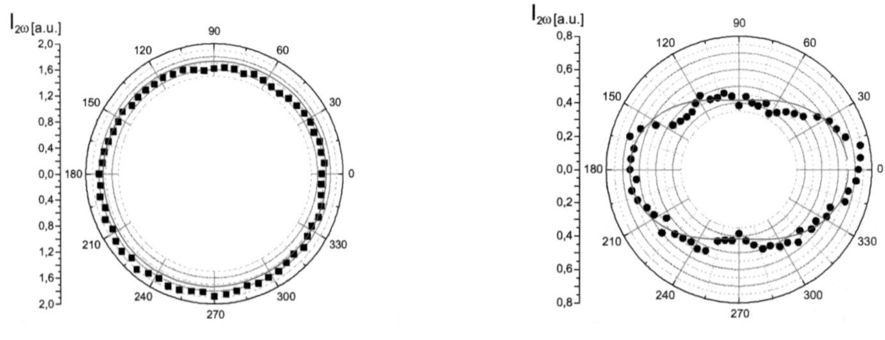

Fig. 5.4. Azimuthal dependencies of the SHG signal detected for a 5CB monolayer adsorbed onto rubbed MMS(a) and MES(b) substrates measured for $p-p$ polarization combinations of the fundamental and the second-harmonic beam.

both rubbed and unrubbed MMS (Fig. 5.4a) and MPP (not shown) coated substrates, the SHG signal shows an isotropic dependence on sample rotation around the surface normal, indicating that there is no detectable anisotropy in the distribution of the 5CB molecules. This is not a surprise as one will not expect an anisotropic interaction of 5CB molecules with the non-polar silane tails randomly distributed in space. From the measured values of the susceptibility components, a tilt angle of $72^o \pm 3^o$ was obtained for the rubbed MMS and MPP. Considering the error bars, this is practically the same as for the unrubbed case.

The third material, MES, behaves differently. In this case the adsorbed LC film does show an anisotropy (Fig. 5.4b): the 5CB molecules follow the rubbing direction. The surface alignment is characterized by $Q_a \approx 0.3$ and $\theta \approx 71^o$. The value of the surface in-plane order parameter is in the lowest range of the usual values for rubbed polymer surfaces: $Q_a \approx 0.3 - 0.5$ [29,35,36]. The observed tilt angle is the same as for the other silanes and comparable with alignment angles of cyano-biphenyl monolayers usually found on substrates that induce a planar macroscopic alignment [15, 22].

By preparing thin cells with rubbed and unrubbed silane-coated substrates, their macroscopic alignment properties could be studied by means of polarized optical microscopy. The cells made with unrubbed plates were inhomogeneous, with only small randomly oriented domains, whereas all the rubbed substrates, including to our surprise the MPP and MMS cases, showed uniform alignment. This behaviour shows that a macroscopic alignment of a nematic LC on a silane coated substrate does not necessarily depend on a uniform alignment of the first monolayer in contact with the substrate. These results are in agreement with the experiments made by Chen and co-workers [15]: they have shown that a rubbed surfactant-coated substrate does not orient the adsorbed first LC monolayer while it does induce a macroscopic alignment in a cell. For the MES case, the rubbing presumably has induced some anisotropy in the surface morphology (grooves) that was sufficient to align the LC molecules.

It is interesting to note that in all cases, comparable tilt angles were found. This implies that the rubbing of silanes does not affect the polar orientation of the 5CB molecules; only the azimuthal orientation is influenced.

5.4 SHG Study of Photo-induced Changes in Poly(vinyl) Cinnamate (PVCN)

As discussed in Chapter 1, photo-polymers that have been unidirectionally modified by exposure to linearly polarized light have the ability to induce alignment of liquid crystals [37–41]. This works presumably via anisotropic van der Waals forces [38], similar to the alignment on conventionally rubbed polymer layers [18,22]. In spite of this similarity, generally there are large differences found between the surface anchoring strengths of these two types of substrates. Rubbed polymers, like polyimides, usually give strong anchoring with an azimuthal surface anchoring energy coefficients $W\varphi \approx 10^{-4}$ Jm^{-2} [42]. In contrast, unidirectionally photo-modified polymers like photoresists and various kinds of azo-polymers give weak surface anchoring with $W\varphi < 10^{-5}$ Jm^{-2} [27, 43, 44]. The origin of this difference is not yet well understood, but must be related to details in the structure of the top layer of the polymer. Therefore, one may gain insight in the alignment mechanism by SHG experiments, as will be shown below.

As example we take poly(vinyl cinnamate) (PVCN), one of the best known photo-polymers for the alignment of liquid crystals, which primarily belongs to the class of photo-crosslinkable polymers, though in addition also photo-isomerization takes place [38,45,46] (see also previous Chapter). The observed maximum value of $W\varphi \sim 10^{-6}$ Jm^{-2} for PVCN is about two orders of magnitude lower than the values known for standard rubbed polyimide surfaces [42]. This difference is surprising, because the observed birefringence of the linearly photo-polymerized (LPP) PVCN films, $\Delta n \approx 0.02$ [38, 43, 44, 47], is only slightly lower than the birefringence of the rubbed polyimide layers [17]. Another unusual feature of the LPP PVCN is also that the dependence of $W\varphi$ on the UV exposure time is about 10 times slower than that of Δn [43, 44]. Such an uncorrelated behaviour of $W\varphi$ and Δn strongly disagrees with theoretical predictions [48, 49], hence it was suggested that the axial selectivity and kinetics of the photo-reaction processes on the surface of the PVCN might be very different from those in the bulk [43].

These apparent contradictions could be resolved by an SHG study of the photo-polimerization process of PVCN and of the alignment of LC molecules on the thus prepared PVCN substrates. Due to the unsaturated carbon-carbon double bounds of the cinnamoyl side groups (see Fig. 1.9), the SHG response (at $\lambda \sim 800$ nm) of a PVCN layer is of the same order of magnitude as the SHG response of an adsorbed nCB film. This means that in this case the substrate contribution $\chi_0^{(2)}$ to the SHG signal (see (5.2)) cannot be ignored. At the same time, this provides a convenient possibility to probe the surface anisotropy of both the substrate and the LC adsorbate.

Effect of Unpolarized UV Light

To characterize photochemical changes of the PVCN and their relation to the SHG signal, the irradiation with unpolarized UV light was first examined. Figure 5.5 shows the dependence of $I_{2\omega}$ for the p polarized SHG as a function of the linear polarization angle δ of the fundamental laser beam for a PVCN sample exposed to different unpolarized UV irradiations. $\delta = 0°$ corresponds to the s polarization. The upper curve corresponds to the measurement before the UV irradiation and the lower two to measurements after 2 min and after 5 minutes of UV exposure, respectively. These graphs show that UV exposure, decreases the magnitude of $I_{2\omega}$ as well as modifies its angular dependence: after 5 min of exposure the value of $I_{2\omega}$ ($\delta = 0°$) is decreased about twice

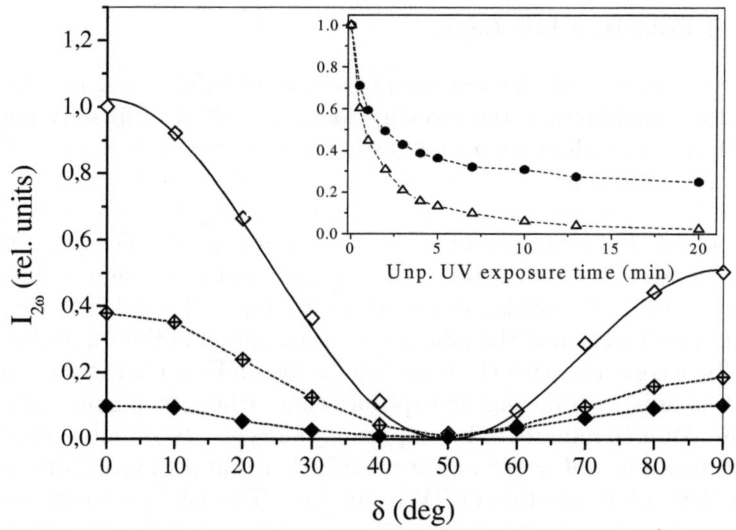

Fig. 5.5. The intensity of p polarized SHG as a function of the polarization direction of the fundamental beam: before UV irradiation (open diamonds), after 2 minutes of unpolarized UV exposure (crossed diamonds), and after five minutes of the unpolarized UV exposure (solid diamonds). The value $\delta = 0°$ corresponds to s input polarization. The solid line is a fit according to reference [50]. Inset: relative decrease of the SHG intensity for $\delta = 0°$ $s-p$ (triangles) and $\delta = 90°$ $p-p$ (dots) as a function of irradiation time. The dashed lines are guide to the eye.

as much as $I_{2\omega}(\delta = 90°)$. This effect can be more clearly noticed in the inset of Fig. 5.5 which shows a relative decrease of $I_{2\omega}(\delta = 0°)$ ($s-p$ signal) and $I_{2\omega}(\delta = 90°)$ ($p-p$ signal) as a function of the UV exposure time.

Comparison between the SHG data and the results of UV absorption spectroscopy reveals a prominent proportionality of the NLO response to the concentration of the *trans*-cinnamoyl side groups $c_{tr}(t)$ [50, 51]. This feature can also be explained by a theoretical analysis based on quantum-chemical calculations of the second order molecular hyperpolarizability β of the cynnamoyl group by using the MOPAC93 program [52]. These calculations showed that the SHG signal from *cis* comformers is about ten times lower than the SHG signal from *trans* comformers, while the SHG contribution from the dimers is negligible.

Therefore the fast initial decrease of the SHG signal for both polarization combinations can be associated with the initial depletion of the *trans*-cinnamoyl groups due to the *trans-cis* isomerization, while the succeeding slower decrease is mainly related to the cross-linking (as depicted in Fig. 1.9). For prolonged exposure times the $s-p$ signal almost vanishes, while the $p-p$ signal approaches a constant value of around 30% of its initial magnitude. This remaining signal can be related to the SHG contribution from the PVCN-glass interface.

Effect of Polarized UV Light

Due to the absence of any remaining bacground SHG signal for the $s-p$ polarization combination, the investigation of the effect of linearly polarized UV light on these photopolymer films is most conveniently done in this geometry. The increasing anisotropy due to the axially selective polymerization is demonstrated by measurements of the azimuthal dependence of $I_{2\omega,s-p}(\Phi)$ for various LP UV exposure times, as shown in Fig. 5.6. The angle $\Phi = 0°$ corresponds to the situation when the direction of the UV light polarization and the s polarization of the optical beams were parallel and is hence associated with the direction of the minimum concentration of the *trans*-cinnamoyl groups. In accordance with the behaviour shown in Fig. 5.5, the average SHG signal decreases with increasing exposure time, while its angular anisotropy increases. After 60 minutes of UV exposure, the $I_{2\omega,s-p}(\Phi)$ exhibits two nearly the same maxima at $\Phi = 90°$ and $\Phi = 270°$, that is at orthogonal orientations to the polarization direction of UV irradiation. The solid lines represent a fit including the nonvanishing susceptibility components that correspond to the C_{2v} symmetry [16]. Thin LC cells prepared with these photoalignment substrates exhibited homogeneous alignment perpendicular to the polarization direction of the UV light, i.e. parallel to the direction of the SHG maxima at $\Phi = 90°$ and $\Phi = 270°$. This confirms that the *trans*-cinnamoyl groups are mostly responsible for the LC alignment.

From the experimental data shown in Fig. 5.5 the average tilt angle of the cinnamoyl groups on the surface of the unexposed PVCN film is determined as $\theta_{tr} = 78° \pm 10°$, which is in good agreement with theoretical models describing photopolimeryzation of PVCN [48, 49]. The ratio χ_{s-p}/χ_{p-p} stayed almost constant during the entire time of the unpolarized UV exposure, suggesting that the value of θ_{tr} is not modified by the photoreactions. Therefore, the observed decrease of $\chi^{(2)}$ during UV irradiation takes place predominantly due to a decrease of the surface density $N_S(t)$ of the *trans*-cinnamoyl groups and not due to their reorientation. This conclusion also explains the observed

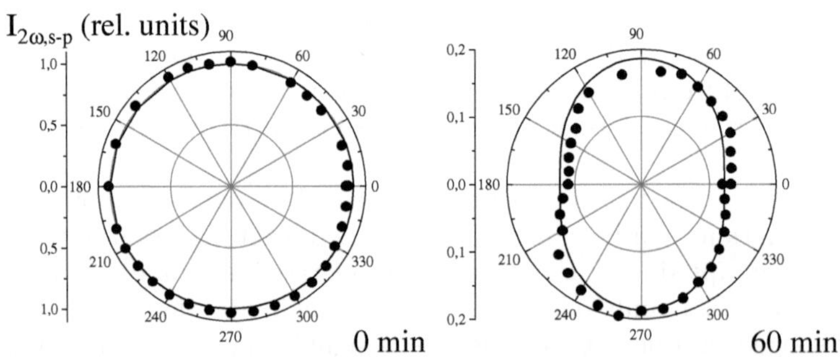

Fig. 5.6. Azimuthal dependencies of the $s-p$ SHG intensity before and after 60 minutes exposure to LP UV irradiation. $\Phi = 0$ corresponds to the UV polarization direction. Solid lines are fits according to reference [50].

strong correlation between $\chi^{(2)}(t)$ and the bulk concentration of the *trans* groups $c_{tr}(t)$. From this it also follows that, contrary to some assumptions given in the literature, the depletion of *trans*-cinnamoyl groups on the surface of the PVCN occurs in a manner very similar to that in the bulk.

Surface Order Parameter

From the fitted susceptibility components, the surface order parameter Q_{tr} can also be obtained using (5.10). Figure 5.7 shows that Q_{tr} monotonously increases with increasing exposure time. The maximum detected value of Q_{tr} is about 0.12, which is in good agreement with the result achieved by linear dichroism measurements [51] and is also very similar to the value of Q_s recently observed by SHG mesaurements on a photo-isomerizable azo-side-chain polymer film [53]. On the other hand, $Q_{tr} = 0.12$ is significantly below the value of the bulk order parameter of the LPP PVCN ($|S| = 0.4$) calculated on the basis of birefringence measurements [43]. The disagreement suggest that, in contrast to the SHG case, the contributions of dimers and *cis*-conformers to the linear optical response are quite important.

Figure 5.7 also shows the time dependence of the relative surface density of the *trans*-cynnamoyl groups $N_S(t)/N_S(t=0)$ during the LP UV exposure, as obtained from the SHG data. It again reveals the presence of two separated photo-depletion processes. From these data we can calculate the dependence of the alignment coefficient $w_{tr} \propto Q_{tr}(N_S)^{1/2}$ of the *trans* cinnamoyl groups on the LP UV exposure time. This parameter is related to that part of the birefringence that is due to the azimuthal anisotropy of the *trans* cinnamoyl groups [54]. The time dependence $w_{tr}(t)$ is shown in Fig. 5.8. For comparison also the result of the measurement of the net bulk induced birefringence $\Delta n(t)$ of the same sample is given. During the initial stage of the LPP process, which is associated with the *trans-cis* isomerization, the $w_{tr}(t)$ and $\Delta n(t)$

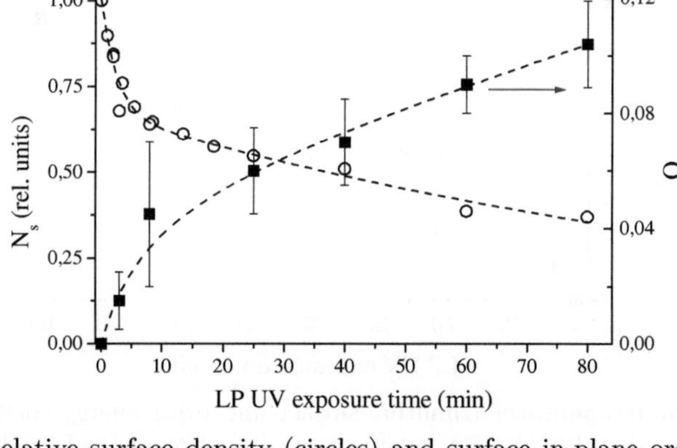

Fig. 5.7. Relative surface density (circles) and surface in-plane order parameter (squares) of trans-cinnamoyl side groups as a function of the LP UV exposure time.

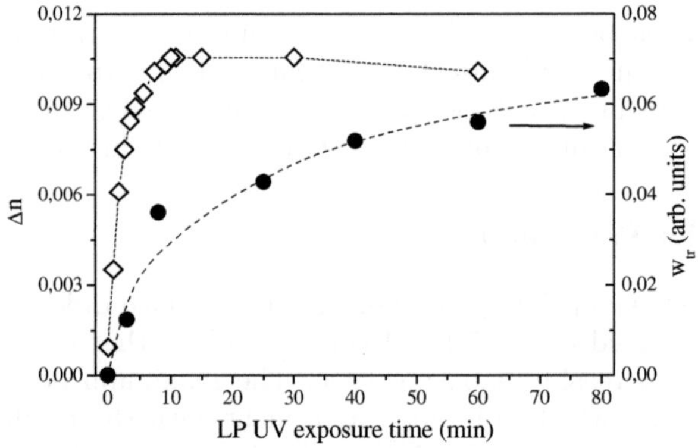

Fig. 5.8. Birefringence of the LPP sample (diamonds) and calculated alignment coefficient of the trans-cinnamoyl groups (dots) as function of the LP UV exposure time. Lines are guides to the eye.

similarly increase with increasing exposure time. After this, during the cross-linking, a profound difference can be noticed. The $\Delta n(t)$ exhibits a plateau and later on a slow decrease - a dependence which is usually reported in the literature [38, 43, 44, 47]. The $w_{tr}(t)$, on the other hand, continues to increase also during the cross-linking. This result suggests that the dimers which are formed by the cross-linking very probably reduce the birefringence $\Delta n(t)$. A possible explanation for this effect is that the dimers are actually oriented along the UV polarization direction and not perpendicularly to it, as is generaly assumed in the literature.

Measurements of the LC azimuthal anchoring energy coefficient $W\varphi$ of the PVCN substrates (Fig. 5.9) were performed by preparing twist cells with

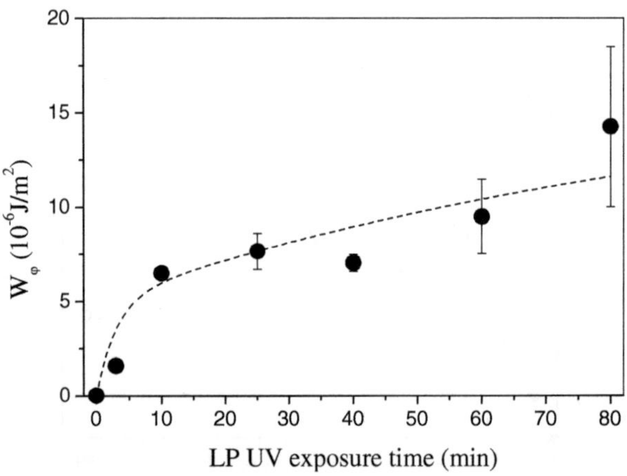

Fig. 5.9. Room temperature azimuthal surface anchoring energy coefficient of the 5CB liquid crystal on LPP PVCN substrate as a function of the LP UV irradiation time. The dashed line is guide to the eye.

a strongly rubbed polyimide layer as a reference surface [55, 56]. One can notice that the observed behaviour of $W\varphi$ on the LP UV exposure time is very similar to the dependence of $w_{tr}(t)$. This correlation supports the earlier conclusion that the anchoring of liquid crystals on the PVCN surface is mainly determined by their interaction with the *trans* conformation of the side groups.

Over-writing of Optically Induced Alignment

One of the interesting features of an optically induced alignment is the possibility for a subsequent realignment of the preferential axis and hence liquid crystal orientation direction. This effects has been recently reported also for PVCN [47]. To characterise the related properties of such PVCN substrates, the effect of two subsequent photo-polymerizations on the azimuthal SHG anizotropy is shown in Fig. 5.10. At first the $I_{2\omega,s-p}(\Phi)$ was measured after 40 min of the LP UV exposure with the UV polarization parallel to the s direction i.e the $\hat{e}_y$axis. Then the UV polarization direction was set parallel to the $\hat{e}_x$ axis and the sample was further exposed for 90 minutes. After this the $I_{2\omega,s-p}(\Phi)$ was measured again. The average value the SHG signal as usually decreases with the additional UV exposure, while the typical unidirectional azimuthal dependence of the $I_{2\omega,s-p}(\Phi)$ is strongly perturbed. A modification of the direction of the photoalignment axis is therefore feasible, but in each step the concentration of the *trans*-cinnamoyl side groups is strongly reduced. As the *trans* groups are found to be the most important factor for the liquid crystal alignment, this means that the anchoring strenght of the substrate will strongly decrease after each realignment.

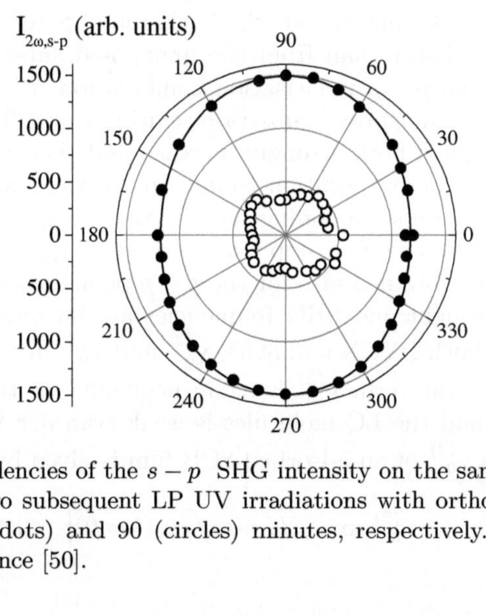

Fig. 5.10. Dependencies of the $s-p$ SHG intensity on the sample rotation angle Φ measured for two subsequent LP UV irradiations with orthogonal polarization directions for 40 (dots) and 90 (circles) minutes, respectively. Solid line is a fit according to reference [50].

To summarize, the SHG results show conclusively that the anchoring of liquid crystals on LPP PVCN is governed by the in-plane order of the *trans* conformation of the side groups. Because the UV light causes trans-cis isomerisation and cross-linking, i.e. leads to a removal of these transgroups instead of a reorientation, the resulting ordering is rather weak and perpendicular to the polarization direction of the UV light. The SHG data allow to probe the alignment coefficient of the *trans*-groups and also show why W_φ is not proportonial to the induced birefringence [16, 50, 51].

5.5 SHG Study of Liquid Crystal Alignment on PVCN

As was already demonstrated in Sect. 5.3 for LC films adsorbed on silane coated surfaces, the intrinsic interface sensitivity of SHG can be further exploited to study the ordering of LC molecules deposited on photo-modified PVCN films [57]. In this way some of the ideas discussed in the previous section can be verified directly.

Figure 5.11 shows the dependence of $I_{2\omega}$ on evaporation time of 8CB on an unexposed PVCN substrate and on one that was exposed to unpolarized UV light for 20 minutes. The unexposed substrate shows a strong increase of the signal in the $s-p$ polarization combination and a much weaker but similar behaviour for the $p-p$ polarization combination. The initial almost linear increase is followed by a plateau and later on by a step like change. Further deposition leads to a gradually decreasing dependence for long evaporation times. The results for the UV exposed substrate are significantly different. From the SHG results discussed in the previous section we know that such exposures lead to a practically complete photo-transformation of the cinnamoyl side groups. Because of this the background SHG signal is about an order of magnitude lower than from the unexposed substrate and appears predominantly in the $p-p$ polarization combination, which is opposite to the situation for the unexposed substrate. During the 8CB evaporation the $s-p$ and $p-p$ signals both strongly increase and then reach a plateau of similar magnitude. The evaporation time to reach the plateau is about ten times longer than the corresponding one for the unexposed substrate. The inset to Fig. 5.11 shows again a step like decrease of $I_{2\omega,s-p}$ for $t > 100$ min. Similar results were obtained also for the $p-p$ polarization combination.

Because the incident and SHG frequencies are far enough from the absorption bands of both PVCN and 8CB, $\chi_0^{(2)}$ and $\chi_a^{(2)}$ in (5.2) are both real. In addition to this, the term $\chi_{int}^{(2)}$ is again negligible, as the interaction between the PVCN and the LC molecules is weak (van der Waals) [38]. With these assumptions $\chi_a^{(2)}$ of an adsorbed 8CB film is given by:

$$\left|\chi_a^{(2)}\right| \propto \left|\sqrt{I_{2\omega}} - \sqrt{I_{2\omega}(t=0)}\right| \tag{5.11}$$

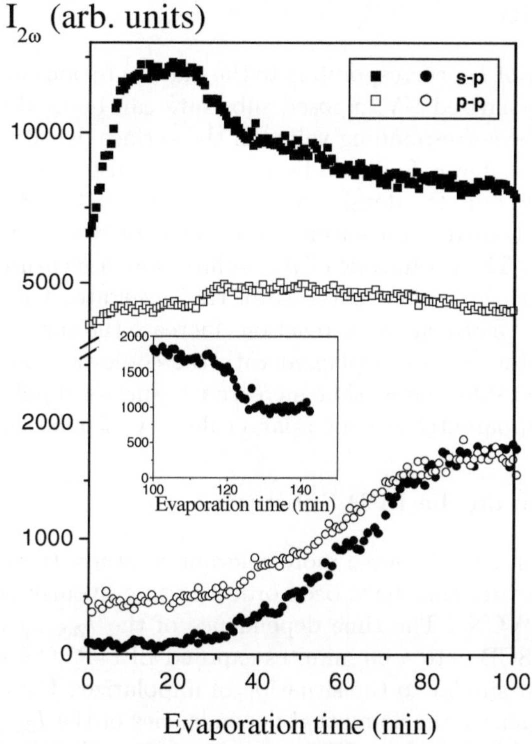

Fig. 5.11. Time dependence of the SHG intensity during evaporation of 8CB on unexposed PVCN substrate (squares) and on PVCN substrate exposed to unpolarized UV light for 20 min (dots). The inset shows the long term behaviour.

where $I_{2\omega}$ is the SHG from the substrate covered with 8CB and $I_{2\omega}(t = 0)$ is the signal from the bare substrate. The $-$ sign of the relative phase of these two contributions follows from the variation of $I_{2\omega}(t)$ during evaporation, that indicates that the contributions from the PVCN substrate and from the 8CB layer add constructively (in-phase).

The first plateau in $I_{2\omega}(t)$ is generally supposed to coincide with the formation of the first complete monolayer of the LC molecules [21]. The analysis of the corresponding SHG data thus provides information on the nonlinear susceptibility $\chi_a^{(2)}$ of the first monolayer and from this the orientation θ_0 of the 8CB molecules can be deduced, yielding $\theta_0 = 65° \pm 2°$ for 8CB on the unexposed and $\theta_0 = 76° \pm 5°$ for 8CB on the UV exposed PVCN substrate. The photo-modification thus appears to increase the average surface tilt angle of the molecules in the adjacent 8CB monolayer. The obtained values of θ_0 are very similar to the polar angles of nCB molecules usually found on substrates that induce a planar alignment [15, 22].

Surface Polarity

The magnitudes of $I_{2\omega}$ corresponding to the plateau regions for the unexposed and for the unpolarized UV exposed substrate can be used to calculate the ratio between the corresponding values of the surface density N_a of adsorbed 8CB molecules that was found to be 1 : 4. As discussed in the introduction, N_a actually gives only the density of the polar oriented 8CB molecules, while the molecules attached to the surface in head to tail "pairs" do not contribute to the SHG [21]. The magnitude of N_a is thus more a measure for the surface polarity than for the surface coverage of the substrate. The SHG data thus suggest that the photo-chemical reactions increase the surface polarity of the PVCN. The latter may be explained with the conformational modifications of the cinnamoyl side chains, that increase the surface dipole moment of the substrate and apparently provide a larger density of polar surface sites.

LC Deposition on the LPP Substrates

For applications it is of course more relevant to study the LC organization on PVCN substrates that have been ordered using linearly polarized UV radiation (LPP-PVCN). The time dependence of the $I_{2\omega,s-p}$ detected during evaporation of 8CB onto a 60 minutes exposed LPP-PVCN substrate exhibited a behaviour similar to the situation of unpolarized UV exposure.

Figure 5.12 shows the azimuthal dependencies of the $I_{2\omega,s-p}$ and $I_{2\omega,p-p}$ of this sample and of the bare LPP substrate, where $\Phi = 0°$ as usually corresponds to the incident s polarization parallel to the direction of the UV polarization (see Fig. 5.1). The results show that the LP UV induced anisotropy is considerably reduced after the 8CB adsorption. This indicates that the SHG

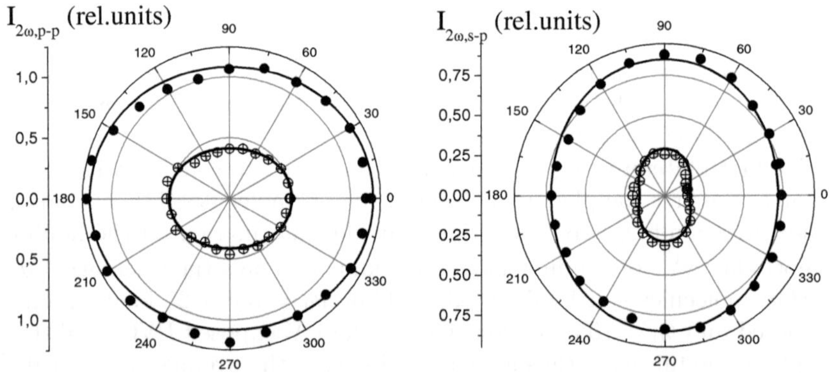

Fig. 5.12. Dependence of $s-p$ and $p-p$ SHG intensity on the rotation angle Φ for bare PVCN substrate exposed to 60 min of LP UV (crossed circles), and the same substrate with the deposited 8CB monolayer (closed circles). Solid lines are fits according to reference [57].

contribution from the adsorbed 8CB film is practically isotropic. From the measurements shown in Fig. 5.12 the in-plane order parameter of the substrate and of the 8CB monolayer can be deduced. The resulting values are $Q_{tr} = 0.15$ and $Q_a \sim 0.02 \pm 0.02$.

The value of $Q_{tr} = 0.15$ of the PVCN is only 2–3 times smaller than the usual values of $Q_s \sim 0.3 - 0.5$ found for rubbed polymer surfaces [19, 29, 35]. This suggests that the weak surface anchoring of liquid crystals on PVCN is not due to its low surface anisotropy, but arises from some other difference in the alignment mechanism. As discussed in the previous section, the *trans*-cinnamoyl side groups are probably most responsible for the LC alignment of PVCN. However, the SHG results have shown that during the LP UV exposure the concentration of these groups is strongly reduced due to their photo-transformation, resulting in a low concentration of relatively well aligned anchoring sites in a matrix of less effective photoproducs. This will lead to very low values of Q_a or, in other words, to almost no ordering in the first LC layer. Note that, as shown in Sect. 5.3.2, even on (rubbed) substrates with a large surface anchoring energy, the first LC monolayer often shows no measurable azimuthal ordering. This actually shows that the surface order parameter Q_s, as derived from SHG experiments, might not be a good measure for the azimuthal surface anchoring W_φ.

5.6 Structure of Multilayer 8CB Films Evaporated onto Solid Substrates

The previous section showed some unusual nonmonotonous time dependencies of the SHG signal detected during the evaporation of 8CB molecules on PVCN substrates (Fig. 11). These features suggest that the growth of multilayer 8CB films took place. In the following we will demonstrate how SHG can be used to get more insight into the corresponding growth process and particularly to resolve the problem of the structure formed above the first monolayer. The results thus obtained appear to be complimentary to and support the ellipsometry study discussed in Chap. 4.

Despite the numerous SHG studies of orientational ordering of nCB molecules evaporated onto various surfaces, the technique was very rarely used to probe films thicker than a monolayer. B. Jérôme at al., for instance, analysed prolonged evaporation of 5CB on mica and reported that the layer-by-layer growth had terminated after formation of the first monolayer [58]. In some other experiments a slow decrease of $I_{2\omega}$ was observed after the plateau region, and was related to the quadrupolar contribution from the multilayer bulk-like film formed on top of the interfacial monolayer. But to generate a comparable quadrupolar SHG response, the films should be relatively thick. This is contrary to the general observation that none of the several attempts to achieve a homogeneous multilayer nCB structure has been successfull up till now, as the adsorbate always forms droplets regardless of the type of the

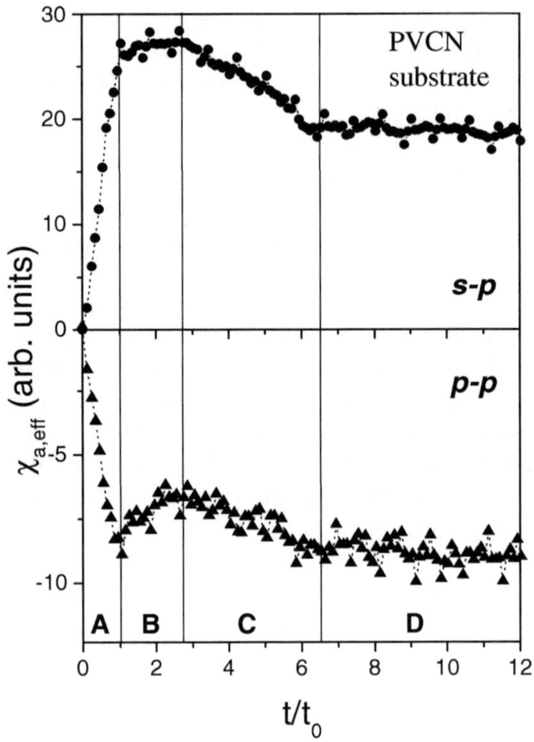

Fig. 5.13. The $s-p$ and $p-p$ components of the nonlinear optical susceptibility of 8CB evaporated onto a PVCN substrate as a function of deposition time. The scale is given relatively with respect to the time necessary to reach the first nearly flat region of the predominant ($s-p$) SHG signal. Vertical lines denote borders between various growth stages.

substrate [59]. It is, however, not clear whether the droplets are superimposed on the bare substrate or on a specific thin interfacial film [60–62].

Figure 5.13 shows a typical dependence of the effective susceptibility of the adsorbed LC layer χ_a as a function of the evaporation time of 8CB on a UV unexposed PVCN substrate. The relative phase of the signals in the $s-p$ and $p-p$ configurations was resolved by measuring the dependence of $I_{2\omega}$ on the incident linear polarization angle δ (similar to the dependencies shown in Fig. 5.5) and indicated that the $s-p$ and $p-p$ components of χ_a have opposite phase during the entire experiment. As the exact growth rate of the 8CB films depends on the temperature, configuration and position of the evaporation source, the time scale is given in units of t_0 corresponding to the starting point of the first nearly flat region of the $s-p$ signal. This plateau is generally associated with completion of the first full monolayer of 8CB molecules on the surface [15, 21, 63]. Typical values of the t_0 at given experimental conditions were about 10 minutes. To simplify the designation, the observed features are devided into four characteristic time intervals, which are denoted in the figure as the regions A, B, C and D. For $t > t_0$ (region A) the magnitude of both susceptibility components increases linearly with time. For $t_0 < t < 2.7t_0$ (region B), the $s-p$ component is almost constant, while the magnitude of the $p-p$ component slightly decreases. The interval of $2.7t_0 < t < 6.5t_0$ (region C) is again related to a nearly linear variation, but now the magnitude of the $s-p$ component decreases, while the magnitude of the $p-p$ component increases. In the last time region (D), corresponding to $t > 6.5t_0$, both susceptibility components remain practically constant, though, for prolonged evaporation a slow continuous decrease is usually observed.

To test whether the observed features are specific for the PVCN substrates, we repeated the SHG measurements by using quartz glass substrates. The result is shown in Fig. 5.14. Though the nonlinear optical response of 8CB films evaporated on quartz plates is higher than in the case of PVCN

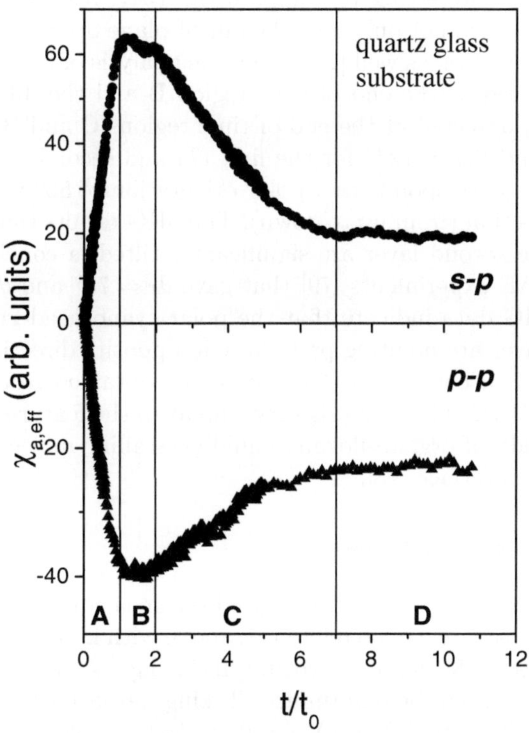

Fig. 5.14. The $s - p$ and $p - p$ components of the effective nonlinear optical susceptibility of 8CB evaporated onto a quartz substrate as a function of deposition time. The data are a superposition of measurements obtained for three different temperatures of the evaporation source. The scale is given relative with respect to the time necessary to reach the first nearly flat region of the predominant $s - p$ SHG signal. Vertical lines denote borders between various growth stages.

substrates, qualitatively the behaviour of the signals is very similar. Only the $p-p$ component shows an opposite behaviour with respect to PVCN within the time regions B and C. The higher absolute values are related to the higher density of molecules on this more polar substrate (see also discussion below).

As discussed in the previous Chap. 4, ellipsometry results indicate that the growth of the first layer takes place in regions A and B, whereas in region C a subsequent bilayer is formed. This is contrary to the usual conclusion from SHG experiments that suggest the monolayer growth to be completed at the starting of the plateau region of the SHG signal, i.e. at the end of time region A. According to the discussion given in the Introduction, this discrepancy can be explained by the fact that SHG probes only the density of the polar oriented part of the adsorbed molecules N_p while the molecules attached to the surface in head-to-tail "pairs" practically do not contribute to the SHG. The "pairs", however, do contribute to the BAE signal and therefore the latter gives a better calibration of the first monolayer formation.

It is therefore interesting to compare the BAE results with those from SHG, as the latter cannot only give the out of plane orientation, but in principle the polar direction as well [19]. From the analysis of the SHG data, using the values detected at the end of time region B and the difference between the values of χ_a detected at the end of time regions C and B (Fig. 5.13), we find $\theta_I = 68°$ and $\theta_{II} = 124°$ for the first (I) and second (II) layer respectively. Here 124° corresponds to a polar orientation of 56° with an inversion of the polar direction (from up to down). The SHG results thus show that the molecules in the second layer are significantly tilted, a conclusion that also follows from BAE experiments [79] that gave $\theta_I = 78°$ and $\theta_{II} = 40°$ Additionally, the SHG data indicate that the polar cyano head-groups ($-$C$\equiv$ N) in the second layer are pointing preferably in opposite direction with respect to the first layer, a conclusion that *cannot* be obtained from linear optical experiments as BAE. This result is very similar to the features recently found by the SHG study of organosiloxane liquid crystalline molecules compressed at the air-water interface [78].

Polarity of 8CB Multilayer

Because in our experiments the SHG signal was measured only relatively with respect to the substrate, i.e. no reference sample with known N_p was available, the exact values of N_p for the two interfacial layers can not be evaluated. However their ratio can be determined. Taking into account the above given values for θ_0 it follows that $N_{pII}/N_{pI} = 0.25$. As the molecular dipole moment of 8CB mesogens ($p_{8CB} \sim 6D$) is parallel to the long molecular axis $\hat{e}_3$, this means that the macroscopic static dipole moment of the second layer is considerably lower than that of the first layer, i.e. $P_{0,II}/P_{0,I} = -0.37$. For quartz glass substrates a similar SHG analysis gives $\theta_I = 74°$ and $\theta_{II} = 114°$, $N_{pII}/N_{pI} = 0.53$ and correspondingly $P_{0,II}/P_{0,I} = -0.78$. The difference between $P_{0,II}$ and $P_{0,I}$ hence depends on the substrate and is, similarly as the molecular orientation [80], very probably determined by the polarity and dielectric susceptibility of the substrate.

The discrepancy between molecular tilt angles of 8CB molecules on PVCN substrates as found from BAE and from SHG measurements can be attributed to different assumptions made in the corresponding data analysis, but it can also be related to the relatively broad molecular angular distributions within the layers. As the two techniques probe different moments of the molecular angular distribution function, a perfect matching of the results would actually be quite surprising. From this point of view a larger discrepancy detected for the second layer ("bilayer") indicates a broader angular distribution of the 8CB molecules within this layer. The broad angular distribution can explain an apparent inconsistency that, despite a considerable tilting, the thickness of the interfacial bilayer is generally found to be equivalent to the thickness of the bilayers of the volume phase. This hypothesis is in agreement with recent results of x-ray absorption spectroscopy (NEXAFS) of the 8CB films evaporated onto unidirectional alignment layers which has shown that the uniaxial nematic order parameter of the contact monolayer is always larger than that of the subsequent film [81]. Application of techniques that probe additional moments of $f(\theta)$, like for instance Raman spectroscopy, would be very helpful to further resolve this problem.

Growth Stages of 8CB

The observations discussed above indicate that during evaporation of 8CB molecules onto solid substrates a variety of growth stages take place. The deposition of the first layer is initially dominated by the adsorption of evaporated molecules onto "polar surface sites". This leads to the preferential orientation of their polar cyano head-groups down towards the surface of the substrate [21]. The increasing density of polarly oriented molecules produces an increasing SHG signal. Because the molecular dipole moment of 8CB mesogens is large compared to the dipole moments of typical surface bonds such as O-H ($\sim$ 1.5 D), C=O ($\sim$ 2.5 D), etc., even their significant tilting leads to an overall dipole of the adsorbate with respect to the substrate.

The intermediate space between the polar surface sites is filled in by the molecules oriented in equal probabilities with their head-groups pointing up or down with respect to the surface, apparently forming head-to-tail "pairs" that contribute only little to the SHG [21,58]. This process continues also after all polar surface sites are occupied and produces minor variations of the signal that take place until the full surface coverage is attained. The SHG results show that on a quartz substrate which, according to its larger signal, provides a larger density of polar surface sites than PVCN, the corresponding nearly flat SHG region is considerably narrower compared to PVCN. The limiting case is in our opinion observed in the SHG experiments performed during compression of 8CB molecules at the air-water interface, which is known to give a close packed polarly oriented monolayer and in which the flat SHG region is completely absent [73, 78].

During the growth of the second layer the adsorbed molecules tend to screen the excess dipole of the underlying structure. For that reason more of them bind to the first layer with their cyano head-groups pointing up than down with respect to the surface. The corresponding second harmonic field has an opposite phase with respect to the first layer and consequently results in a decrease of the net SHG response. This continues until the overall surface dipole is compensated and layer-by-layer growth of the film terminates. After the second layer is completed, any additional deposition leads to formation of bulk droplets sitting on top of the interfacial film.

This scenario supports the frequently stated hypothesis that dipolar interactions play a principal role in determining the wetting characteristics of nCB molecules on solid substrates [82]. The associated strong cohesion is presumably the reason that nCB liquid crystals, contrary to many other smectogenic materials, are not able to wet any surface studied [59]. An interesting open problem that arises in relation with this is, how an analogous process of compensation of surface dipole moment takes place in thick nCB cells. Here, this compensation may as well extend over several molecular layers and hence produce a dipolar SHG signal comparable to the dipolar SHG response from the first monolayer and to the bulk quadrupolar SHG response.

5.7 Conclusion

In this Chapter we have shown how the nonlinear optical technique of Second Harmonic Generation (SHG) can be used to probe the molecular organization of liquid crystal molecules at various LC-substrate interfaces. As an optical, non evasive technique, SHG can be used to in-situ probe the evolution of, for example, the effect of polarized UV radiation on photopolymers on the formation of the first layers at a solid-LC interface. Though SHG does not have molecular resolution like STM, it can provide complementary information like molecular tilt angles. A future challenge would be to directly combine the strong points of an optical technique like SHG with the unsurpassed molecular resolution of scanning probe methods. In that way, we may also be able to solve some of the unanswered questions of this Chapter like: Why does SHG show a perfectly isotropic distribution of the first LC layer on a rubbed silane substrate that subsequently makes a very well aligned LC cell?

References

1. F.R. Aussenegg, A. Leitner and M.E. Lippitsch *Surface studies with lasers(ed.)*, Springer Verlag, Berlin 1983.
2. P. Halevi, *Photonic Probes of Surfaces* (Nort-Holland, Amsterdam, 1995).
3. D. Bedeaux, J. Vlieger, *Optical Properties of Surfaces* (World Scientific, Singapore, 2002).
4. T. F. Heinz, in *Nonlinear Surface Electromagnetic Phenomena*, edited by H. E. Ponath, G. I. Stegeman (North-Holland, Amsterdam, 1991).
5. J. Hohlfeld, Th. Rasing, Appl. Phys. B **74**, 615 (2002).
6. Z. Chen, Y. R. Shen, G. A. Somorjai, Annu. Rev. Phys. Chem. **53**, 437 (2002).
7. L. Marrucci, D. Paparo, G. Cerrone , C. de Lisio C, E. Santamato, S. Solimeno, S. Ardizzone, P. Quagliotto, Opt. Laser. Eng. **37**, 601 (2002).
8. Y. R. Shen, IEEE J. Sel. Top. Quant. **6**, 1375 (2000).
9. E. Matthias, F. Trager, Appl. Phys. B **68**, 287 (1999).
10. J. F. McGilp, J. Phys. D Appl. Phys. **29**, 1812 (1996).
11. Y. R. Shen, Annu. Rev. Phys. Chem. **40**, 327 (1989).
12. Th. Rasing, G. Berkovic, Y. R. Shen, S. G. Grubb, M. W. Kim, Chem. Phys. Lett. **130**, 1 (1986).
13. G. Berkovic, Th. Rasing, Y. R. Shen, J. Opt. Soc. Am. B **4**, 452 (1987).
14. Y.R. Shen *The principles of Nonlinear Optics* (Wiley press., 1984).
15. M. B. Feller, W. Chen. Y. R. Shen, Phys. Rev. A **43**, 6778 (1991).
16. I. Drevenšek Olenik, M. W. Kim, A. Rastegar, Th. Rasing, Appl. Phys. B **68**, 599 (1999).
17. N. A. J. M. van Aerle, A. J. W. Tol, Macromolecules **27**, 6520 (1994).
18. J. A. Castelano, Mol. Cryst. Liq. Cryst. **94**, 33 (1983).
19. Y. R. Shen, Liq. Cryst. **5**, 635 (1989).
20. W. Chen, M. B. Feller, Y. R. Shen, Phys. Rev. Lett. **63**, 2665 (1989).
21. C. S. Mullin, P. Guyot-Sionnest, Y. R. Shen, Phys. Rev. A **39**, 3745 (1989).
22. M. Barmentlo, N. A. J. M. van Aerle, R. W. J. Hollering, J. P. M. Damen, J. Appl. Phys. **71**, 4799 (1992).
23. N. A. J. van Aerle, M. Barmentlo, R. W. J. Hollering, J. Appl. Phys. **74**, 3111 (1993).
24. D. Johannsmann, H. Zhou, P. Sonderkaer, H. Wierenga, B. O. Myrvold, Y. R. Shen, Phys. Rev. E 48, 1889 (1993).
25. X. Zhuang, L. Marrucci, Y. R. Shen, Phys. Rev. Lett. **73**, 1513 (1994).
26. T. Sakai, K. Shirota, T. Yamada, H. Hoshi, K. Ishikawa, H. Takezoe, A. Fukuda, Jpn. J. Appl. Phys. **35**, 3971 (1996).
27. V. P. Vorflusev, H. -S. Kitzerow, V. G. Chigrinov, Appl. Phys. A **64**, 615 (1997).
28. J. R. Dennis, V. Vogel, J. Appl. Phys. **83**, 5195 (1998).
29. R. Meister, B. Jérôme, Macromolecules **32**, 480 (1999).
30. B. Jerome and Y. R. Shen, Phys. Rev. E **48**, 4556 (1993).
31. J. G. Yoo, H. Hoshi, T. Sakai, B. Park, K. Ishikawa, H. Takezoe, Y. S. Lee, J. Appl. Phys. **84**, 4079 (1998).
32. P. Guyot-Sionnest, H. Hsiung, Y. R. Shen, Phys. Rev. Lett. **57**, 2963 (1986).
33. S. Soria, D. Schuhmacher, G. Marowsky, R. Barberi, F. Ciuchi, S. Paus, Th. Rasing, J. Nonlinear Opt. Phys.&Mat. 10, 133 (2001).
34. S. Soria, D. Schuhmacher, G. Marowsky, R. Barberi, F. Ciuchi, Th. Rasing, Mol. Cryst. Liq. Cryst. **372**, 291 (2002).

35. T. Sakai, J. G. Yoo, Y. Kinoshita, K. Ishikawa, H. Takezoe, A. Fukuda, T. Nihira, H. Endo, Appl. Phys. Lett. **71**, 2274 (1997).
36. X. Wei, X. Zhuang, S. C. Hong, T. Goto, Y. R. Shen, Phys. Rev. Lett. **82**, 4256 (1999).
37. W. M. Gibbons, P. J. Shannon, S. T. Sun, B. J. Swetlin, Nature **351**, 49 (1991).
38. M. Schadt, K. Schmitt, V. Kozinkov, V. Chigrinov, Jpn. J. Appl. Phys. **31**, 2155 (1992).
39. T. Ya. Marusii, Yu. A. Reznikov, Mol. Mat. **3**, 161 (1993).
40. P. J. Shannon, W. M. Gibbons, S. T. Sun, Nature **368**, 532 (1994).
41. M. Schadt, H. Seiberle, A. Schuster, Nature **381**, 212 (1996).
42. E. S. Lee, P. Vetter, T. Miyashita, T. Uchida, Jpn. J. Appl. Phys. **32**, L1339 (1993).
43. G. P. Bryan-Brown, I. C. Sage, Liq. Cryst. **20**, 825 (1996).
44. X. T. Li, D. H. Pei, S. Kobayashi, Y. Iimura, Jpn. J. Appl. Phys. **36**, L432 (1997).
45. T. Ya. Marusii, Yu. A. Reznikov, Mol. Mater. **3**, 161 (1993).
46. P. L. Egerton, E. Pitts, A. Reiser, Macromolecules **14**, 95 (1981).
47. R. Yamaguchi, A. Sato, S. Sato, Jpn. J. Appl. Phys., Part 2 **36**, L432 (1997).
48. H. G. Galabova, D. W. Allender, J. Chen, Phys. Rev. E **55**, 1672 (1997).
49. A. Th. Ionescu, R. Barberi, M. Giocondo, M. Iovane, A. L. Alexe-Ionescu, Phys. Rev. E **58**, 1967 (1998).
50. I. Drevenšek Olenik, M. W. Kim, A. Rastegar, Th. Rasing, Phys. Rev. E **60**, 3120 (1999).
51. N. Klopčar, I. Drevenšek-Olenik, M. Čopič, M. W. Kim, A. Rastegar, Th. Rasing, Mol. Cryst. Liq. Cryst. **368**, 395 (2001).
52. J. J. P. Stewart, *Mopac 93.00 Manual* (Fujitsu Limited, Tokyo, 1993).
53. B. Park, H. S. Kim, J. Y. Bae, J. G. Lee, H. S. Woo, S. H. Han, J. W. Wu, M. Kakimoto, H. Takezoe, Appl. Phys. B: Lasers Opt. **B66**, 445 (1998).
54. W. H. de Jeu, *Physical Properties of Liquid-Crystalline Materials*, edited by G. W. Gray (Gordon and Breach, Philadelphia, PA, 1980).
55. V. P. Vorflusev, H. S. Kitzerow, V. G. Chigrinov, Jpn. J. Appl. Phys. **34**, L1137 (1995).
56. M. W. Kim, A. Rastegar, I. Drevenšek Olenik, M. W. Kim, Th. Rasing, J. Appl. Phys. **90**, 3332 (2001).
57. I. Drevenšek Olenik, M. W. Kim, A. Rastegar, Th. Rasing, Phys. Rev. E **61**, R3310 (2000).
58. B. Jérôme, Y. R. Shen, Phys. Rev. E **48**, 4556 (1993).
59. M. Bardosova, R. H. Tredgold, Mol. Cryst. Liq. Cryst. **355**, 289 (2001).
60. B. M. Ocko, Phys. Rev. Lett. **64**, 2160 (1990).
61. E. Olbrich, O. Marinov, D. Davidov, Phys. Rev. E **48**, 2713 (1993).
62. M. Woolley, R. H. Tredgold, P. Hodge, Langmuir **11**, 683 (1995).
63. M. Barmentlo, F. R. Hoekstra, N. P. Willard, R. W. J. Hollering, Phys. Rev. A **43**, 5740 (1991).
64. J. Daillant, G. Zalczer, J. J. Benattar, Phys. Rev. A **46**, R6158 (1992).
65. E. G. Bortchagovsky and L. N. Tarakhan, Phys. Rev. B **47**, 2431 (1993).
66. M. P. Valignat, S. Villette, J. Li, R. Barberi, R. Bartolino, E. Dubois-Violette, and A. M. Cazabat, Phys. Rev. Lett. **77**, 1994 (1996).
67. R. Barberi, N. Scaramuzza, V. Formoso, M. P. Valignat, R. Bartolino, A. M. Cazabat, Europhys. Lett. **34**, 349 (1996).

68. O. Ou Ramdane, P. Auroy, and P. Silberzan, Phys. Rev. Lett. **80**, 5141 (1998).
69. F. Vandenbrouck, S. Bardon, M. P. Valignat, A. M. Cazabat, Phys. Rev. Lett. **81**, 610 (1998).
70. S. Bardon, R. Ober, M. P. Valignat, F. Vandenbrouck, A. M. Cazabat, J. Daillant, Phys. Rev. E **59**, 6808 (1999).
71. L. Xu, M. Salmeron, S. Bardon, Phys. Rev. Lett. **84**, 1519 (2000).
72. R. Lucht and Ch. Bahr, Phys. Rev. Lett. **85**, 4080 (2000).
73. J. Xue, C. S. Jung, and M. W. Kim, Phys. Rev. Lett. **69**, 474 (1992).
74. M. C. Friedenberg, G. G. Fuller, C. W. Frank, and C. R. Robertson, Langmuir **10**, 1251 (1994).
75. M. N. G. de Mul and J. A. Mann, Jr., Langmuir **10**, 2311 (1994).
76. P. Schmitz and H. Gruler, Europhys. Lett. **29**, 451 (1995).
77. J. Fang, C. M. Knobler, H. Yokoyama, Physica A **244**, 91 (1997).
78. M. Harke, M. Ibn-Elhaj, H. Möhwald, and H. Motschmann, Phys. Rev. E **57**, 1806 (1998).
79. I. Drevenšek Olenik, K. Kočevar, I. Muševič, Th. Rasing, Eur. Phys. J. E **11**, 169 (2003).
80. S. G. Grubb, M. W. Kim, Th. Rasing, Y. R. Shen, Langmuir **4**, 452 (1988).
81. K. Weiss, C. Wöll, D. Johannsmann, J. Chem. Phys. **113**, 11297 (2000).
82. M. Ibn-Elhaj, H. Riegler and H. Möhwald, M. Schwendler, and C. A. Helm, Phys. Rev. E **56**, 1844 (1997).

6 Liquid Crystal Alignment on Surfaces with Orientational Molecular Order: A Microscopic Model Derived from Soft X-ray Absorption Spectroscopy

Jan Lüning and Mahesh G. Samant

6.1 Introduction

To investigate the microscopic origin of liquid crystal alignment on rubbed polymer surfaces several novel experimental techniques have been developed, which allow probing of the liquid crystal-polymer interface on a molecular level. The understanding of the alignment mechanism presented in this Chapter is based on the detection of the molecular orientation at rubbed polymer surfaces by surface sensitive, polarization dependent soft x-ray absorption spectroscopy resolving the near edge x-ray absorption fine structure (NEXAFS) of the absorption coefficient. NEXAFS spectroscopy is a powerful experimental technique for the investigation of ordering phenomena in thin films and at surfaces with element specificity, chemical selectivity, and sensitivity to the presence of orientational order. Its sampling depth can be tuned from the outermost surface layer only to a few ten nanometers deep into the film bulk. Also, this technique is especially powerful for the investigation of ordering phenomena in the absence of long-range order, as often the case in polymers, due to the local nature of the x-ray absorption process.

These capabilities of NEXAFS spectroscopy enabled us to obtain experimental results, which led us to develop a model explaining the liquid crystal alignment on rubbed polymer surfaces. This model suggests that liquid crystal alignment only requires a statistically significant preferential bond orientation at the polymer surface without the necessity of regions with crystalline or quasi-crystalline order. The model is capable of explaining the in-plane alignment direction for polymer films with alignment parallel (e.g., polyimide) as well as perpendicular to the rubbing direction (e.g., polystyrene). In addition, it correctly predicts the presence and the direction of the pretilt angle observed for polymer films with liquid crystal alignment parallel to the rubbing direction, and the general absence of a liquid crystal pretilt for polymer films with perpendicular alignment. Furthermore, the applicability of the model is not restricted to rubbed polymer surfaces, as it also correctly predicts the alignment direction and the absence or the presence/direction of a liquid crystal pretilt angle for polymer surfaces treated with polarized UV or ion beam irradiation.

In fact it was the understanding within the model of liquid crystal alignment on ion beam irradiated polymer surfaces, which inspired the development of an entirely new liquid crystal alignment material that has recently successfully been demonstrated by IBM (see also Chap. 9).

6.2 Principle and Properties of NEXAFS Spectroscopy

In this Section we will introduce the principles and experimental details of NEXAFS spectroscopy. The focus of the discussion will be on showing that this technique has all the capabilities required to address the origin of liquid crystal alignment on rubbed polymer surfaces.

In x-ray absorption spectroscopy one observes the absorption of an incoming x-ray photon by the excitation of an electron into an unoccupied state of the material's electronic structure[1] as illustrated in Fig. 6.1A. While valence electrons are excited by optical photons, x-rays excite due to their considerably higher photon energy more tightly bound core electrons. Due to their high binding energy these core electrons do not contribute to the chemical bonds. Their orbitals are therefore atomic-like, localized at a specific atom. The following capabilities of x-ray absorption spectroscopy are a consequence of the participation of the localized core electron in the photon absorption process. First of all, x-ray absorption spectroscopy is element specific, since each element has different core electron binding energies. The elemental composition of a material of interest can therefore directly be obtained quantitatively from a coarse spectrum. This is demonstrated for polyimide in Fig. 6.1B.

Furthermore, when the near edge x-ray absorption fine structure (NEXAFS) is resolved (Fig. 6.1C), information about the chemical composition is obtained. The reason for this is twofold. Firstly, different neighboring atoms cause different chemical shifts of the core electron binding energy [2]. Secondly, since the core electron is localized, the electronic structure is probed locally at the site of the absorbing atom. Hence, the occurring resonances in the absorption fine structure reflect the unoccupied orbitals of the chemical bonds, which the absorbing atoms have. Thus, the chemical structure can be deduced from the resonances present in the NEXAFS spectrum. The assignment of the resonances in the carbon K NEXAFS spectrum of polyimide to carbon atoms of the different functional groups as indicated by the filled circles in Fig. 6.1C is taken from the literature [3, 4].

Finally, NEXAFS spectroscopy is sensitive to the presence of charge[2] ordering. As shown in Fig. 6.1D for polyimide, the actual shape of a NEXAFS spectrum can strongly depend on the orientation of the electric field vector

[1] X-ray absorption spectra of molecular materials can be well-interpreted within the single electron picture presented here, which neglects electron correlation effects. For a discussion of correlation effects see, e.g., reference [1].

[2] Equally important is the sensitivity to spin ordering referred to as x-ray magnetic linear/circular dichroism [5].

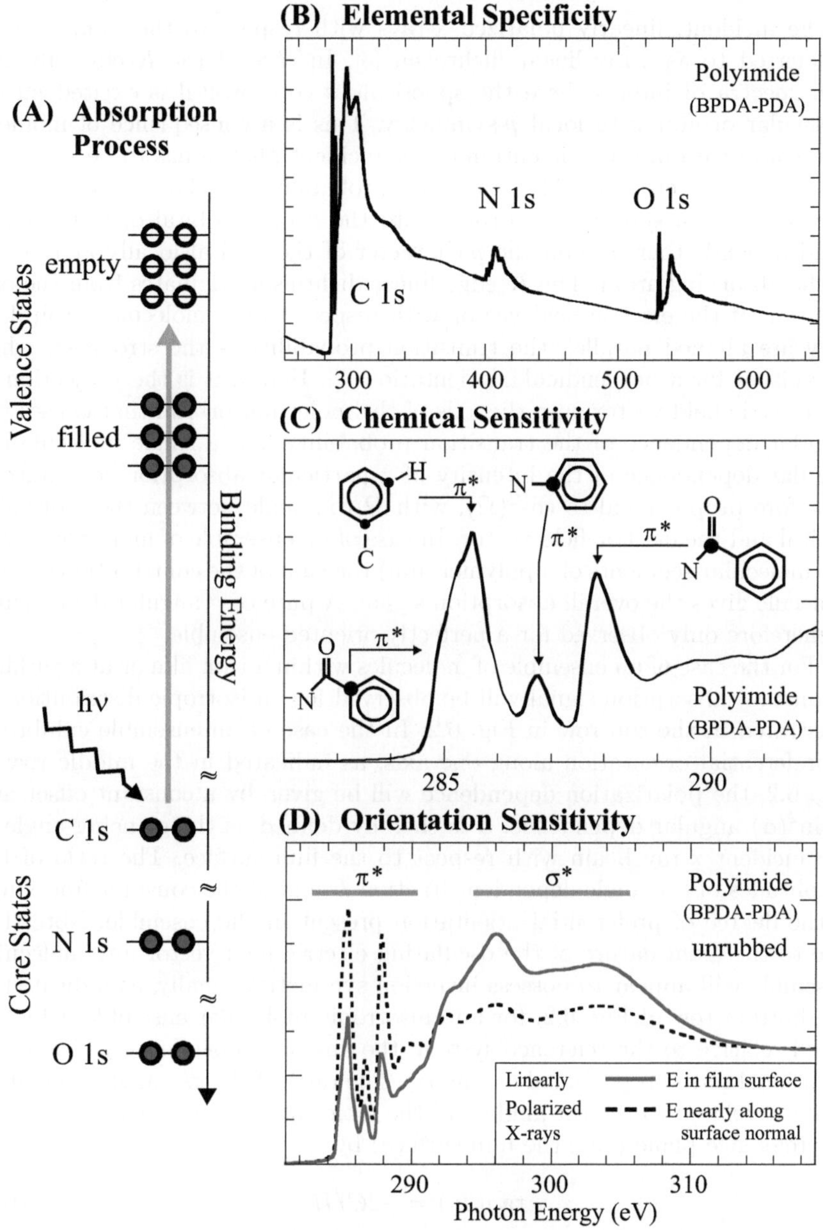

Fig. 6.1. The principle of x-ray absorption spectroscopy (A), in which the incident photon is absorbed by the excitation of a core electron into an unoccupied valence state. Element specificity (B), sensitivity to the local chemical environment (C), and the sensitivity to the presence of orientational order (D) result from the core electron being localized at a specific atom in an atomic-like orbital.polyimide

of the incident, linearly polarized x-rays with respect to the sample. This is referred to as x-ray linear dichroism [6]. In case of the K-edge absorption spectra of interest here the spherical $1s$ core orbital is excited into a molecular orbital with local p-symmetry. This is a consequence of momentum conservation, the momentum of the incident photon has to be absorbed by the excited electron. The transition probability for the excitation of an electron from a spherical $1s$ orbital into the σ or π orbital of a molecular bond depends therefore on the p-character of the orbital locally at the site of the absorbing atom. The K-edge linear dichroism originates from the orientation of the electric field vector with respect to the molecular orbital. If they are aligned parallel, the transition probability is the strongest, while it vanishes for a perpendicular orientation [7]. Hence, it is the projection of the electric field vector onto the axis of the molecular orbital that causes the angular dependence of the transition probability. For a single molecule the angular dependence of the intensity of a particular absorption resonance is therefore proportional to $\cos^2(\Omega)$, with Ω the angle between the molecular orbital and the electric field vector. In case of an ensemble of molecules (or of the molecular segments of a polymer film) the sum of the contribution of each molecule gives the overall absorption signal. A pure $\cos^2$ angular dependence is therefore only observed for a perfectly oriented ensemble.

For the case of an ensemble of molecules within a thin film or at a surface, a constant absorption signal will be observed for an isotropic distribution as illustrated in the top row in Fig. 6.2. In the case of an ensemble exhibiting a preferential orientation along one axis, as indicated in the middle row of Fig. 6.2, the polarization dependence will be given by a constant offset and a $\sin^2(\alpha)$ angular dependence. α is thereby defined as the glancing angle of the incident x-ray beam with respect to the film surface. The ratio of the amplitudes of the angle dependent fraction B and of the constant fraction A is the degree of preferential orientation present in the ensemble. Note that due to the axial nature of the oscillating electric field vector any molecular ensemble will appear to possess inversion symmetry. Finally, as indicated in the bottom row of Fig. 6.2, for an anisotropic molecular ensemble, which is tilted relative to the reference system, the angular dependence will have an additional term proportional to $\sin(2\alpha)$. The ratio of the two angle dependent terms is related to the tilt angle γ of the molecular ensemble with respect to the reference plane (e.g., the film surface) by

$$\tan(2\gamma) = -2C/B\,, \tag{6.1}$$

with C the proportionality constant of the $\sin(2\alpha)$ term. Although the above discussion is carried out for a 2-dimensional molecular distribution, it also holds true for a 3-dimensional ensemble, because of the axial nature of the oscillating electric field vector. Furthermore, it applies to resonances associated with both, σ and π orbitals. The only difference between the polarization dependence of σ and π orbitals of a given molecular ensemble is that they have

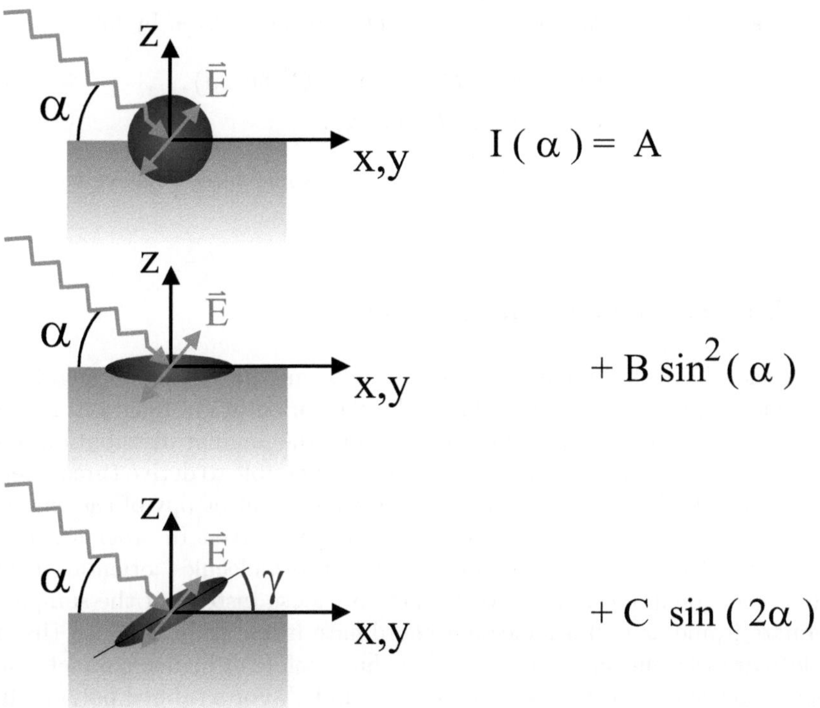

Fig. 6.2. Relationship between preferential orientation within an ensemble of molecules (represented by the dark areas with respect to the film surface) and the dependence of the experimentally observed intensity (of a related absorption resonance) on the x-ray incidence angle α.

the opposite orientation, which is due to the orthogonal orientation of σ and π orbitals. This is demonstrated by the polarization dependence of the $\sigma^\star$ and $\pi^\star$ resonances in the spectra of unrubbed polyimide shown in Fig. 6.1D.

In summary, the angular dependence of an absorption peak intensity within a plane for an arbitrarily oriented molecular ensemble is given by

$$I(\alpha) = A + B \cdot \sin^2(\alpha) + C \cdot \sin(2\alpha) \ . \tag{6.2}$$

For a rubbed polymer surface we note that there are two principal, orthogonal planes perpendicular to the film surface. These are the planes oriented parallel and perpendicular to the rubbing direction (see Fig. 6.3A and B). The first one, which we will refer to as parallel plane, clearly possesses mirror symmetry, while the mirror symmetry of the plane perpendicular to the rubbing direction, referred to as perpendicular plane, may be broken by the directional nature of the rubbing process. Hence, the most general expression for the polarization dependence on rotation of the electric field vector within the parallel ($I^{\parallel}$) and the perpendicular ($I^{\perp}$) plane is given by [3]

$$I^{\parallel}(\alpha) = A^{\parallel} + B^{\parallel}\sin^2(\alpha) + C^{\parallel}\sin(2\alpha)\,, \qquad (6.3)$$
$$I^{\perp}(\alpha) = A^{\perp} + B^{\perp}\sin^2(\alpha)\,,$$

with the glancing angle α of the incident x-rays measured relative to the film surface.

6.3 Molecular Orientation Factors

From the above discussion it is clear that from the NEXAFS angular dependence one cannot derive the molecular orientation function $F(\alpha, \Psi)$, which contains the precise information about the angular distribution of the molecules within the ensemble. It is, however, possible to derive three orientation factors f_i that characterize the orientational anisotropy of the ensemble by their relative alignment along three orthogonal axes. In other words, the orientation factors f_i are the projection of the molecules' orientation onto three orthogonal axes.[3] It is convenient to choose instead of the sample coordinate frame (x,y,z) a molecular coordinate frame (x',y',z') such that the molecular distribution is symmetric, i.e., has twofold or higher symmetry with respect to the three orthogonal axes x', y', and z'. For a rubbed polymer film, as discussed above, we expect twofold symmetry about the y axis, while the directional rubbing process may lead to a distinction of the $+x$ from the $-x$ direction. Consequently, the molecular distribution may be asymmetric about the x and z axis of the sample frame. However, by rotating the sample frame as indicated in Fig. 6.3C about the y axis by the angle γ introduced in Fig. 6.2 we obtain the desired symmetric molecular frame $(x', y' = y, z')$. According to (6.1) and (6.3) the angle γ is thereby given by

$$\tan(2\gamma) = -2C^{\parallel}/B^{\parallel}\,. \qquad (6.4)$$

The orthogonal orientation factors $(f_{x'}, f_{y'}, f_{z'})$ are then simply the projection of the molecular distribution function onto the orthogonal axes of the molecular frame, and for a normalized molecular distribution function $f_{x'} + f_{y'} + f_{z'} = 1$.

In the absence of any mirror symmetry breaking by the rubbing process, the molecular and the sample frame are identical. In this case, the experimental intensities for 100% linearly polarized x-rays with the electric field vector oscillating along the x, y, and z axes are directly proportional to the

[3] This technique has been used before for the characterization of the orientation of liquid crystal molecules [8, 9].

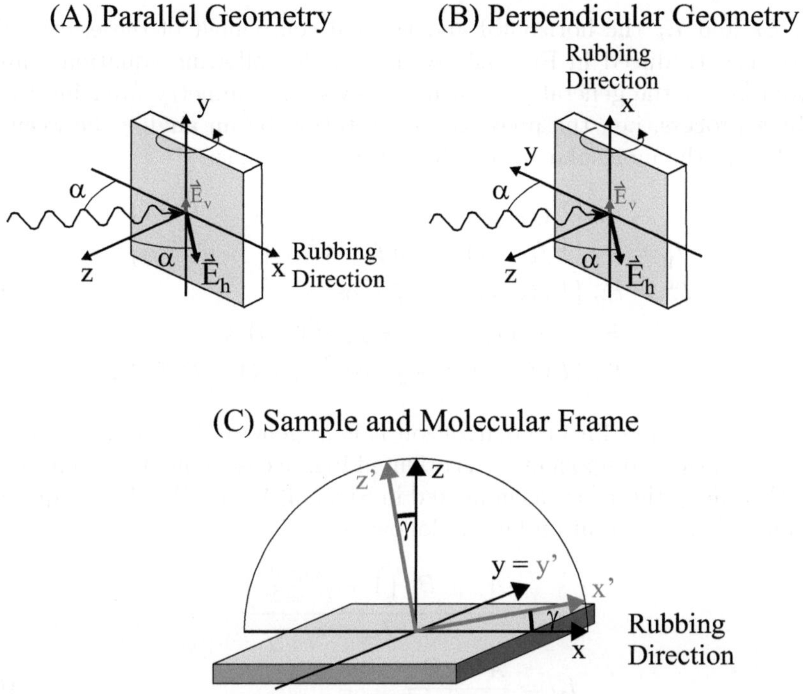

Fig. 6.3. Sketch of the experimental geometries to illustrate for elliptically polarized x-rays the orientation of the major horizontal $\mathbf{E}_h$ and minor vertical $\mathbf{E}_v$ component of the electrical field vector for parallel (A) and perpendicular (B) orientation of the film surface with respect to the rubbing direction ($+x$ direction). (C) Relationship between the sample frame (x,y,z) and the molecular frame (x',y',z'), in which the molecular distribution of polymer segments at the film surface has at least twofold symmetry about all three orthogonal axes.

orientation factors:

$$\begin{aligned} I_x &= I_{tot} f_{x'} \\ I_y &= I_{tot} f_{y'} \\ I_z &= I_{tot} f_{z'} \end{aligned} \tag{6.5}$$

I_{tot} is the total integrated NEXAFS intensity $I_{tot} = I_x + I_y + I_z$. One important implication of this is that for linearly polarized x-rays the total intensity can always be obtained by measurements along three orthogonal directions, independent of the relative orientation of the sample coordination system.

In general, X-ray sources have a finite degree of linear polarization, and we define the polarization parameter P to characterize the intensity or energy density of the electric field in the synchrotron orbit plane. Hence, it is

$$P = \frac{|E_h^2|}{|E_h^2| + |E_v^2|}, \tag{6.6}$$

with E_h and E_v the horizontal and vertical component of the electric field vector as introduced in Fig. 6.3. With this the following equations can be derived [3] for the general case, which allows for symmetry breaking by the rubbing process, i.e., the presence of a rotation by an angle γ between the sample and the molecular frame about the $y = y'$ axis:

$$\begin{aligned} I_x &= I_{tot}\{P(f_{x'}\cos^2\gamma + f_{z'}\sin^2\gamma) + (1-P)f_{y'}\} \\ I_y &= I_{tot}\{Pf_{y'} + (1-P)(f_{x'}\cos^2\gamma + f_{z'}\sin^2\gamma)\} \\ I_z^{\perp} &= I_{tot}\{P(f_{x'}\sin^2\gamma + f_{z'}\cos^2\gamma) \\ &\quad +(1-P)(f_{x'}\cos^2\gamma + f_{z'}\sin^2\gamma)\} \\ I_z^{\parallel} &= I_{tot}\{P(f_{x'}\sin^2\gamma + f_{z'}\cos^2\gamma) + (1-P)f_{y'}\}\ . \end{aligned} \tag{6.7}$$

Note that for finite linear polarization it is in general $I_z^{\parallel} \neq I_z^{\perp}$, because the smaller elliptical component E_v is oriented in one case along the y axis and in the other along the x axis as indicated in Fig. 6.3(A) and (B). From equations 6.3 and 6.5 the orientation factors follow as:

$$\begin{aligned} f_{x'} &= \frac{A^{\perp} + B^{\parallel}\left(1 + \frac{\sin^2\gamma}{P\cos 2\gamma}\right)}{I_{tot}} \\ f_{y'} &= \frac{A^{\parallel} + B^{\perp}}{I_{tot}} \\ f_{z'} &= \frac{A^{\perp} + B^{\parallel}\left(1 - \frac{\cos^2\gamma}{P\cos 2\gamma}\right)}{I_{tot}}\ . \end{aligned} \tag{6.8}$$

Interestingly, these equations imply that the polarization factor can be determined from such measurements:

$$P = \frac{B^{\perp} - B^{\parallel}}{A^{\parallel} - A^{\perp} + B^{\perp} - B^{\parallel}}\ . \tag{6.9}$$

6.4 Experimental Setup for NEXAFS Spectroscopy

The setup for a polarization dependent, surface sensitive soft x-ray absorption experiment is sketched in Fig. 6.4A. The sample is mounted onto a rotatable holder with a vertical rotation axis lying in the sample surface and a second rotation axis along the sample normal. The latter allows rotation of the sample such that the major, horizontal component of the electric field vector rotates within the plane parallel (Fig. 6.3A) or perpendicular (Fig. 6.3B) to the rubbing direction, when the sample is rotated around the vertical rotation axis.

The absorption signal of the sample is obtained by recording simultaneously two electron yield signals, which originate from the decay of the core electron vacancies created in the absorption process. This decay occurs within

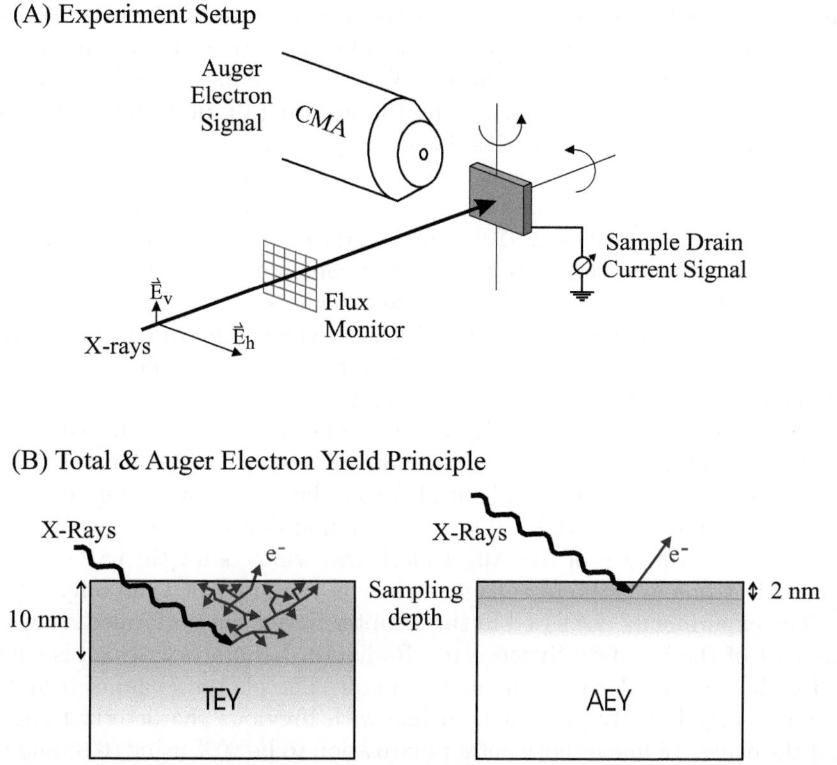

Fig. 6.4. (A) Experimental set-up for collecting polarization-dependent x-ray absorption spectra by monitoring simultaneously the total (TEY) and Auger electron yield (AEY) via the sample drain current and the electron energy analyzer count rate, respectively. The sample can be rotated around its surface normal and a vertical rotation axis allowing for arbitrary orientation relative to the electric field vector of the incident, elliptically polarized x-rays. (B) Illustration of the electron scattering cascade (left side) following the Auger decay of a core electron vacancy. The mean free path length of the low energy secondary electrons determines the $1/e$ sampling depth of the TEY signal. If only Auger electrons are monitored (right side), the sampling depth is determined by the short mean free path length of the Auger electrons in matter.

the soft x-ray energy region primarily via the Auger process. In this the core electron vacancy is filled by a shallower bound electron and the excess energy is transfered to another electron, the so-called Auger electron, making it a high kinetic energy electron traveling through the material. As illustrated on the left hand side of Fig. 6.4B, due to the short mean free path of electrons in matter the Auger electron will scatter with other valence electrons. In this way it will distribute its kinetic energy, and thus the information of the occurrence of an absorption event, to a bunch of other valence electrons, of which some might eventually penetrate the sample surface. The total electron yield signal (TEY) is then simply obtained by measuring the total number of electrons escaping the sample surface. This can be done either by collecting the emitted electrons or by measuring the resulting drain current through a wire connecting the otherwise insulated sample to ground. Since x-rays have a much longer penetration length in matter than electrons, the number of electrons escaping the sample surface[4] is in good approximation proportional to the absorption coefficient [10], which determines the number of photons absorbed within the thin subsurface layer contributing to the TEY signal. The TEY 1/e sampling depth, i.e., the thickness of the subsurface layer, from which 67% of the yield signal originates, is given by the average length of the electron scattering cascade. This is typically on the order of 5 - 10 nm for the carbon K-edge in carbonaceous materials.

When using an electron energy analyzer one can selectively measure only the number of electrons escaping the sample surface, which have a kinetic energy matching the one of the initial Auger electrons. This is referred to as Auger Electron Yield (AEY). The 1/e sampling depth is then given by the mean free path length of the Auger electrons, which is for the carbon KVV Auger electrons in carbonaceous materials on the order of 1 nm only [11].

The experiments discussed in this chapter have been performed at beamline 10-1 of the Stanford Synchrotron Radiation Laboratory, which is equipped with a spherical grating monochromator. The photon beam of beamline 10-1 is nearly linearly polarized. In line with previous characterizations we find the degree of linear, horizontal polarization to be 80% using (6.9) and the present data. The energy resolution $E/\Delta E$ was set to 3000, which is significantly better than the inherent broadening present in the discussed polymer spectra. With the help of an 80% transmissive gold grid, the intensity of the incident x-rays was monitored simultaneously with the electron energy analyzer count rate and the sample drain current. These two sample signals were normalized by the gold grid signal to obtain the AEY and TEY spectra, which are proportional to the absorption coefficient within the about 2 nm (AEY) and 10 nm (TEY) thin outermost surface layer. From these yield spectra a pre-edge background was subtracted and the edge jump far above the carbon K edge (340 – 380 eV) was arbitrarily scaled to unity. This procedure discussed in detail elsewhere [6] produces spectra corresponding to the same number of sampled carbon atoms, which enables the comparison of spectra recorded with different x-ray incidence geometries. Peak intensity differences in such a comparison therefore indicate the ratio of carbon atoms with specific orientations of their corresponding molecular orbitals within the sampled layer. All spectra shown in this chapter have been treated this way.

[4]Note that due to the strong electron scattering the majority of the electrons escaping the sample surface have a low kinetic energy. This is the origin of the intense inelastic tail in x-ray photoelectron spectroscopy spectra.

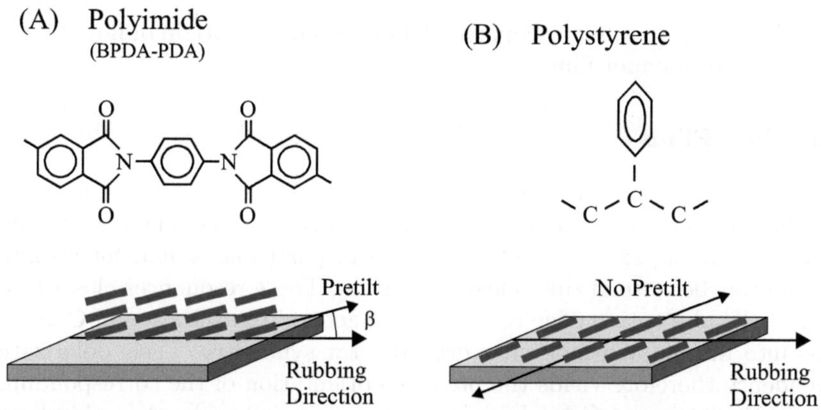

Fig. 6.5. On rubbed polyimide films (A) liquid crystals orient parallel to the rubbing direction with an upwards tilt with respect to the rubbing direction. Note that this pretilt angle β is defined as the average out-of-plane tilt angle of the liquid crystal rods in the bulk of the liquid crystal ensemble. This may differ from the angle of the first monolayer, for which β is indicated here for simplicity. On rubbed polystyrene (B) liquid crystals orient perpendicular to the rubbing direction without any out-of-plane tilt angle.

The outcome of the normalization procedure is shown by the two spectra in Fig. 6.1D, which are plotted over an extended photon energy range.

6.5 Polarization Dependent NEXAFS Spectra of Polyimide and Polystyrene Surfaces

Rubbed polyimide and polystyrene films show very different alignment properties for liquid crystals as illustrated in Fig. 6.5A and B. Polyimide films show the intuitively better understandable alignment of the liquid crystals parallel to the rubbing direction [12]. In addition the liquid crystals exhibit an upwards tilt angle relative to the rubbed polyimide film surface. This is the technologically important liquid crystal pretilt angle β. This is experimentally determined as the average tilt angle of the liquid crystal molecules in the liquid crystal bulk. On rubbed polystyrene films, on the other hand, liquid crystals align perpendicular to the rubbing direction without any pretilt angle relative to the film surface [13]. We will discuss in this section how NEXAFS spectroscopy can be applied to study these rubbed polymer film surfaces[5]. The presented experimental results will allow us to understand the different alignment behavior of polyimide and polystyrene. This understand-

[5]Rubbed polymer surfaces have been investigated also by others using NEXAFS spectroscopy. See, e.g., references [4, 14–16].

ing will guide us to a microscopic model describing the origin of liquid crystal alignment on polymer films.

Unrubbed Films

The carbon K-edge absorption spectra of polymers exhibit a rich fine structure, like the one of polyimide shown in Fig. 6.1C and D. The lowest-energy resonances in the 280 – 290 eV range are in particular suited for an intensity analysis because of their clear separation. These resonances closer to the onset of the absorption edge correspond in general to excitation of C 1s electrons into unoccupied molecular orbitals of π symmetry. Their polarization dependence therefore yields the preferred orientation of the corresponding π system. Excitation of C 1s electrons into unoccupied molecular orbitals of σ symmetry give mostly rise to resonances at higher photon energies. In principle, these resonances can also be used to study molecular orientation, but a quantitative analysis is often hampered by their generally larger width and the close excitation energies of $\sigma^\star$ resonances of different molecular bonds. It is, however, important to note that $\sigma^\star$ and $\pi^\star$ resonances associated with the same molecular bond, or more generally, the same molecular group, exhibit the same polarization dependence, albeit reversed due to their orthogonal orientation. This is clearly demonstrated by the angular dependence of the $\pi^\star$ and $\sigma^\star$ resonances of polyimide in Fig. 6.1D. It is therefore legitimate to only analyze the angular dependence of $\pi^\star$ resonances, which can experimentally be done more accurately. From the derived anisotropy of the π system, the anisotropy of the corresponding σ system follows directly due to their orthogonality.

For the two polymers of interest here, polyimide and polystyrene, we note that their overall orientation is well-characterized by the orientation of the π system, since all of the carbon atoms in polyimide –and the vast majority in polystyrene– contribute to the π system (see the molecular structures sketched in Fig. 6.6).

The spectra shown in Fig. 6.1D have been recorded on an unrubbed polyimide film. The strong polarization dependence of the $\pi^\star$ resonances (and the related, reversed one of the $\sigma^\star$ resonances) indicates that the polymer chains have a strong preferential orientation within the sampled subsurface layer. The origin of this in- versus out-of-plane anisotropy is the presence of the film surface, driven by steric effects and the minimization of the surface energy. Hence, it is quite common to find such an in- versus out-of-plane anisotropy in the surface region of many polyimide films as well as polystyrene films, especially, after annealing [3,17].

In case of the BPDA-PDA polyimide film of Fig. 6.1D the intensity of the $\pi^\star$ resonances is enhanced for the electric field vector oriented nearly perpendicular to the film surface. Hence, the π orbitals are preferentially oriented perpendicular to the film surface. This implies that the phenyl rings and the $C = O$ bonds are preferentially oriented parallel to the film surface.

Correspondingly, the $\sigma^\star$ resonances are enhanced for the electric field vector oriented parallel to the film surface. For polystyrene surfaces, on the other hand, the phenyl rings are preferentially oriented perpendicular to the film surface, such that the polystyrene polymer chains run parallel to the film surface [3].

Rubbed Films

The absorption spectra of rubbed[6] polyimide and polystyrene are shown in the left and right hand side of Fig. 6.6. The spectra have been recorded with the electric field vector aligned parallel to the rubbing direction (solid black, $E \parallel x$), in-plane perpendicular to the rubbing direction (dashed, $E \parallel y$), and nearly perpendicular[7] to the film surface (solid gray, $E \parallel z$). The upper and the lower panels show the AEY and TEY spectra, sampling the top 2 nm and 10 nm of the rubbed film surfaces, respectively. The plotted energy range is chosen such that the spectral region of the $\pi^\star$ resonances is shown enlarged, on which our further analysis will concentrate for the reasons discussed before.

For both films the spectra indicate a strong polarization dependence, in fact, one notices that each orientation of the electric field vector corresponds to a distinct spectrum. Therefore, the molecular distribution is anisotropic with respect to the film surface as well as within the film surface itself. One further notices that the polarization dependence is more pronounced for the more surface sensitive AEY data than for the deeper into the film sampling TEY data. This shows that the redistribution of the molecular arrangement introduced by the rubbing process decreases in strength with increasing distance from the film surface. Clearly, this is in line with our expectation based on the nature of the rubbing process itself, in which the fibers of the rubbing tool pull on the molecular chains at the film surface. Assuming as a first approximation an exponential decrease of the rubbing effect on the molecular arrangement one can estimate from the known AEY and TEY sampling depths the "$1/e$ alignment depth" of the rubbing process to be on the order of 10 nm. [15] Of course, this alignment depth is material specific and also depends strongly on the particular rubbing conditions. It is, however, interesting to note that under the rubbing conditions applied here, which correspond to the ones used for manufacturing of liquid crystal displays, the effect of the rubbing process reaches significantly below the film surface.

Looking in detail at the polarization dependence of the polyimide spectra shown on the left hand side of Fig. 6.6, we find that the $\pi^\star$ resonances show the

[6] The films were rubbed at room temperature with a velour cloth under a load of 2 g/cm^2 over a distance of 300 cm at a speed of 1 cm/s [15, 18].

[7] *Nearly perpendicular* refers to the electric field vector oriented 20° off the z axis, since experimentally it cannot be aligned perpendicular to the film surface. For this qualitative discussion, the small difference between $I_z^{\parallel}$ and $I_z^{\perp}$ (6.7) is neglected, but it has to be taken into account for the quantitative analysis presented later.

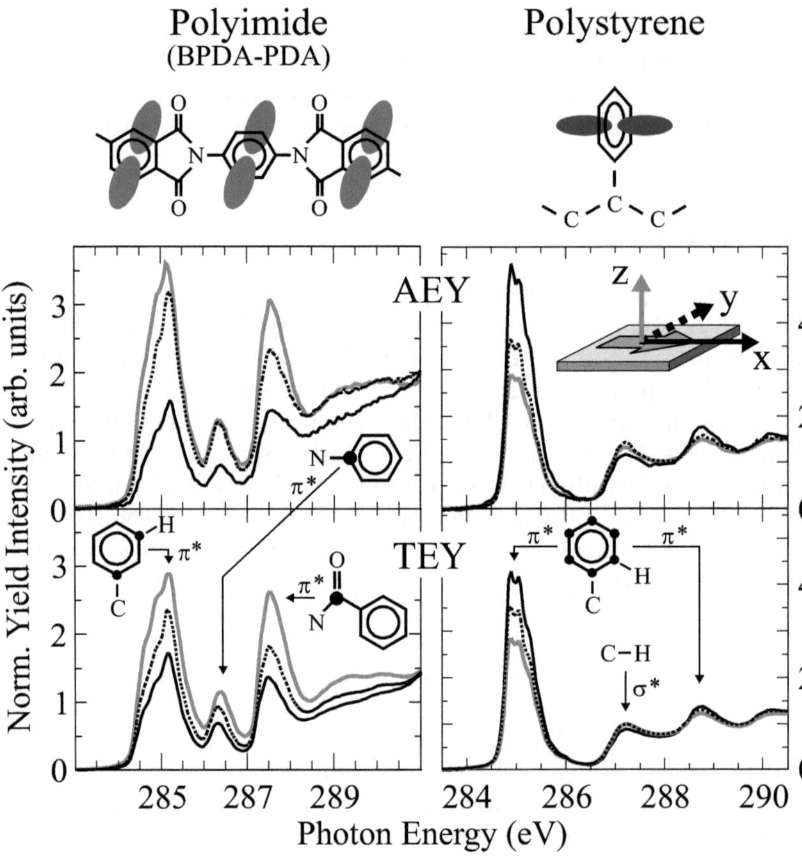

Fig. 6.6. Polarization dependent NEXAFS spectra of rubbed BPDA-PDA polyimide (left) and polystyrene (right) recorded by AEY (top) and TEY (bottom) detection with the electric field vector oriented parallel to the rubbing direction x (solid black), and perpendicular to the rubbing direction in-plane along y (dashed black) and 20° of the surface normal z (solid gray). The assignment of the resonances to the carbon atoms indicated by filled circles is taken from the literature [3].

strongest intensity for the electric field vector aligned nearly perpendicular to the film surface (solid gray, $E \parallel z$). This indicates that as in unrubbed polyimide films the majority of the polyimide rings is oriented parallel to the film surface. Ring segments oriented perpendicular to the film surface are preferentially oriented parallel to the rubbing direction. This is indicated by the much stronger intensity of the $\pi^\star$ resonances for the electric field oriented perpendicularly to the rubbing direction (dashed, $E \parallel y$) than for parallel alignment (solid black, $E \parallel x$). In summary, these spectra indicate that the polyimide chains are preferentially oriented along the rubbing direction, with the chain segments more parallel than perpendicular to the film surface.

The polarization dependence observed for the $\pi^\star$ resonance of polystyrene shows the reversed order. As for the unrubbed polystyrene film [3] the weakest $\pi^\star$ intensity is found for the electric field vector nearly perpendicular to the film surface (solid gray, $E \parallel z$). This indicates that the majority of the polystyrene phenyl rings is oriented perpendicularly to the film surface. Within the film surface, however, we find a significant redistribution of phenyl rings due to the rubbing process. The strongest $\pi^\star$ intensity is observed for the electric field vector aligned parallel to the rubbing direction (solid black, $E \parallel x$). Hence, the majority of the π orbitals is oriented parallel to the rubbing direction, which corresponds to the majority of the phenyl rings being oriented with their plane perpendicular to the rubbing direction. In summary the observed polarization dependence indicates that the rubbing process preferentially orients the polystyrene chains within the film surface along the rubbing direction.

In conclusion we note that for both polymers the rubbing process aligns the polymer chains preferentially along the rubbing direction. The observation of opposite polarization dependences for polyimide and polystyrene reflects the orthogonal orientation of their π systems with respect to the chain directions. While the π system is oriented parallel to the polystyrene chain direction, it is oriented perpendicular to the polyimide polymer chain.

Orientation Factors of Rubbed Films

To quantitatively analyze the molecular distribution within the rubbed polymer surfaces we have recorded absorption spectra for a series of electric field vector orientations within the two principal planes perpendicular to the film surface. These are the previously introduced planes parallel (x-z) and perpendicular (y-z) to the rubbing direction. The polarization dependence is summarized by the intensity of the $\pi^\star$ resonance in the normalized AEY spectra, which are plotted versus the photon incidence angle α as solid and open symbols in Fig. 6.7. The solid lines are the result of a fit to the data using (6.3).

Leaving the already discussed overall opposite sense of the polarization dependence for the moment aside, we note a general similarity in the polarization dependence of the two polymers. In both cases we find the stronger polarization dependence in the plane parallel to the rubbing direction (filled symbols). And the most distinct $\pi^\star$ intensity is obtained for the electric field vector oriented parallel to the rubbing direction ($E \parallel x$, 90° x-ray incidence angle). This indicates that for both polymer surfaces the major anisotropy is along the rubbing direction. Within the plane perpendicular to the rubbing direction the polarization dependence is much weaker, which reflects a much smaller anisotropy. Of particular importance is the asymmetry of the polarization dependence within the plane parallel to the rubbing direction with respect to the $\alpha = 0°$ incidence angle (orientation of the electric field vector along the surface normal). These are indicated by the small arrows,

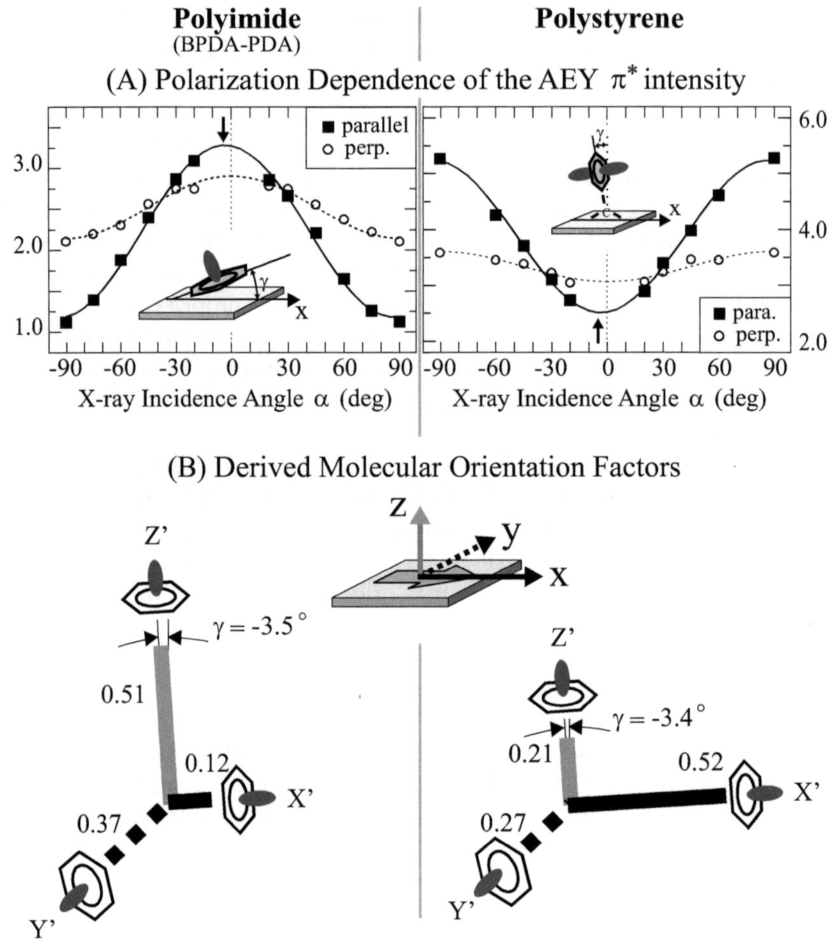

Fig. 6.7. (A) Polarization dependence of rubbed BPDA-PDA polyimide (left) and polystyrene (right). Note that the angular dependence for the electric field vector rotated within the y-z plane perpendicular to the rubbing direction (open circles) is symmetric about the $\alpha = 0°$ x-ray incidence angle (E parallel to the surface normal), while within the x-z plane parallel to the rubbing direction (solid squares) the symmetry point of the angular dependence is shifted to negative angles. This indicates the preferential upwards/backwards tilt of the phenyl ring planes with respect to the rubbing direction, as indicated in the Fig. insets. (B) Illustration of the molecular distribution factors ($f_{x'}$,$f_{y'}$,$f_{z'}$) and the tilt angle γ as defined in (6.8) and (6.4), which are obtained by fitting of the polarization dependence (solid lines in (A)) using (6.3).

positioned at the center of the respective fits (solid lines). For both films the center of the curves is shifted to negative incidence angles. As discussed before, this corresponds to a breaking of the symmetry of the molecular distribution along the rubbing direction within the sample frame. The phenyl ring planes possess a preferential upwards/backwards tilt with respect to the rubbing direction as illustrated in the insets of Fig. 6.7A. The polarization dependence within the plane perpendicular to the rubbing direction, on the other hand, shows a symmetric dependence with respect to the $\alpha = 0°$ incidence angle. In line with our expectation, this implies a symmetric molecular distribution in the plane perpendicular to the rubbing direction.

From the fitted intensity dependence of the π^* resonances, using the relationships given in (6.8), the orientation factors $f_{x'}$, $f_{y'}$, and $f_{z'}$ are derived, which give the fraction of ring segments at the film surface oriented with their π system along the x', y', and z' axes of the molecular frame. The derived factors are illustrated graphically at the bottom of Fig. 6.7. For polyimide we find that within the molecular frame only a small minority of about 12% of the ring segments is oriented with their π orbital parallel to the rubbing direction. With 51% the majority of the π orbitals is pointing along the surface normal, and the remaining 37% of the polymer segments have their π orbital oriented in-plane perpendicular to the rubbing direction. Due to the orthogonal relationship between the π orbital and the molecular plane these numbers state that only 12% of the chain segments are oriented perpendicular to the rubbing direction, while the majority of 51% is oriented parallel to the film surface, and the remaining 37% are oriented along the rubbing direction with their plane turned perpendicular to the film surface. The molecular frame, within which the molecular distribution is symmetric along all three orthogonal axes, is tilted by $-3.5°$ counter-clockwise about the y axis of the sample frame. Due to this tilt, the average polyimide ring, and thus the average polyimide polymer chain segment, has within the sample frame an upwards tilt with respect to the rubbing direction ($+x$ direction).

For polystyrene, the molecular frame is also tilted counter-clockwise about the y axis of the sample frame, with $-3.4°$ by a similar amount. On the other hand, the molecular orientation factors of polystyrene indicate that in the molecular frame with 52% the vast majority of phenyl rings is oriented perpendicular to the rubbing direction. With 27% and 21% an about equal number of phenyl rings is oriented along the rubbing direction and parallel to the sample surface, respectively. Hence, the average phenyl ring in polystyrene is oriented perpendicular to the rubbing direction and slightly backwards tilted. This implies, that as in the case of polyimide, the average polystyrene polymer chain segment is oriented along the rubbing direction pointing slightly upwards.

Origin of the Molecular Reorientation in the Rubbing Process

Within the simple picture illustrated in Fig. 6.8 one can understand how the rubbing process causes the observed preferential molecular orientation. The tips of the rubbing cloth fibers can be thought of to pull on the surface segments of the polymer chains into the rubbing direction ($+x$ direction), which increases the number of chain segments oriented towards this direction. The associated segment motion is illustrated in Fig. 6.8 for two representative polymer chains of polyimide (left side) and polystyrene (right side). For a polymer chain running within the film surface (top row) this increases for both polymers the number of chain segments oriented along the rubbing direction ($+x$ direction) relative to those oriented along the perpendicular direction (y-axis). Since for polyimide,[8] the π orbital is oriented perpendicular to the chain segments, this leads to an in-plane anisotropy with the π system preferentially oriented perpendicular to the rubbing direction, as it is observed by NEXAFS spectroscopy. For polystyrene, on the other hand, because of the parallel orientation of the π system along the chain segments, this leads to an in-plane anisotropy with the π system preferentially oriented along the rubbing direction, again in line with our observation.

To understand the origin of the observed out-of-plane anisotropy, polymer chains running from within the bulk of the film into the film surface are shown in the bottom row of Fig. 6.8. Segments with their π system within the x-z plane are shown, because only their redistribution can be responsible for the observed tilt of the molecular distribution within this plane. Pulling of the rubbing fibers into the $+x$ direction will increase the number of segments oriented along the x axis relative to those perpendicular to the film surface. In line with the observation, this leads for polyimide to an anisotropy with the π system preferentially oriented perpendicular to the film surface, while for polystyrene it increases the π fraction pointing along the rubbing direction. As indicated in the figure, pulling into the $+x$ direction of a chain running from within the film into the film surface leads for polyimide segments to a preferred upwards tilt of the chain segments with respect to the rubbing direction. Hence, the π system is expected to show within the x-z plane an average backwards tilt with respect to the surface normal. This is the observed counter-clockwise rotation of the molecular distribution around the y axis of the sample frame.

The reverse effect is predicted for the phenyl rings and the π system of polystyrene. The bending of the polymer chains into the rubbing direction is expected to lead to a preferential backwards tilt of the phenyl rings with respect to the surface normal, which gives the π system an average upwards

[8] For unrubbed polyimide the majority of chain segments is oriented parallel to the film surface with their π system pointing along the surface normal. Pulling on such a chain, however, affects the π system distribution only, if the polymer chain is "rotated" perpendicular to the film surface, in which case the number of segments oriented parallel and perpendicular to the rubbing direction will increase.

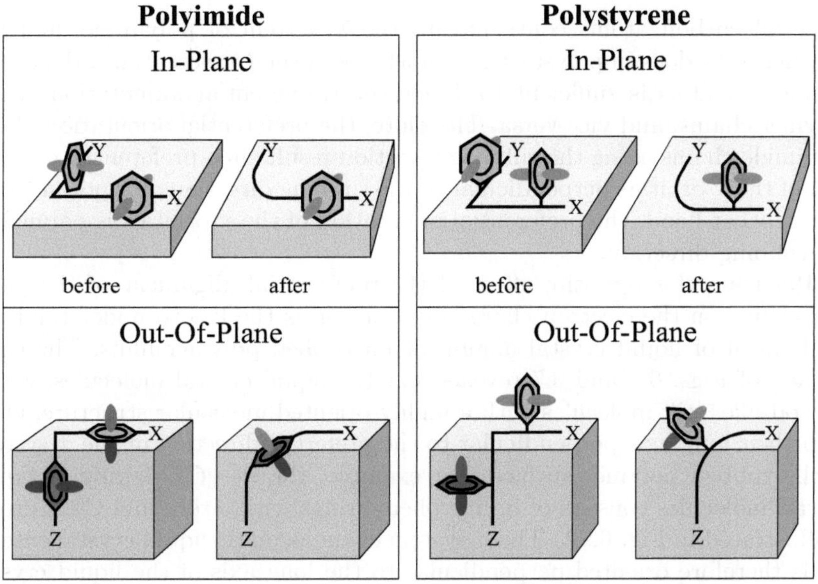

Fig. 6.8. The effect of rubbing on the molecular distribution of polymer segments at the film surface is illustrated to explain the origin of the observed charge anisotropies, which are characterized by the distribution factors illustrated in Fig. 6.7B.

tilt with respect to the rubbing direction along the $+x$ direction. Again, this is what we observed. Hence, for both polymers we understand the origin of the molecular distribution anisotropies and of the occurring counter-clockwise rotation of the molecular distributions around the y axis of the sample frame.

6.6 Molecular Anisotropy and Liquid Crystal Alignment on Rubbed Polymer Surfaces

Summarizing the discussion of the previous Sections we can state that rubbing of a polymer surface results in a preferential reorientation of the polymer chains into the rubbing direction. As a consequence of the directional pull on chains running perpendicular to the film surface, the inversion symmetry of the molecular distribution along the rubbing direction is broken. Since in molecular materials like polymers the distribution of the electron charge is oriented with respect to the molecular structure, the preferential orientation of the polymer chains gives rise to an anisotropic charge distribution. Both, the σ and the π part of the electron charge, exhibit an anisotropy. For a given molecular unit the anisotropy of the σ and π orbitals are orthogonal due to their orthogonal orientation. The orientation of the anisotropic charge distribution thereby depends on the molecular structure of the polymer chain.

Since all carbon atoms contribute to the π system of polyimide, and the vast majority does in polystyrene, knowledge of the (preferential) orientation of the π orbitals is sufficient to derive the (preferential) orientation of the polymer chains, and vice versa. Therefore, the preferential orientation of the polyimide chains along the rubbing direction results in a preferential orientation of the π orbitals perpendicular to the rubbing direction. For polystyrene, on the other hand, the preferential orientation of the π orbitals is parallel to the rubbing direction.

Realizing this opposite effect of the preferential alignment of the polymer chains on the electron charge distribution is the key to understand the mechanism of liquid crystal alignment on rubbed polymer films. The comparison of Figs. 6.5 and 6.7 reveals that the liquid crystal molecules, which are rod-like, stiff molecules with a highly oriented molecular structure, align with their long axis perpendicular to the preferred direction of the π system of the rubbed polymer surface. For example, the $n - CB$ family of liquid crystal molecules consists of in line phenyl rings with a terminal C≡N group as illustrated in Fig. 6.9A. The π system of the nematic liquid crystal ensemble is therefore oriented perpendicular to the long axis of the liquid crystal molecules (Fig. 6.9B). We therefore observe empirically that the liquid crystal molecules align on the rubbed polymer surface in that direction, which corresponds to a parallel alignment of the anisotropic charge clouds of the liquid crystal and the rubbed surface: For polyimide (Fig. 6.9C) the liquid crystal molecules are oriented parallel to the rubbing direction and their π orbitals are therefore aligned parallel to the preferential orientation of the π system within the film surface. This parallel orientation of the anisotropic charge clouds, however, shall not imply the presence of a chemical π bond. It merely means that for the attractive interaction between the liquid crystal and the rubbed polymer surface a parallel orientation of the π orbitals of the liquid crystal molecules and the rubbed surface is energetically preferred over a perpendicular orientation. One may thus picture the liquid crystal alignment on the rubbed polymer surface as the liquid crystal rods being guided by the preferred orientation of the polymer ring planes at the film surface. The microscopic origin of the liquid crystal pretilt is the out-of-plane tilt angle γ of the polyimide phenyl ring planes, depicted in the inset of Fig. 6.7.

For polystyrene, on the other hand, optimal parallel alignment of the respective π systems is achieved for the liquid crystal oriented perpendicular to the rubbing direction (Fig. 6.9D). In this orientation, the backwards tilt of the polystyrene phenyl rings with respect to the rubbing direction does not cause any out-of-plane directional asymmetry along the perpendicular axis of the liquid crystal orientation. Consequently no pretilt of the oriented liquid crystals is expected, as it is indeed observed.

In summary, the NEXAFS results led to a model, which identifies as origin of liquid crystal alignment on rubbed polymer surfaces the presence

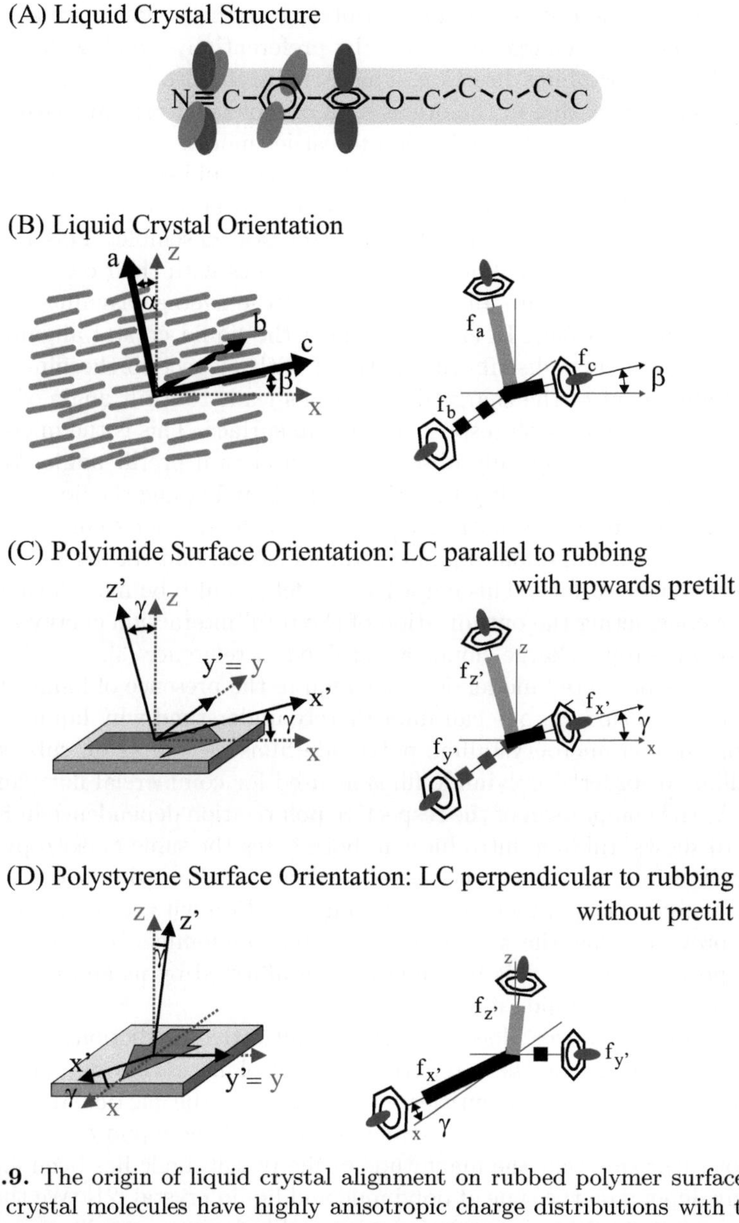

Fig. 6.9. The origin of liquid crystal alignment on rubbed polymer surfaces: (A) Liquid crystal molecules have highly anisotropic charge distributions with their π system preferentially oriented perpendicular to their long axis, which gives an ensemble of oriented (nematic) liquid crystals an asymmetric charge distribution (B). This can be characterized by molecular orientation factors (f_a,f_b,f_c) describing the preferential orientation of the π system. For the experimentally observed alignment directions the anisotropic charge distribution of the liquid crystal is oriented parallel to the one of the rubbed polyimide (C) and polystyrene (D) surface, which optimizes their interaction energy [3].

of an average, macroscopic charge anisotropy at the polymer surface. Rubbing introduces this charge anisotropy by preferentially aligning the polymer chains along the rubbing direction. This structural anisotropy translates into a charge anisotropy due to the orientation of molecular orbitals with respect to their molecular axis. In distinction to earlier models [19–25], the presence of patches with crystalline-like local order at the rubbed polymer surface is not required. The average anisotropy has only to be statistically significant so that it can be "sensed" by the liquid crystal ensemble. This ensemble then orients such that the liquid crystal molecules with their own asymmetric charge distribution are aligned parallel to the average asymmetry of the rubbed polymer surface. In case that along the liquid crystal alignment direction the molecular distribution is tilted with respect to the film surface, parallel alignment of the charge distributions leads to a tilt angle of the liquid crystal molecules with respect to the film surface. This is the microscopic origin of the macroscopically observed liquid crystal pretilt angle. We note that the model uses implicitly that the interaction aligning the liquid crystals parallel to each other (nematic ordering) is much stronger than the interaction between the individual liquid crystal molecules and the segments of the rubbed polymer surface. This empirical model can also be more formally derived by considering the optimization of the total interaction energy between the two anisotropic charge clouds as outlined in reference [3].

Since the presented model does not require the presence of long-range order at the film surface, one can immediately understand why liquid crystals align on rubbed microcrystalline polyimide films as well as on rubbed noncrystalline, disordered polyimide films as used for commercial flat panel displays. As the comparison of the respective polarization dependence in Fig. 6.7 and 6.10 shows, rubbing introduces in both cases the same anisotropy in the charge distribution. In fact, the model predicts for all polyimide derivates an in-plane alignment parallel to the rubbing direction with an upwards pretilt angle, provided that the π system is oriented predominantly perpendicular to the polymer chain. This has indeed been affirmed by us and others for a wide variety of polyimide films.

The model, however, does not make quantitative predictions for the size of the pretilt angle or the strength of the in-plane anchoring energy. The absolute value of these quantities will depend on the particular molecular structure of both, the rubbed polymer surface and the liquid crystal. Hence, it is not surprising that the magnitude of the pretilt angle has been found to strongly depend on the type of polyimide and liquid crystal [21]. We therefore only expect a qualitative or relative relationship between pretilt angle β and anchoring energy of the liquid crystal and molecular tilt angle γ and in-plane anisotropy of the rubbed polymer surface. For a quantitative analysis of the pretilt angle one would also need to consider steric (excluded volume) interactions [9, 26, 27], which favor a liquid crystal alignment parallel to the film surface over a homeotropic alignment, i.e., perpendicular to the film surface. Hence, we expect the liquid crystal bulk pretilt angle β to be smaller than the molecular tilt angle γ of the polymer segments at the film surface. This also explains the in-plane alignment for polystyrene, although the charge anisotropy of the film surface would favor weakly a homeotropic alignment (with 27% versus 21% of the segments aligned in the respective directions).

We further note that the proposed model does not take the topographic structure explicitly into account. Grooves in the polymer surface caused by the rubbing fibers, however, may give rise to a macroscopic anisotropic topography without significantly disturbing the molecular distribution due to the different length scales on which these phenomena occur. For polymer films with their π system along the polymer chains as in polystyrene, for which our model predicts liquid crystal alignment perpendicular to the rubbing direction, this may lead to a competing interaction. Hence, depending on the relative strength of these two alignment forces, this may lead to either a parallel or a perpendicular alignment of the liquid crystals.

We can now also address the case of rubbed polymer films without any π system such as polyethylene. Our model readily explains the observed [28] liquid crystal alignment along the rubbing direction, since the preferential orientation of the chain segments leads to the largest charge density along the rubbing direction. The quadrupolar molecular interaction then favors an alignment of the largest liquid crystal charge density, which is along the long axis of the liquid crystal, parallel to the rubbing direction. Our model also explains the origin of liquid crystal alignment on polymer surfaces, which have been irradiated by polarized UV radiation [29–31]. For example, as was shown in Chap. 5, irradiation of PVCN with polarized UV light decreases the charge density along the polarization direction and consequently leads to liquid crystal orientation perpendicular to the polarization direction. The anisotropy is believed to be created in such systems by directional bond breaking or bond formation, e.g. crosslinking. More generally, based on the model derived from our experimental NEXAFS spectroscopy results, one may predict that any method creating a statistically significant surface bond anisotropy should be applicable for liquid crystal alignment. One would expect that covalent materials, owing to the directional nature of the bonds, are particularly suited as alignment materials. A good example is graphite (see Chap. 7). As we will discuss in the following these predictions have indeed led us to develop a new alignment layer material for liquid crystal displays.

6.7 NEXAFS Study of Ion Beam Irradiated Polymer Surfaces

In the late 90's it had been observed that liquid crystals align on polyimide films irradiated with a low energy (50 – 200 eV) directional ion beam [32,33]. The technologically important aspect of this observation is that ion beam irradiation is a non-contact technique, which is performed in a clean environment and which at these low ion beam energies only modifies the film surface. Hence, it offers to overcome the inherent problems associated with the rubbing process, most importantly contamination of the polymer film surface (rubbing debris) and damage of the underlying electronic circuit (local charge built-up in the rubbing process of the non-conducting polymer layer). These studies demonstrated that high-quality and reliable displays can be built on the basis of ion beam irradiated polyimide films. The liquid crystals thereby align along the direction of the ion beam irradiation, as illustrated in the top part of Fig. 6.10. We note that for both, rubbing and ion beam irradiation, the in-plane alignment of the liquid crystals is parallel to the direction of the surface treatment. The ion beam irradiated polyimide surface also gives rise to a pretilt angle of the liquid crystals, which is essential for display quality. This pretilt angle, however, points for ion beam irradiated polyimide downwards into the film surface, which is opposite to the direction observed on rubbed polyimide surfaces.

The results of a polarization dependent NEXAFS study on a rubbed and an ion beam irradiated disordered polyimide film[9] is shown in Fig. 6.10. As before, the angle dependence of the $\pi^\star$ intensity of the surface sensitive AEY data is shown . One notices on the first glance that both angular dependences have the same general orientation. In particular, within the film surfaces a much smaller intensity is observed for the electric field vector oriented along the respective rubbing and ion beam directions ($\alpha = 90°$, solid squares) than for perpendicular orientation ($\alpha = 90°$, open circles). This indicates that on both surfaces a charge anisotropy is introduced with a preferential in-plane orientation of the π system perpendicular to the direction of the applied treatment. Hence, our model predicts for both surfaces a liquid crystal alignment parallel to the direction of the surface treatment, which is in line with the experimental observation.

For the asymmetry of the angular dependence within the x-z plane parallel to the treatment direction (solid symbols), one notices an opposite shift of the curves with respect to the $\alpha = 0°$ incidence angle. This indicates that the molecular orientation is tilted into opposite directions (see (6.3)). Hence,

[9]IBM-X polyimide is the proprietary polymer used in IBM's flat panel displays. This polyimide is disordered/amorphous, while the BPDA-PDA polyimide film discussed before (Fig. 6.7) has microcrystalline domains. We note that rubbing introduces in both film surfaces the same charge anisotropy, independently of the presence/absence of locally ordered domains.

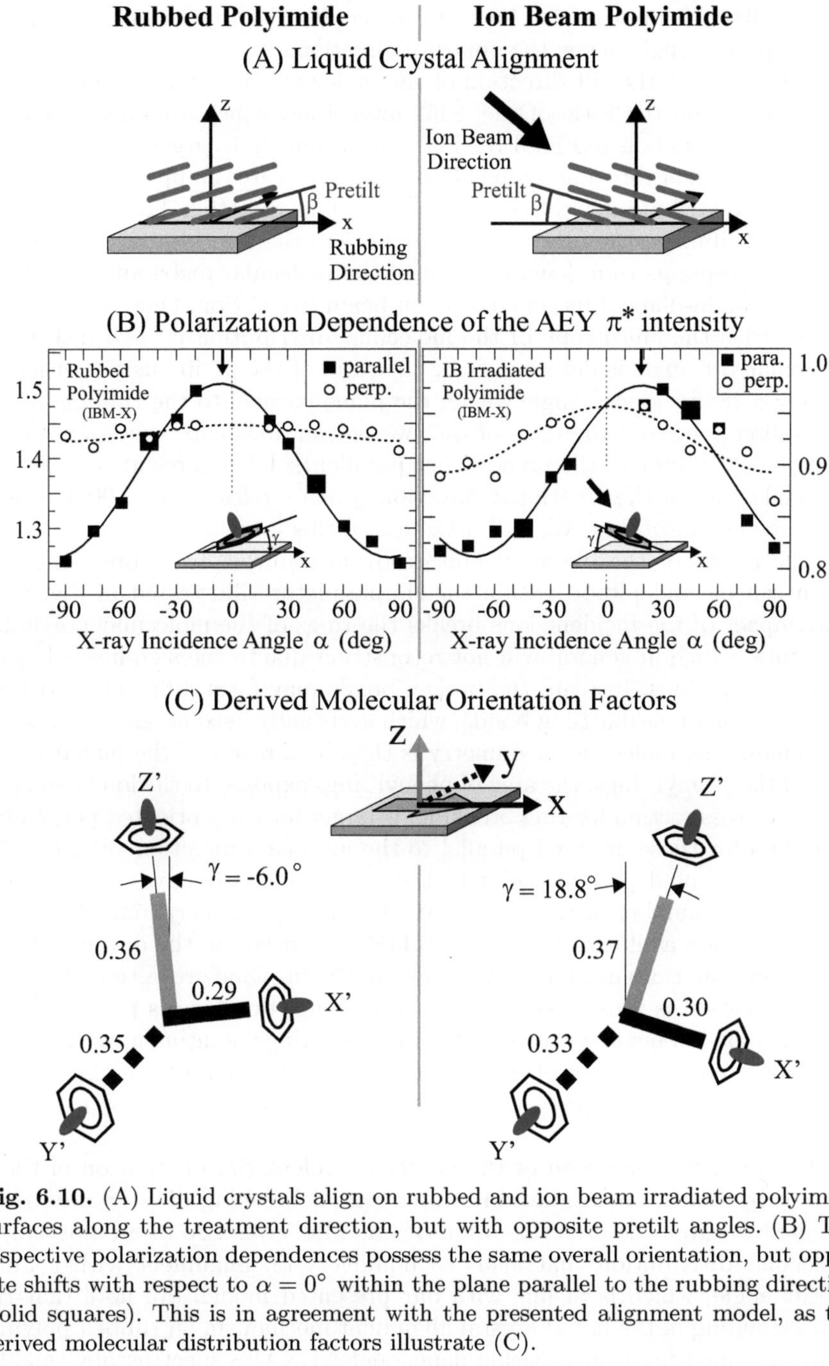

Fig. 6.10. (A) Liquid crystals align on rubbed and ion beam irradiated polyimide surfaces along the treatment direction, but with opposite pretilt angles. (B) The respective polarization dependences possess the same overall orientation, but opposite shifts with respect to $\alpha = 0°$ within the plane parallel to the rubbing direction (solid squares). This is in agreement with the presented alignment model, as the derived molecular distribution factors illustrate (C).

again in line with our model, both the molecular distributions and the liquid crystal pretilt angle, show the opposite direction.

We note that the tilt direction of the molecular distribution can directly be derived from the ratio of the $\pm 45°$ intensities, which are emphasized by the enlarged symbols in Fig. 6.10. The $-45°$ intensity is larger than the $+45°$ intensity for the rubbed polyimide film, which indicates an upwards tilt of the molecular distribution with respect to the rubbing direction. For the ion beam irradiated film, on the other hand, one notices the opposite intensity ratio, which corresponds to a downward tilt of the molecular distribution with respect to the in-plane direction of the ion beam irradiation. One can therefore characterize the anisotropy of the molecular distribution by two ratios: the molecular tilt angle is determined by the ratio of the $\pi^\star$ intensities observed for $\alpha = \pm 45°$ incidence angle within the plane parallel to the rubbing or ion beam direction. And the ratio of the two normal incidence spectra with the electric field vector in the plane either parallel (solid squares at $\alpha = 90°$) or perpendicular to the treatment direction (open circles at $\alpha = 90°$) reveals the in-plane anisotropy. We will make use of this later.

The origin of the observed anisotropy introduced by a directional ion beam can be understood within the simple picture illustrated in Fig. 6.11. The impact of the incident ions breaks the rings of the molecular polyimide structure, which in general will not reconstruct due to the significant kinetic energy of the ions. Instead, the broken bonds may form differently oriented bonds or remain as dangling bonds, which eventually become saturated by air molecules. The molecular asymmetry is thereby a result of the planar structure of the phenyl rings: the area of phenyl rings exposed to the ion beam, and thus the cross section for ring breaking, is larger for rings oriented perpendicular than for those oriented parallel to the ion beam direction (Fig. 6.11A). For rings oriented perpendicular to the film surface this gives rise to a decrease of the number of rings oriented with their plane perpendicular to the beam direction as illustrated in Fig. 6.11B. The origin of the downwards tilt of the molecular orientation with respect to the in-plane projection of the ion beam direction is illustrated in Fig. 6.11C. Since downwards pointing rings intersect the ion beam with a smaller area than rings pointing upwards, more upwards pointing rings will be broken. The distribution of remaining rings therefore points downwards, which is opposite to the tilt direction caused by rubbing.

In summary, the origin of the observed preferential orientation of the π system by ion beam irradiation is preferential breaking of polyimide rings oriented towards the directional ion beam. This gives rise to an anisotropic molecular distribution that leads to liquid crystal alignment with a finite pretilt angle, which is in line with our presented model. We note that the understanding of the liquid crystal alignment mechanism on rubbed polymer films, obtained from polarization dependent NEXAFS spectroscopy, enables us to also understand the origin of liquid crystal alignment on ion beam

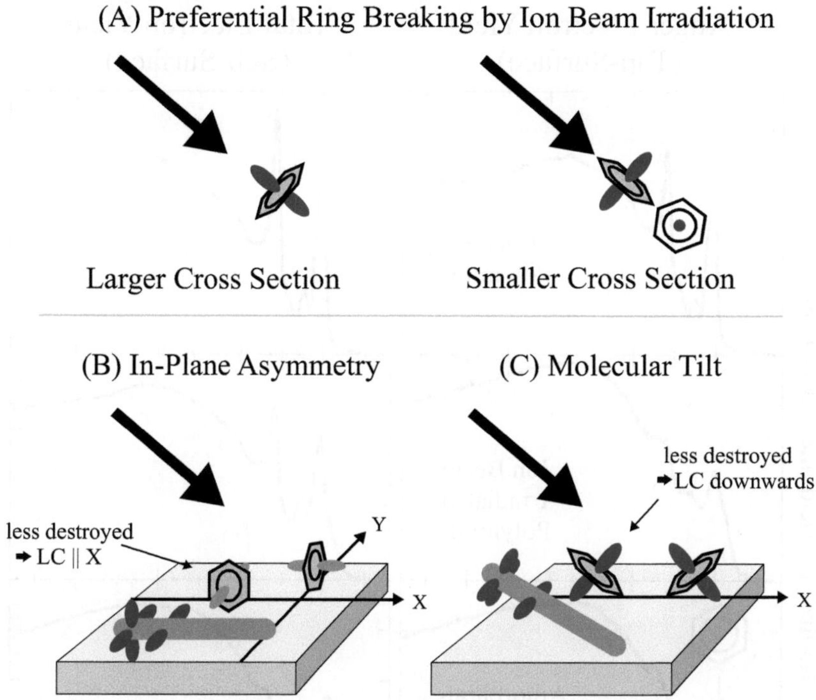

Fig. 6.11. (A) Directional ion beam irradiation can introduce molecular anisotropy, since the cross section for bond breaking depends for planar molecular groups on the relative orientation. (B) On polyimide this gives rise to the observed charge asymmetry, characterized by the distribution factors illustrated in Fig. 6.10C.

irradiated polymer films. In addition, the NEXAFS spectra of the ion beam irradiated polyimide film reveal an even more important information that will be discussed in the next section.

6.8 Replacing the Polymer: Ion Beam Irradiated Amorphous Carbon

In Fig. 6.12 we show the $\pi^\star$ region of the more surface sensitive AEY (left column) and the deeper into the film sampling TEY spectra (right column) of a rubbed (top row) and an ion beam irradiated (middle row) polyimide film. Comparing the spectra one clearly notices that the ion beam irradiation[10] leads to significant ring breaking at the film surface. While the characteristic fine structure of polyimide is still prominent in the deeper below the film

[10]Ion beam parameter are: 75 eV, 4×10^{15} Ar atoms/cm^2, 15° angle of incidence with respect to the surface normal.

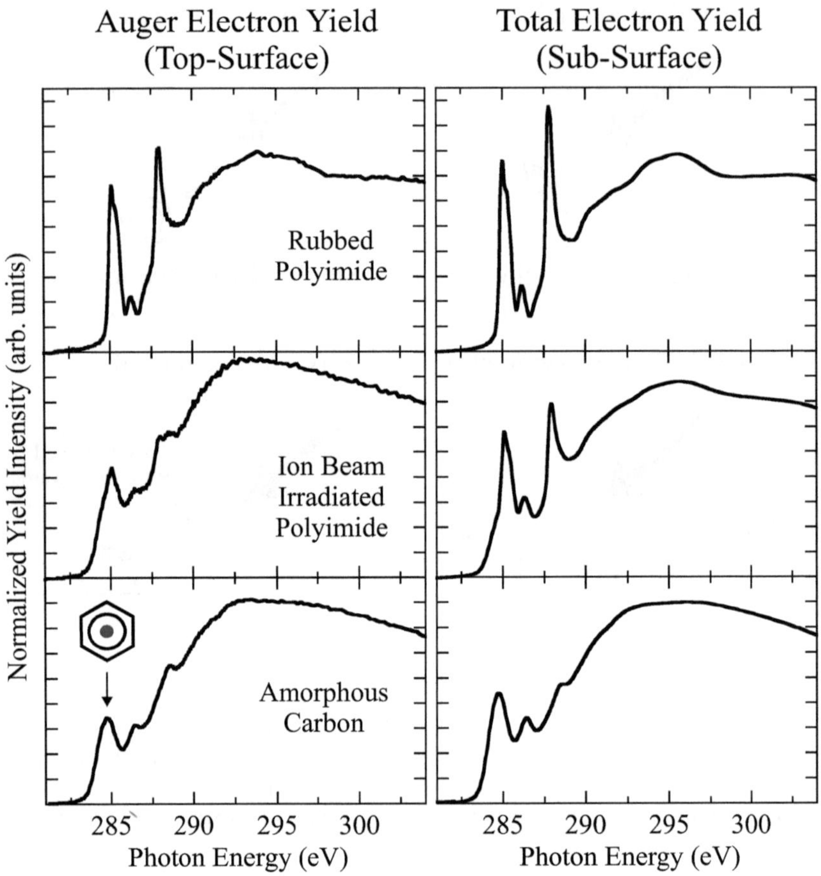

Fig. 6.12. Auger (left) and total electron yield (right) NEXAFS spectra of rubbed polyimide (top), ion beam irradiated polyimide (middle), and amorphous carbon (bottom). The comparison reveals the presence of a layer of amorphous carbon at the surface of the ion beam irradiated polyimide film.

surface sampling TEY spectrum, it is basically absent in the more surface sensitive AEY spectrum, which shows a distinct fine structure. The molecular origin of this shape is revealed by the comparison of the AEY spectrum of the ion beam treated polyimide film with the spectrum in the bottom row, which has been recorded on an amorphous carbon film. The similarity of these two spectra indicate that the ion beam irradiation destroys completely the molecular structure of the polyimide polymer in the film surface leading to the formation of a layer of amorphous carbon at the film surface with a thickness of about 2 – 5 nm. This thickness estimate follows from the comparison of the TEY and AEY spectra and their known sampling depths, and it is also in agreement with Monte Carlo simulations of the applied ion beam irradiation.

The observation of a layer of amorphous carbon at the film surface, in combination with the observation of liquid crystal alignment on this surface, suggested the breakthrough idea to replace the polyimide polymer film with an amorphous carbon layer [34]. The essential requirement for liquid crystal alignment (as stated by our model), namely the presence of an anisotropic distribution of directional bonds, can be fulfilled by an ion beam irradiated amorphous carbon layer. This is demonstrated by the presence of the resonance associated with π orbitals at 285 eV in the absorption spectrum of amorphous carbon (bottom of Fig. 6.12). Its presence indicates that amorphous carbon contains unsaturated sp_2 and sp hybridized carbon atoms. While sp_3 hybridization does not lead to any anisotropy, the directional nature of carbon double and triple bonds formed by sp_2 and sp hybridized carbon atoms can lead to a breaking of the isotropy of the molecular distribution. It therefore mainly remains the question whether a statistically significant anisotropy in these carbon bonds can be achieved by ion beam irradiation of an amorphous carbon layer.

That this should indeed be the case is demonstrated in Fig. 6.13, in which the AEY spectra of rubbed polyimide, ion beam irradiated polyimide, and ion beam irradiated amorphous carbon are shown. Plotted are the four essential spectra necessary to characterize the anisotropy of the molecular distribution as discussed before. The comparison of the $\pi^\star$ intensities in the two in-plane geometries (left side) shows for all three cases the same orientation. Hence, our model predicts in-plane alignment of the liquid crystals along the respective rubbing and ion beam direction in all three cases. We note that not only the overall spectral shape of the ion beam irradiated polyimide and amorphous carbon film is very similar, but also the intensity of the $\pi^\star$ resonances and their dichroism is of comparable strength. The in-plane anisotropy of the molecular distribution at the surface of the ion beam irradiated amorphous carbon film is therefore expected to give rise to a sufficient in-plane anchoring energy for liquid crystal alignment along the ion beam direction.

Comparing the spectra in the right panel, which characterize the molecular tilt angle, one finds the same polarization dependence for the two ion beam irradiated materials, which is opposite to the one of the rubbed polyimide film. Hence, a downwards liquid crystal pretilt angle is expected for both ion beam treated surfaces. Again, since the overall shape and the $\pi^\star$ intensities and their dichroism is comparable for the two ion beam irradiated films, liquid crystals are expected to exhibit a technologically sufficient pretilt angle on an ion beam irradiated amorphous carbon layer.

In line with these expectations, liquid crystal alignment has indeed been observed on ion beam irradiated amorphous carbon layers deposited either by sputtering or CVD [35]. In fact, also a wide variety of other materials showed liquid crystal alignment upon ion beam irradiation, for example, SiN_x, hydro-

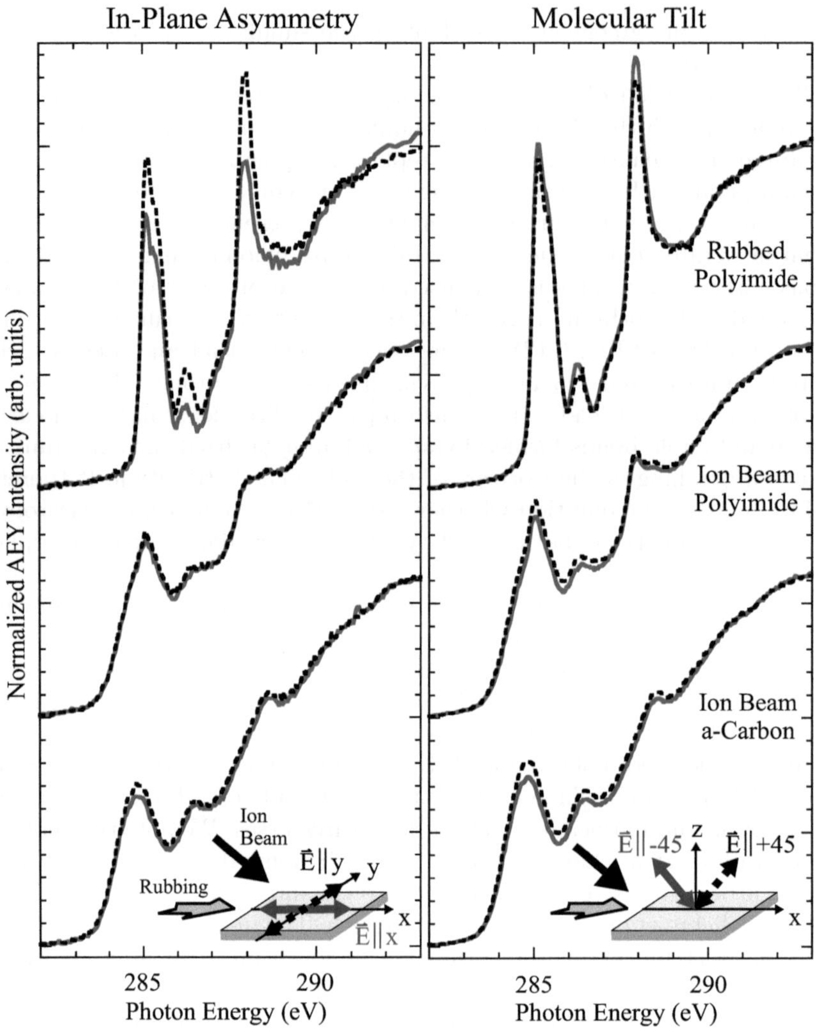

Fig. 6.13. Auger electron yield carbon K-edge NEXAFS spectra of rubbed polyimide (top), ion beam irradiated polyimide (middle), and ion beam irradiated amorphous carbon (bottom) for the indicated geometries. The intensity ratios of the plotted $\pi^\star$ resonances characterize the in-plane asymmetry (left column) and the molecular tilt direction (right column).

genated amorphous silicon, SiC, SiO_2, glass, Al_2O_3, CeO_2, SnO_2, $ZnTiO_2$, and $InTiO_2$ [35]. Bearing our model for the mechanism of liquid crystal alignment in mind this is not totally surprising, since in principle any material with oriented bonds, which can be preferentially broken to introduce an average anisotropic charge distribution, should be suitable for liquid crystal alignment. One should, however, bear in mind that in one or the other of the above cases not the ion beam irradiated material itself, but a carbonaceous surface contamination may have contributed or given rise to the observed liquid crystal alignment.

The ion beam irradiation offers a wide variety of parameters for optimization of the alignment properties of the irradiated surface. Among these are the kind of ion used, their kinetic energy, the exposure dose, the gas used for backfilling of the vacuum chamber (which will saturate any remaining dangling bond), and of course the incident angle of the ion beam. This incident angle will certainly have an influence on the tilt of the molecular distribution with respect to the in-plane projection of the ion beam. For both "extreme" geometries, the ion beam either oriented perpendicular or parallel to the surface, no tilt of the molecular distribution is expected. The reason for this is that whether a phenyl ring (or any other directional bond) points with a given angle upwards or downwards, it will have in both cases the same area exposed to the incident ion beam. Hence, we expect the molecular tilt angle γ, i.e., the tilt of the polymer segments at the film surface, to depend on the ion beam incidence angle Θ. In particular, we expect that there is an optimum ion beam incidence angle Θ, for which the the molecular tilt angle γ is the largest. This implies that the liquid crystal pretilt angle β, predicted by our model to originate from the tilt of the molecular distribution at the film surface, should also depend on the ion beam incidence angle Θ.

The angular dependence of the liquid crystal pretilt angle β and of γ, the molecular tilt angle of the chain segments at the polymer film surface, are plotted in Fig. 6.14 for ion beam irradiated[11] polyimide (top) and amorphous carbon (bottom). The agreement shown by the two angular dependences is in botch cases excellent, which lends further, strong support for our model. We note that the value of the molecular tilt angle γ exceeds that of the liquid crystal, which is expected due to the steric (excluded volume) interactions discussed before. For this angle range, however, we observe a linear relation between the molecular tilt angle γ and the pretilt angle β of the liquid crystal. We also remind that the liquid crystal pretilt angle β is the average pretilt angle of the liquid crystal ensemble, which is dominated by the "bulk" of the liquid crystal. That the liquid crystals in close contact with the ion beam irradiated surface may possess a larger pretilt angle than the rest of the liquid crystal can not be excluded. This could, however, be investigated with the nonlinear optical techniques described in Chap. 5.

[11] Ion beam doses of 75 eV and 3.8×10^{15} Ar atoms/cm^2 were used for polyimide and 200 eV and 4.5×10^{15} Ar atoms/cm^2 for amorphous carbon.

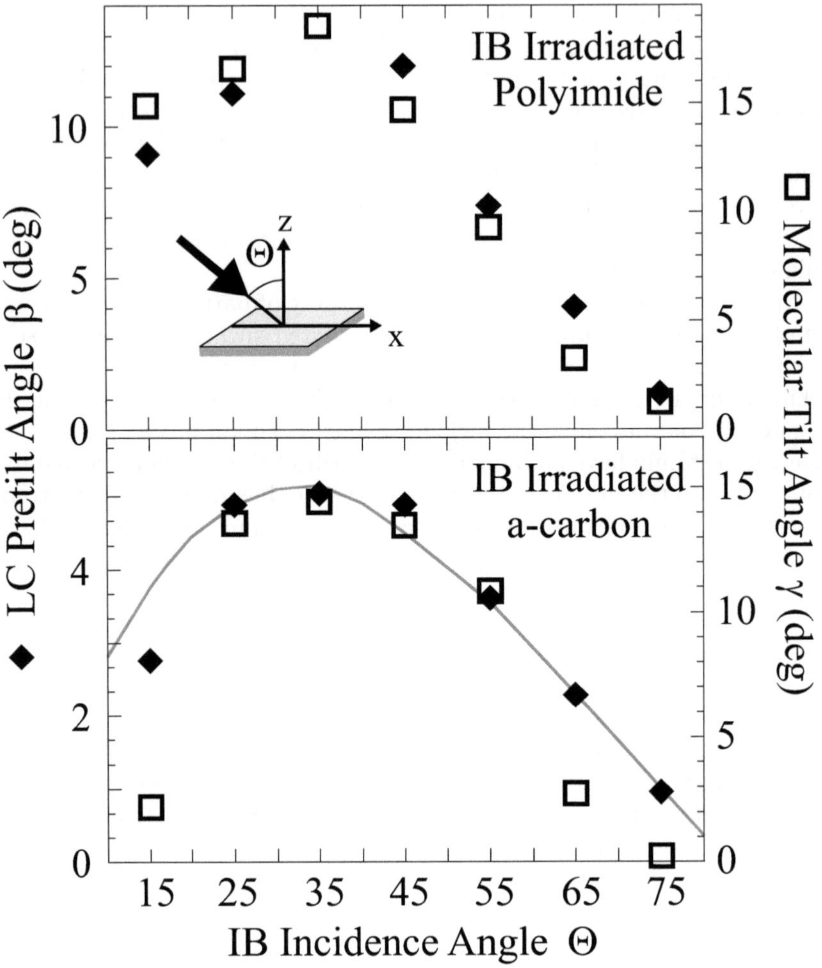

Fig. 6.14. Ion beam incidence angular dependence of the liquid crystal pretilt angle β and the molecular tilt angle γ of the polymer segment distribution at the film surface for polyimide (top) and amorphous carbon (bottom). As predicted by the alignment model the liquid crystal pretilt angle β follows the molecular tilt angle γ. The line is a fit to $\gamma(\Theta)$ using a model that assumes finite, but different cross sections for breaking of phenyl rings oriented along or perpendicular to the ion beam direction [35].

Conclusions

In this Chapter we have demonstrated how the surface sensitive technique of polarization dependent, soft x-ray absorption spectroscopy resolving the near edge x-ray absorption fine structure (NEXAFS) can be used to investigate and understand the origin of liquid crystal alignment on rubbed polymer surfaces. These measurements reveal an anisotropic distribution of the molecular segments of the polymer chains at the rubbed surfaces. Both, the out-of-plane distribution and the in-plane distribution with respect to the rubbing direction, are found to be anisotropic. This structural anisotropy goes along with a charge anisotropy due to the orientation of the molecular orbitals with respect to the molecular bonds. The liquid crystal molecules possess also a highly anisotropic charge distribution due to their rod-like shape. Since for parallel alignment of these two anisotropic charge distributions their attractive interaction energy is optimized, the alignment direction of the liquid crystal molecules is determined by the orientation of the surface charge anisotropy. It is important to note that a statistically significant average anisotropy is sufficient for this, no crystalline-like long-range order is required. This model explains straight forwardly the occurrence of liquid crystal alignment parallel and perpendicular to the rubbing direction as observed for polyimide (disordered and microcrystalline) and polystyrene, respectively, as well as the respective presence and absence of a liquid crystal pretilt angle. Furthermore, the model also explains the liquid crystal alignment observed for polymer surfaces treated by polarized UV and ion beam irradiation.

Our study of ion beam irradiated polyimide films revealed that in this case the aligning surface consists of an amorphous carbon layer, which is caused by degradation of the molecular polymer structure in the ion beam. This observation led to the idea to replace the polyimide film with an amorphous carbon film. Indeed, an ion beam irradiated amorphous carbon layer appears to align liquid crystals as well as an ion beam irradiated polymer film. These results convinced IBM to develop a new manufacturing technology for liquid crystal displays, which successfully produced displays of highest quality.

Acknowledgment

The experiments reported here have been carried out at the Stanford Synchrotron Radiation Laboratory (SSRL), which is supported by the Department of Energy, Office of Basic Energy Science. We would like to express our thanks in particular to Jeff Moore and Curtis Troxel for their support. Furthermore, we acknowledge the many contributions to this work of our colleagues from the IBM Research Division in Almaden and Yorktown, USA, and the IBM Display Unit, Japan.

References

1. F. de Groot, J. Elec. Spectrosc. Relat. Phenom. **67**, 529 (1994).
2. K. Siegbahn, *ESCA; Atomic, Molecular, and Solid State Structure Studied by means of Electron Spectroscopy*, Almqvist & Wiksells, Uppsala, 1967.
3. J. Stöhr and M.G. Samant, J. Elec. Spectrosc. Relat. Phenom. **98**, 189 (1999).
4. I. Mori, T. Araki, H. Ishii, Y. Ouchi, K. Seki, and K. Kondo, J. Elec. Spectrosc. Relat. Phenom. **78**, 371 (1996).
5. J. Stöhr, H.A. Padmore, S. Anders, T. Stammler, and M.R. Scheinfein, Surface Review and Letters **5**, 1297 (1998).
6. J. Stöhr, *NEXAFS Spectroscopy*, Springer Series in Science, Vol. 25, Heidelberg, 1992.
7. J. Sakurai, *Advanced Quantum Mechanics*, Addison-Wesley, London, 1967.
8. E.W. Thulstrup and J. Michl, *Elementary Polarization Spectroscroscopy*, VCH Publishers, Inc., New York, 1989.
9. P.G. De Gennes and J. Prost, *The Physics of Liquid Crystals*, Clarendon Press, Oxford, 1993.
10. B.T. Thole, G. van der Laan, J.C. Fuggle, G.A. Sawatzky, R.C. Karnatak, and J.M. Esteva, Phys. Rev. B **32** 5107 (1985).
11. S. Tanuma, C. Powell, and D. Penn, Surface and Interface Analysis **21**, 165 (1994).
12. N. van Aerle, Liquid Crystals **17**, 585 (1994).
13. S. Ishihara, H. Wakemoto, K. Nakazima, and Y. Matsuo, Liquid Crystals **4**, 669 (1989).
14. Y. Ouchi, I. Mori, M. Sei, E. Ito, T. Araki, H. Ishii, K. Seki, and K. Kondo, Physica B **209**, 407 (1995).
15. M.G. Samant, J. Stöhr, H.R. Brown, T.P. Russell, J.M. Sands, and S.K. Kumar, Macromolecules **29**, 8334 (1996).
16. K. Weiss, C. Wöll, E. Böhm, B. Fiebranz, G. Forstmann, B. Peng, V. Scheumann, , and D. Johannsmann, Macromolecules **31**, 1930 (1998).
17. J. Stöhr, M.G. Samant, A. Cossy-Favre, J. Díaz, Y. Momoi, S. Odahara, and T. Nagata, Macromolecules **31**, 1942 (1998).
18. Y. Liu, T.P. Russell, M.G. Samant, J. Stöhr, H.R. Brown, A. Cossy-Favre, and J. Diaz, Macromolecules **30**, 7768 (1997).
19. J.M. Geary, J.W. Goodby, A.R. Kmetz, and J.S. Patel, J. Appl. Phys. **62**, 4100 (1987).
20. S. Kobayashi and Y. Iimura, Proc. SPIE, Liquid Crystal Materials, Devices, and Applications III, edited by Ranganathan Shashidhar, **2175**, 122 (1994).
21. K.W. Lee, S.H. Paek, A. Lien, C. Durning, and H. Fukuro, Macromolecules **29**, 8894 (1996).
22. K.-W. Lee, S. Paek, A. Lien, C. Durning, and H. Fukuro, in K.L. Mittal and K.-W. Lee, editors, *Polymer Surfaces and Interfaces: Characterization, Modification and Application*, VSP, The Netherlands, 1996.
23. H. Mada and T. Sonoda, Jpn. J. Appl. Phys. Part 2 Lett. **32**, L1245 (1993).
24. M.F. Toney, T.P. Russell, J.A. Logan, H. Kikuchi, J.M. Sands, and S.K. Kumar, Nature **374**, 709 (1995).
25. X. Zhuang, D. Wilk, L. Marrucci, and Y.R. Shen, Phys. Rev. Lett. **75**, 2144 (1995).
26. K. Okano, N. Matsuura, and S. Kobayashi, Jpn. J. Appl. Phys. **21**, L109 (1982).

27. K. Okano, Jpn. J. Appl. Phys. Part 2 Lett. **22**, L343 (1983).
28. H. Aoyama, Y. Yamazaki, N. Matsuura, H. Mada, and S. Kobayashi, Mol. Cryst. Liq. Cryst. **72**, 127 (1981).
29. W.M. Gibbons, P.J. Shannon, S.T. Sun, and B.J. Swetlin, Nature **351**, 49 (1991).
30. P.J. Shannon, W.M. Gibbons, and S.T. Sun, Nature **368**, 532 (1994).
31. M. Schadt, H. Seiberle, and A. Schuster, Nature **381**, 212 (1996).
32. P. Chaudhari, J.A. Lacey, S.C.A. Lien, and J.L. Speidell, Jpn. J. Appl. Phys. Part 2 Lett. **37**, L55 (1998).
33. S.C.A. Lien, P. Chaudhari, J.A. Lacey, R.A. John, and J.L. Speidell, IBM J. Res. Develop. **42**, 537 (1998).
34. J. Stöhr, M.G. Samant, J. Lüning, A.C. Callegari, P. Chaudhari, J.P. Doyle, J.A. Lacey, S.A. Lien, S. Purushothaman, J.L. Speidell, Science **292**, 2299 (2001).
35. P. Chaudhari, J. Lacey, J. Doyle, E. Galligan, S.C.A. Lien, A. Callegari, G. Hougham, N.D. Lang, P.S. Andry, R. John, K.H. Yang, M.H. Lu, C. Cai, J. Speidell, S. Purushothaman, J. Ritsko, M.G. Samant, J. Stöhr, Y. Nakagawa, Y. Katoh, Y. Saitoh, K. Sakai, H. Satoh, S. Odahara, H. Nakano, J. Nakagaki, and Y. Shiota, Nature **411**, 56 (2001).

7 Scanning Probe Microscopy Studies of Liquid Crystal Interfaces

Theo Rasing and Jan Gerritsen

The invention and development of Scanning Tunneling Microscopy (STM) by Binnig and Rohrer et al [1] has revolutionized surface science and has started a new era in nanoscience and technology. Scanning a sharp tip over the nano-object under study allows for investigations of many material properties with atomic resolution. After the initial breakthrough of the STM in the mid-80s, when the STM was employed to map atomic positions (more precisely: electron densities), a large number of Scanning Probe Microscopy methods (SPM) have been developed, based on the scanning of objects by using specific forces between the tip and sample (magnetic, chemical, frictional, etc.), combinations with optical methods (Near-field Scanning Optical Microscopy (NSOM), Scanning Tunneling Luminescence (STL), etc.) and the active manipulation of nano-objects by means of the interaction with the 'tip'. That way, atoms and molecules can be manipulated at will and chemical reactions can be catalyzed in order to arrive at artificial nanostructures with unique functionalities. Some good overviews of the state of the art of SPM can be found in the book by Magonov and Wangbo [2] and in the series by Wiesendanger and Güntherodt [3].

However, although atomic resolution can regularly be obtained, this is usually only true in the most ideal circumstances, like on ultraclean samples in ultra high vacuum. The desire to study molecular structures in more realistic environments, like under ambient conditions, in high pressure or in liquids has led to the developments of new generations of microscopes with increasing complexity. In this Chapter we will discuss some or our recent approaches in both the observation and manipulation of liquid crystal molecules and polymeric surfaces, respectively.

7.1 STM Investigations of the Ordering of 8CB Molecules on Graphite

The development of Scanning Probe techniques allows to study liquid crystal molecules and their ordering on and interaction with surfaces with an unprecedented resolution. In this part we will describe how detailed information down to the atomic scale can be obtained by STM studies of 8CB molecules on graphite.

We have developed and built a simple, rigid and low cost microscope [4] that allows to investigate the organization of molecules in ambient, liquid and even gel-like [5] environments. We have chosen to use a simple analogue feedback system and an analogue I-O board [6] in a personal computer. To avoid noise problems with the D-A converter we control the lateral tip velocity instead of the lateral tip position. By integrating a voltage [7] (tip velocity) over a certain time we generate the desired lateral tip positions. In this way we get a more accurate control of the tip position as well as more time to acquire and average the incoming data. The resulting pocket-size microscope (see Fig. 7.1) is so reliable and user friendly that several copies are routinely used for student classes.

The first observations of liquid crystal molecules by STM were already reported in the late eighties [8, 13], showing structural features such as aromatic rings and alkyl chains. Except graphite, also molybdenum disulphide appeared to be a perfect substrate for these studies, and very detailed molecular information could be obtained [9, 10]. An example is shown in Fig. 7.2.

Without electronic structure calculations [11], one can qualitatively understand these observations by realizing that the liquid crystal molecules placed between the tip and the surface counter electrode, will modify the tunneling process. Contrast can then be obtained between less and more conducting parts of the molecules under study. For the nCB type of LC molecules for example, the π orbitals of the bi-phenyl rings will lead to a higher con-

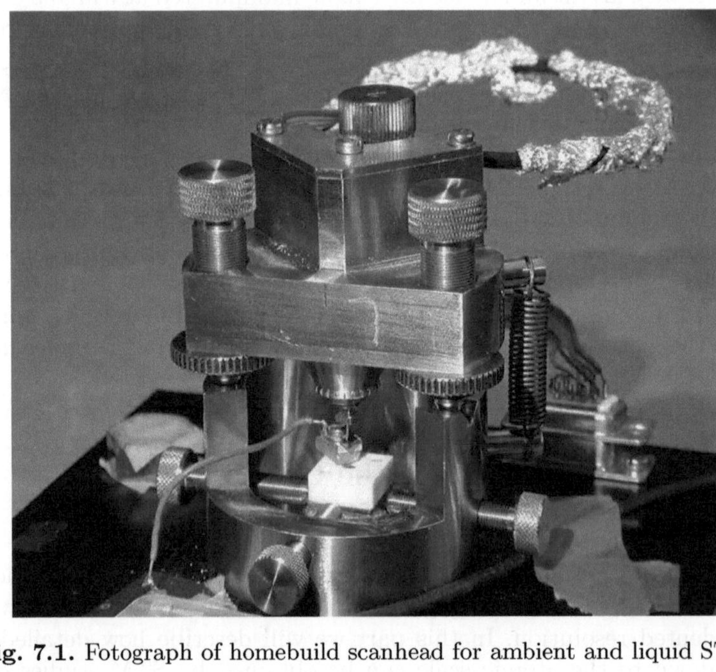

Fig. 7.1. Fotograph of homebuild scanhead for ambient and liquid STM.

Fig. 7.2. STM picture (raw data) of 8CB on HOPG measured at room temperature. Framesize is 30 nm × 30 nm, $V_{\text{sample}} = 870\,\text{mV}$, $i_{\text{bias}} = 20\,\text{pA}$.

ductivity than the much more localized charges on the alkyl chains, resulting in a brighter contrast for the first and a darker for the latter.

For the experiments we used our home build STM described above. As a substrate we used freshly cleaved pyrolytic graphite (HOPG). The samples were prepared by putting a drop of 8CB on the HOPG substrate, heat it above the clearing point and let it cool down. The images were taken with the tip in the droplet. Figure 7.2 shows a constant current image of 8CB on HOPG at room temperature, i.e. at about 20 degrees below the nematic-isotropic phase transition. Because of the strong periodicity in this structure, the raw data (Fig. 7.2) allow to apply correlation averaging techniques [15] without loss or unintentional additon of information. The result is shown in Fig. 7.3. The image in Fig. 7.3 clearly indicates that the 8CB molecules form long range, crystalline like order at the graphite/liquid crystal interface. This means that the crystalline graphite surface not only induces orientational but also positional order in this liquid crystalline material. Figure 7.4 shows a quasi "3-dimensional" enlargement of these results, with superimposed atomic scale models of 8CB molecules, indicating that the bright structures in Figs. 7.2 and 7.3 indeed correspond to the bi-phenyl rings, whereas the darker areas correspond to the alkyl tails.

These atomic scale models of 8CB, consisting of a bi-phenyl group with a CN attached on one side and an alkyl chain on the other, also indicate that

Fig. 7.3. Correlation averaged picture of Fig. 7.2, showing rows of antiparallel ordered 8CB molecules.

the phenyl rings and the alkyl chain cannot lie in a plane. Instead, the two phenyl rings are twisted with respect to each other, whereas the stretched out alkyl chain forms an angle with these two phenyl rings. As a result the phenyl groups do not lie flat but rather are tilted with respect to the surface [13].

Comparing the width of the bright and dark parts with the atomic scale model of 8CB, one realizes that the 8CB on graphite forms a double row structure of anti-parallel ordered molecules. This indicates that apart from the strong interaction with the surface, leading to a crystalline-like ordering, the liquid crystal molecules still experience strong interactions between themselves, leading to this antiparallel (smectic-like) ordering. However, the structure of 8CB on the graphite surface is different from that in the bulk, where the head groups and tails are expected to be more fully interdigitated.

Figure 7.3 also shows a clear superstructure that is spontaneously formed at this 8CB-graphite interface. This superstructure with a periodiciy of 4 molecules can be understood from the competition between the molecule-molecule and the molecule-substrate interaction. The stable configuration of normal alkyl chains C_nH_{2n+2} is obtained when their carbon frameworks adapt the all-trans conformation. These chains can be in perfect registry with the graphite lattice (mismatch in carbon-carbon distance is only 2%) [14], as is also demonstrated in Fig. 7.5 for stearic acid molecules on graphite. The width of a bi-phenyl ring (0.59 nm) is however larger than the width of

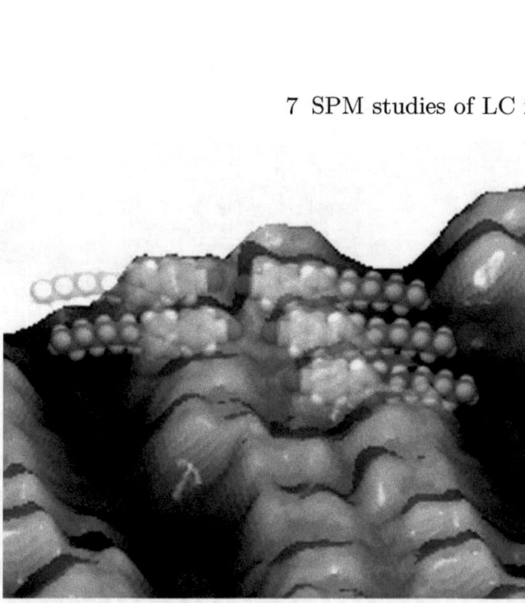

Fig. 7.4. Enlargement of detail of Fig.2, including atomistic model of the 8CB molecules.

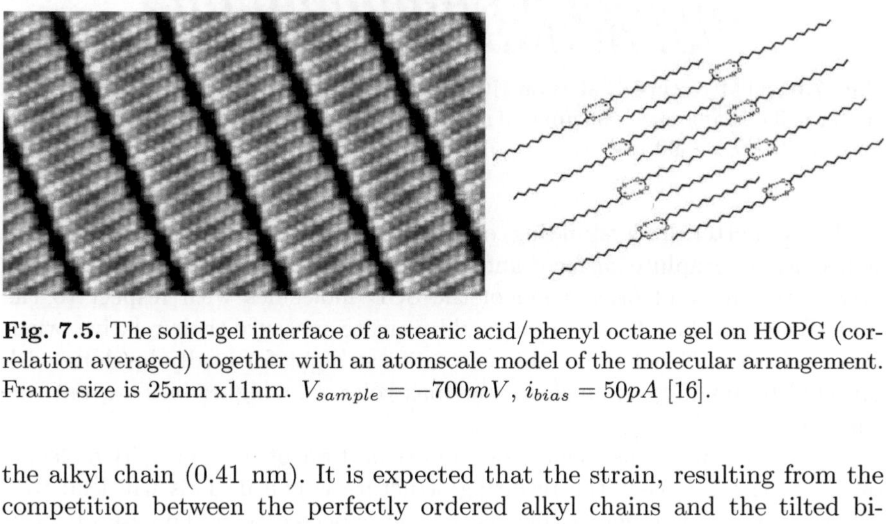

Fig. 7.5. The solid-gel interface of a stearic acid/phenyl octane gel on HOPG (correlation averaged) together with an atomscale model of the molecular arrangement. Frame size is 25nm x11nm. $V_{sample} = -700mV$, $i_{bias} = 50pA$ [16].

the alkyl chain (0.41 nm). It is expected that the strain, resulting from the competition between the perfectly ordered alkyl chains and the tilted bi-phenyl rings, is responsible for this superstructure formation. The fact that this superstructure is indeed related to this competition and thus also to the relative sizes of the bi-phenyl and alkyl part is supported by the fact that similar structures are observed for 10CB, but in that case with a periodicity of 5 molecules [13]. Apparently, the longer alkyl chain can maintain the ordering longer in this case.

The STM image of stearic acid in Fig. 7.5 indeed shows perfect ordering of alkyl chains over a very large area. More examples of this kind of molecular structures revealed by STM imaging and comparisons with electron structure calculations can be found in [2].

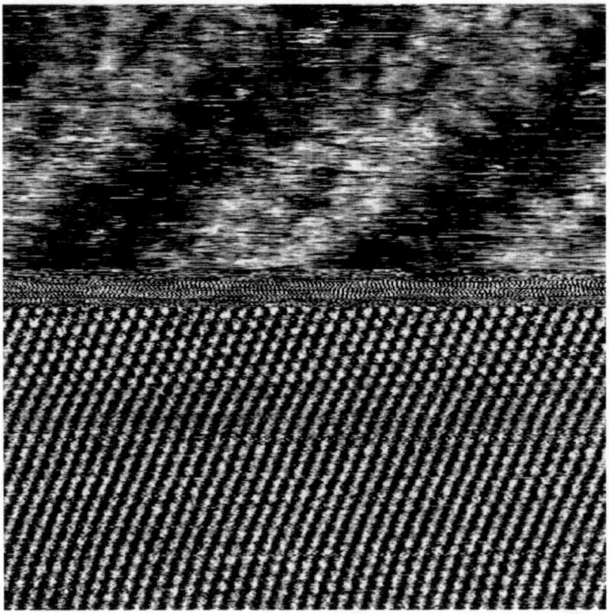

Fig. 7.6. STM picture of 8CB on HOPG (top half) and of the graphite substrate (lower half). Frame size is 30 nm×30 nm, $i_{bias} = 70$ pA. Upperhalf: $V_{bias} = 900$ mV, lower half: $V_{bias} = 90$ mV [17].

For a better understanding of the structure and organization of LC molecules on graphite surfaces and to confirm the hypotheses about the relative structure and orientation of the 8CB molecules with respect to the underlying graphite lattice, one would want to measure their relative orientation directly. Fortunately, this can quite straight forwardly be done with an STM by making an image at low bias voltage or higher bias current, see Fig. 7.6.

An interesting consequence of the orientation of a bend 8CB molecule on a substrate, is that the resulting structure is chiral. This will lead to 6 different types of domains (instead of the expected 3 in the case of an a-chiral situation). Figure 7.7 shows a beautiful example of 5 of these possible configurations.

Apart from these nCB examples, there have also been reports of STM images of ferro-electric liquid crystals on graphite surfaces [19–21]. The interesting conclusion from these results, in particular from those of [21], is that the enchanced tunneling efficiency of the benzene rings cannot simply be explained via the highest occupied or lowest unoccupied molecular orbitals.

Concluding this section, STM has been proven to be a very powerfull tool to obtain structural information of LC molecules on surfaces with molecular and even atomic resolution, though there are still open questions to the ac-

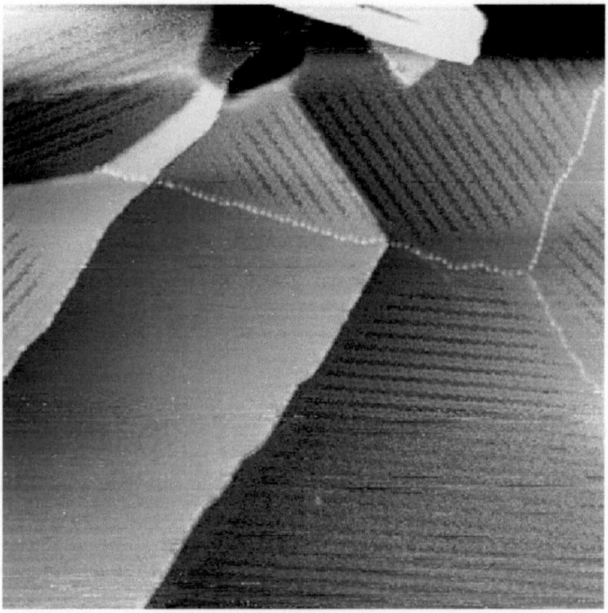

Fig. 7.7. STM picture of different domains of 8CB on HOPG. Frame size: 100nm x 100nm, $V_{bias} = 1030$ mV, $i_{bias} + 10$ pA [18].

tual molecular arrangements for the presented examples. Nevertheless, one should realize that perfectly flat graphite surfaces are quite different from a standard rubbed or otherwise aligned polymer substrate. A real challenge is to obtain similarly detailed information about the molecular arrangement on such relevant surfaces. Fortunately, the development of Atomic Force Microscopy is impressive and already allows to get very detailed information, though not yet at quite the resolution obtained by STM [12]

7.2 Alignment Layers by AFM Micro- and Nano-patterning

Miha Škarabot, Igor Muševič and Theo Rasing

7.2.1 Introduction

In their Science paper in 1994, Ruetschi et al. [22] introduced the idea of using an AFM to create liquid crystal alignment by simply "scratching" in a regular manner a thin nylon layer, deposited on a glass substrate. To their surprise, this scratching, later also known as micro-patterning, nano-patterning or nano-rubbing, created a perfect alignment in a twisted nematic liquid crystal cell, assembled from nano-rubbed substrates. Prior to this discovery, a lot

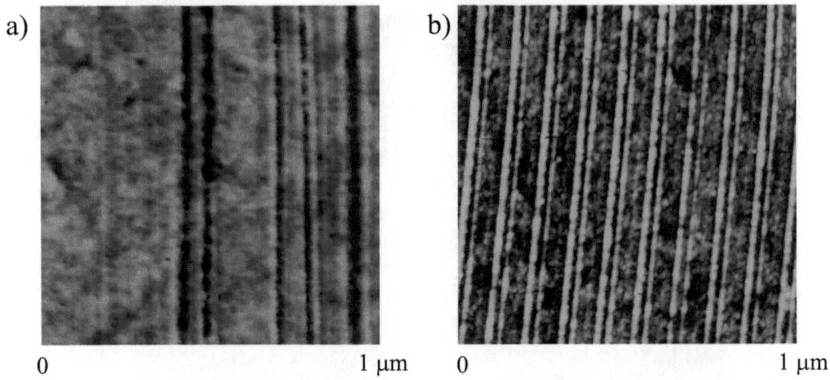

Fig. 7.8. Tapping mode images of a polyimide surface, oriented by (a) ordinary rubbing with a piece of soft cloth, and, (b) AFM rubbing, performed by scanning the AFM tip across the surface in a regular manner, rubbing force is 0.8×10^{-6}N.

of work had been done on the modification of polymer surfaces using AFM, with pioneering work by several authors in 1992 [23–25]. In these experiments, it was observed that an AFM tip creates grooves in a soft polymer surface, which appear very similar to those, produced by rubbing polymer alignment layers for LCD's. A comparison of surface corrugations produced by ordinary rubbing and AFM rubbing is shown in Fig. 7.8.

In their study, Ruetschi et al. [22] systematically analyzed the orienting effects of nylon surfaces, modified by a sharp AFM tip, scanned across the surface at various loading forces, scanning speeds and separations between subsequent lines. They clearly observed, that at relatively large force loads of $\approx 10^{-7}$ N, the AFM tip produces grooves in a nylon surface, that are typically $20 - 30$ nm wide and have a depth of $1 - 2$ nm. These grooves showed most efficient orienting properties at relatively large force loads $\geq 10^{-6}$ N and relatively small line separation of 50 nm, where the anchoring strength was found independent of the scanning speed. In the following paper [26], the same group of authors presented the first examples of nanopatterned surfaces, forming optical gratings and other microscale objects with interesting optical geometries.

Pidduck et al. [27] confirmed similar findings on nanorubbed, alkyl branched polyimide. Optically diffractive elements were successfully produced by nanorubbing polyimide surfaces using large forces of $\geq 10^{-7}$ N and small line separation ≤ 200 nm. They have shown for the first time, that the polymer surface can be over-patterned, where the liquid crystalline molecules always follow the last scanning direction. Contrary to the observations of Ruetschi, practically no sign of topographic modifications of the polyimide surface could be observed. This indicated polymer chain alignment by shear-yielding within a surface layer of a polymer and was a driving question for subsequent experiments. These were performed by several groups, concentrating on the dependence of the anchoring strength coefficients on rubbing parameters [28], depression of the isotropic-nematic transition temperature in nanorubbed regions [29], nanopatterned grey scale and polarization gratings [30, 31], studying mechanism of liquid crystal alignment on sub-micron patterned polymer sufaces [32, 33] and even bare ITO surfaces [34].

Of particular importance is here a series of very original experiments, that has led to patterned surfaces that show bistability [35–37] multistability [38] and very large pretilt angles [39]. The underlying idea is to create a surface orienting pattern, that causes frustration of the nematic liquid crystal ordering. The lowering of the elastic energy with respect to the frustration, dictated by the surface corrugations, results either in several equivalent surface orientations, or leads to the escape of the director away from the surface, resulting in a very high pretilt of ≈ 40 degrees [39].

7.2.2 AFM Patterning

In the experiments with AFM patterned orienting surfaces, usually only one plate is patterned with the AFM tip, whereas the other one is rubbed and assembled in a twisted nematic liquid crystal cell together with the first one. In a typical AFM nanorubbing experiment, silicon nitride cantilevers with a typical tip radius of $20 - 50$ nm are used. The tip is scanned across the polymer surface at a speed of 0.1 mm/s - 1 mm/s, whereas the applied force is of the order $10^{-8} - 10^{-6}$ N. Figure 7.9 shows an example of the resulting surface corrugations on a polyimide surface, imaged with a tapping mode AFM.

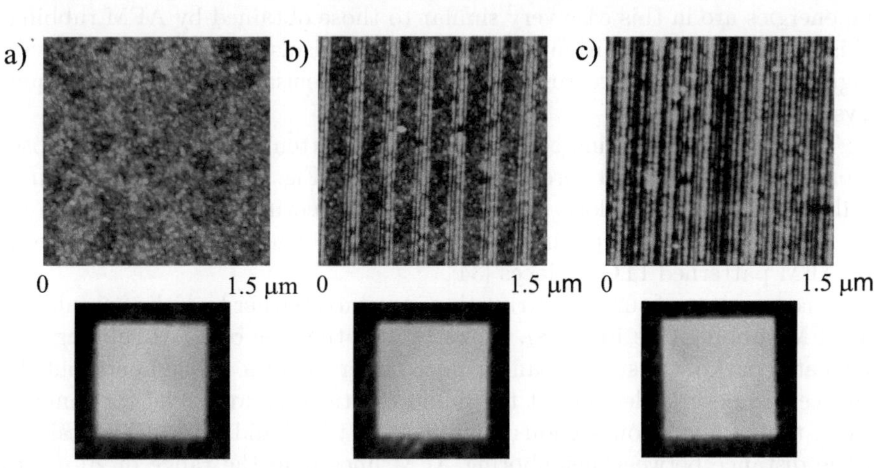

Fig. 7.9. Tapping mode images of 80 nm polyimide films, rubbed using AFM. (a) scan force 0.2×10^{-6} N. (b) scan force 0.8×10^{-6} N. (c) scan force 1.5×10^{-6} N. Optical micrograph of twisted nematic cells, formed by AFM patterned substrate.

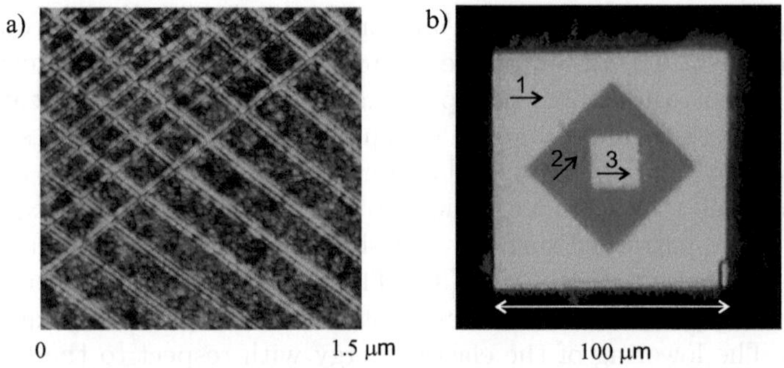

Fig. 7.10. (a) tapping mode image of a polyimide layer patterned in crossed directions. (b) micrograph of a patterned twisted nematic pixel under crossed polarizers. Arrows indicate the scan directions and the number shows the scan order.

It is clearly seen that surface corrugations (grooves) are not visible for weak scanning forces ($\leq$ 200 nN) and they become visible at large applied forces. The depth of grooves increases with the applied force. Surprisingly, all surfaces presented in Fig. 7.9 showed the same quality of alignment in twisted nematic cells, as well as a nearly identical azimuthal surface anchoring energy. These observations suggest that the existence of topographical grooves are not the only mechanism that is responsible for the LC alignment. An alignment of the polymer chains during the AFM patterning should be considered as well. On the other hand, one has to consider complementary experiments of Behdani et al. [40], where surface grooves have been produced into an untreated polyimide surface by laser ablation. The azimuthal anchoring energies are in this case very similar to those obtained by AFM rubbing. This leads to the conclusion, that both mechanisms, that is polymer chain alignment and surface corrugations, play an essentially equal role in liquid crystal alignment.

The AFM nanorubbing can be easily overwritten by performing a subsequent scan in a different direction, as shown in Fig. 7.10. This behaviour is well known in normal cloth rubbing, where the entire surface is affected by the rubbing process. Surprisingly, such overwrite behaviour was also observed for AFM patterned ITO surfaces [34].

A comparison of surface corrugations produced either by ordinary rubbing or AFM rubbing (see Fig. 7.8), shows that in the case of AFM rubbing, the untreated part of the surface can be much larger than the rubbed part, but the surface will nevertheless orient the liquid crystal very well. The experiments show, that a good homogenous alignment of the liquid crystal is obtained, if the distance between neighboring AFM lines is in the range of 20 nm to 1 μm. This is much larger than the typical width of a single AFM line, which is of the order of 20 nm. Figure 7.11 shows three examples of AFM rubbing with different line-to-line separations. For line separations below 1 μm, a homogeneous surface alignment is obtained. It should be noted that this is

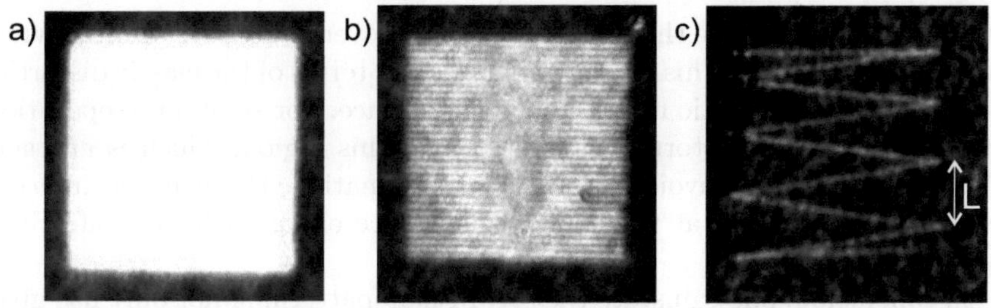

Fig. 7.11. Polarizing microscope micrographs of AFM rubbed surfaces with different line separation. (a) line separation 1 μm, (b) 4 μm and (c) 32 μm.

not an artifact due to diffraction limitations of the polarizing microscope, as domains of reverse twist in 90 degrees twisted nematic cells are observed, indicating a homogeneous surface structure.

When the distance between the AFM lines is increased, it is possible to observe under the polarizing microscope single twist nematic lines, originating from a single AFM rubbing line. An example is shown in Fig. 7.11, where one can clearly observe a very thin twisted nematic region above a single AFM line [33].

The azimuthal surface anchoring energy of AFM rubbed surfaces has been measured as a function of the distance between the lines, as shown in Fig. 7.12. For large line separation, the anchoring drops to zero, as it should, since no direction is preferred on an un-oriented polymer surface. With decreasing line separation, the effective anchoring strength first increases inversely with the line separation. When the line distance is smaller than 300 nm, the surface anchoring energy saturates. It is interesting that

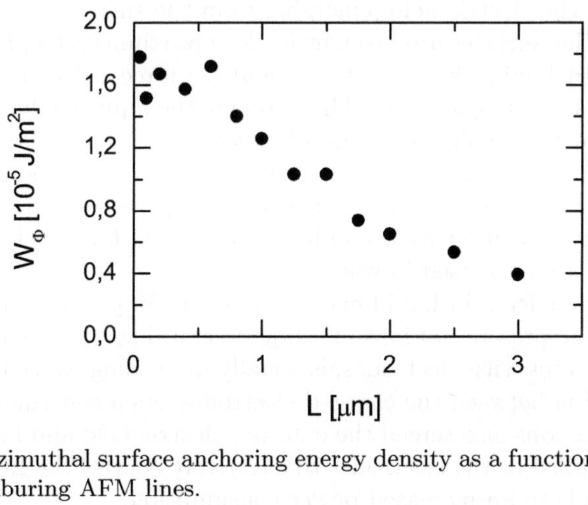

Fig. 7.12. Azimuthal surface anchoring energy density as a function of separation between neighbouring AFM lines.

saturation occurs before the surface is densely covered with AFM lines, which are ten times thinner. This can be understood in terms of the elastic distortion of the interfacial nematic region above the surface. For small line separation, a very large elastic distortion is expected in this region, which is energetically unfavorable. To avoid unfavorable deformation, the nematic interface becomes homogeneous at the expense of surface energy, which is effectively lower.

In conclusion, one can state that AFM nanopatterning not only has given us more insight into the mechanisms of LC alignment, but it also offers exciting new possibilities for tailoring LC surface interactions. With the current development of multiprobe arrays, real applications of this approach may come within the realm of real possibilities.

7.3 Surface Charge and Electric Field at Liquid Crystal Interfaces Observed by AFM

Klemen Kočevar and Igor Muševič

7.3.1 Introduction

When solid surfaces are in contact with a liquid, they may become charged either because of: *(i)* the dissociation of surface chemical groups into the liquid, or *(ii)* the selective adsorption of ions from the liquid onto the surface [42]. In both cases, a surface electric field is developed due to the charged surfaces, and the electric field extends over a Debye screening length into the bulk liquid. This is due to the presence of charged ions in the solution that redistribute themselves due to the electric field and form a double layer, which screens the electric field, emerging from the surfaces.

If two similar surfaces are brought in close proximity, the presence of ions in the confined liquid gives rise to a repulsive force between the confining surfaces that is of entropic origin. This is due to the repulsion between charged ions that have to remain in the gap between the bodies, as demanded by the electroneutrality condition [42]. In water colloid dispersions, the surface charge density may be very large. The resulting electrostatic force is quite strong and may be used as a stabilizing repulsive force between colloidal particles that are immersed in water [43].

The electric effects in liquid crystals are very important in the LCD industry [44]. The presence of ions in a liquid crystal sample, confined in a LC cell between plates with electrodes, is usually disturbing, since it gives rise to ionic conduction between the charged electrodes. As a consequence, power is consumed. The ions also screen the external electric field and higher voltages have to be applied to the electrodes to effectively control the LC orientation, which also leads to an increased power consumption.

Furthermore, it has been argued for a long time that charging of surfaces and the formation of an electric double layer across the liquid crystal-solid interface could have a significant influence on the anisotropic anchoring properties of surfaces [45–49]. The field affects the liquid crystal orientation in the vicinity of the surface and gives a non-local contribution to the surface anchoring, due to the long range of the electrostatic interaction. The non-locality can be observed as a dependence of the anchoring strength on the separation between the surfaces, that confine the liquid crystal [45, 50]. Recently [51], it has been pointed out that the electrostatic interaction could be used for an efficient stabilization of nematic colloidal dispersions with interesting optical properties.

In spite of the obvious technological importance, surface charging and surface electric field have only recently been directly observed in liquid crystals [52] using AFM in the force spectroscopy mode. In this section, we briefly describe the experiments, that enabled a direct measurement of the surface charge and the Debye screening length of the liquid crystal-solid interface.

7.3.2 Electrostatic Force Between Surfaces in Liquids

The surface that is in contact with a liquid may become charged either due to surface ionic adsorption or dissociation. In both cases, the final surface charge is balanced by an equal and oppositely charged region of the counterions. This region is further divided into the "inner", usually transiently bound ionic layer (Stern or Helmholtz layer) and the "outer", diffuse counterion layer or "atmosphere". We are interested in the force, that arises when two identically ionized surfaces are brought in close proximity, so that the diffuse regions of the counterions start to overlap. Although it is called "the electrostatic" force between charged surfaces, the actual force *is of an entropic, and not of electrostatic origin* [42]. The reason for this is that, when bringing two surfaces closer together, the density of counterions that repel each other has to increase in the gap due to the electroneutrality condition.

The calculation of the force between charged surfaces in liquids containing ions is in general a very complicated task. Different effects due to the solvent-ion interaction or the structure of the solvent have to be taken into account to describe the force completely [53]. For this reason, the mean field approximation is employed, that is derived using a continuum approach, neglecting the structure of the ionic solution [42] as well as the inhomogeneous distribution of the charge on the surface [54]. The mean-field approach is valid only when the charged ions are dilute and do not considerably affect the structure of the dielectric medium [42, 53].

We will illustrate how the charged ions redistribute close to the charged surface and how this affects the force. To simplify the discussion, we assume that only a single ionic species is present in the solution. This ionic species represents the counterions to the surface adsorbed charge and the number of the counterions in the solution is exactly the same as the number of adsorbed ions on the surface (electroneutrality). We set the coordinate system so, that confining surfaces are perpendicular to the x axis and located at $x = \pm D/2$. The counterions of density $\rho(x)$ and valence z distribute in the region of the surface electric field according to the Boltzmann's distribution $\rho(x) = \rho_0 \cdot \exp[-ze_0 U(x)/k_B T]$, where e_0 is the fundamental charge, $U(x)$ is the electric potential, k_B is the Boltzmann constant and T is the temperature. As a result of the charge distribution, an electric field is created, which is described by the Poisson equation $\rho(x) = -\epsilon\epsilon_0 \nabla^2 U(x)$, where ϵ is the dielectric constant of the liquid. Together, these two equations are combined into the Poison-Boltzmann (PB) equation for the ionic distribution

$$\frac{d^2U}{dx^2} = -\frac{ze_0\rho_0}{\epsilon\epsilon_0} \cdot \sinh\left(-\frac{ze_0U(x)}{k_BT}\right) . \tag{7.1}$$

The PB equation determines the electric potential $U(x)$, the electric field $E = \mathrm{d}U/\mathrm{d}x$, and the density of (counter)ions $\rho(x)$ at any point. It is a second-order, nonlinear differential equation that cannot be solved analytically in general. The boundary conditions for the PB equation follow from the symmetry of the problem and the overall electroneutrality. When the surfaces are located at $x = \pm D/2$, where D is the surface separation, then the electric field at the midpoint is zero, $E = \mathrm{d}U/\mathrm{d}x|_{x=0} = 0$ and $U(x) = U(-x)$. The surface charge density is $\sigma = \epsilon\epsilon_0 \mathrm{d}U/\mathrm{d}x$ at $x = D/2$, which follows from the neutrality condition. It should be pointed out, that both the surface field and the surface charge density depend on the surface separation, which is known as the charge regulation [55]. For simplicity, the corresponding nonlinear boundary conditions are often replaced either by a *constant charge or constant potential*, depending on the nature of the interfaces.

When the electric potential U is small, $\sinh(x) \approx x$ in (7.1) to first order, and we obtain the linearized form of the PB equation, which is known as the Debye-Hückel (DH) equation:

$$\frac{\mathrm{d}^2U}{\mathrm{d}x^2} = \kappa^2 U . \tag{7.2}$$

Here $\kappa^{-1} = \lambda_D$ is the Debye screening length. $\kappa^2 = \rho_0 z^2 e_0^2/(\epsilon\epsilon_0 k_B T)$, where ρ_0 is the ionic density in the limit of large surface separations. The Debye-Hückel equation can be readily integrated,

$$U(x) = U_s \frac{\cosh(\kappa x)}{\cosh(\kappa D/2)} \tag{7.3}$$

where U_s is the surface electric potential, whereas the surface charge density is

$$\sigma = \epsilon\epsilon_0 \kappa U_s \tanh(\kappa D/2) . \tag{7.4}$$

In the DH limit, the electric potential therefore decreases approximately exponentially away from the surface, with the Debye screening length being the characteristic decay length of the electric field.

In liquid crystals, being poor solvents, a low concentration of dissociated ions is expected, as well as relatively low surface charge densities, so that the DH approximation can be used. In this case, the free energy per unit surface area $\mathcal{F}$ due to the electrostatic interaction between two parallel plates at the separation D is [42]

$$\mathcal{F} \approx \frac{2\sigma^2 \lambda_D}{\epsilon\epsilon_0} e^{-D\lambda_D} \; . \tag{7.5}$$

Once the free energy of the interaction between two parallel plates is known, the net force between the sphere and the flat surface can easily be calculated using the Derjaguin approximation [42, 56]. This approximation connects the force between the sphere and the flat surface with the free energy of the interaction, $F = 2\pi R\mathcal{F}$, which is justified as long as the range of the interaction and the separation between the interacting surfaces are both much smaller than the radius of the sphere. Since the range of the interaction is $\approx 100\,\mathrm{nm}$, this is much smaller than the radius of the sphere that is $\approx 10\,\mu\mathrm{m}$ and the use of the Derjaguin approximation is justified. The force F between the sphere of radius R and a flat surface then is

$$F = \frac{4\pi R\sigma^2 \lambda_D}{\epsilon\epsilon_0} e^{-D/\lambda_D} \; . \tag{7.6}$$

In this equation, we have assumed equal surface charge densities on both interacting surfaces. By measuring the separation dependence of the electrostatic force, one can directly determine the Debye screening length λ_D. Simultaneously, the surface charge density σ can be determined from the amplitude of the measured force.

7.3.3 AFM Observation of Force Due to Charged Interfaces

The electrostatic force between two glass surfaces, immersed in an isotropic liquid crystal, has been measured with a commercial Nanoscope III AFM (Digital Instruments). It was equipped with a heating stage to control the temperature of the liquid crystal, as described elsewhere [57]. The AFM was used in a force spectroscopy mode, where a surface of a sample is periodically approached and retracted to the probe of the AFM and simultaneously the force on the AFM probe is monitored. We have used flat plates of either LaSF glass or sapphire, attached to the piezo scanner of the AFM and a small sphere made of BK7 glass, that was attached to the commercial AFM cantilever with an elastic constant of $k = 0.1\,\mathrm{N/m}$. The radius of the sphere was typically $10\,\mu\mathrm{m}$. All glass surfaces were thoroughly cleaned in an ultrasonic detergent bath, carefully rinsed with distilled water and acetone and finally cleaned with an RF oxygen plasma. After the cleaning, a monolayer of DMOAP (N,N-dimethyl-N-octadecyl-3-aminopropyltrimethoxysilyl chloride) was deposited onto the glass surfaces to assure good homeotropic orientation of the liquid

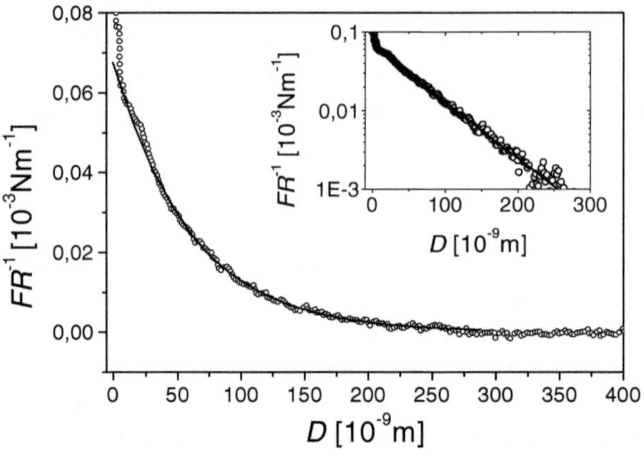

Fig. 7.13. The separation dependence of the normalized force FR^{-1}, between the silanated glass sphere ($R = 9\,\mu\text{m}$) and silanated LaSF glass plate in 8CB at $T - T_{NI} = 7.1\,\text{K}$. The solid line is a fit to an electrostatic repulsive force given by (7.6). The fit parameters are: $\sigma = 1 \times 10^{-4}(1 \pm 0.07)\,\text{Asm}^{-2}$, $\lambda_D = 73(1 \pm 0.07)\,10^{-9}\text{m}$. Inset shows the same data in a log-lin scale, to emphasize the exponential dependence of the force on the separation.

crystal. Nematic liquid crystals 5CB (4-cyano-4'-n-pentylbiphenyl) and 8CB (4-cyano-4'-n-octylbiphenyl) were used in the experiments.

A typical force measurement that indicates the presence of the electrostatic force, is presented in Fig. 7.13. Here, the force between a silanated $10\,\mu$m glass sphere and a flat surface of a silanated glass plate is measured in the isotropic phase of 8CB at a temperature 7.1 K above the phase transition temperature into the nematic phase (T_{NI}). A very strong repulsive force can be observed. As one can see from the inset, this force decreases exponentially with increasing separation and can be detected even at a separation of 300 nm. At smaller separations of ≈ 20 nm, we have observed an attractive force, which is a result of the capillary condensation of the partially ordered isotropic liquid crystal into the developed nematic phase, as already reported [58]. The exponentially decaying repulsive force showed no temperature dependence in a wide range of temperatures above the nematic to isotropic phase transition temperature.

Similar results were observed in the isotropic phase of 5CB. A typical measurement of the force between a silanated glass sphere and a silanated LaSF glass plate immersed in 5CB, is presented in Fig. 7.14.

As can be seen on Figs. 7.13 and 7.14, the observed repulsive force can be perfectly fitted to an exponentially decaying function, with a decay length of $70 - 100$ nm. The only exponentially decaying force with such a long range and no significant temperature dependence, is the electrostatic force.

The measured electrostatic forces could in all cases be very well fitted to (7.6). The results of the analysis, which include Debye screening lengths

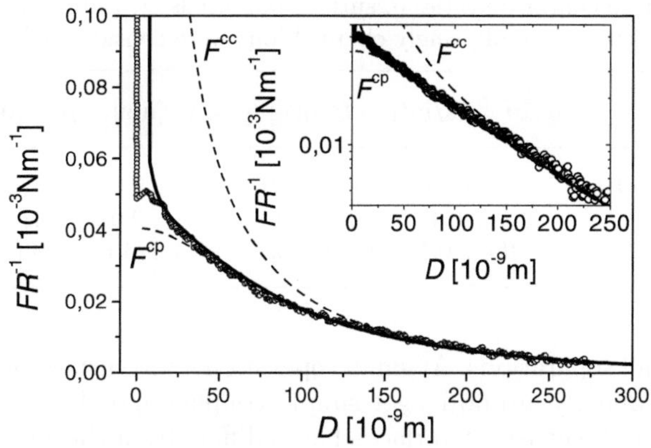

Fig. 7.14. The separation dependence of the normalized force FR^{-1}, between the silanated glass sphere ($R = 10\,\mu$m) and silanated LaSF glass plate in 5CB at $T - T_{NI} = 11$ K (open circles). The dashed lines represents the numerical solution of the PB equation. The upper dashed curve is calculated in a constant surface charge density approximation (F^{cc}) with $\sigma = 10^{-4}$ Asm^{-2} and the lower dashed line is calculated in a constant surface potential approximation (F^{cp}) with $\Psi_0 = 80$ mV. The full line is the weighted superposition of both curves $F = pF^{cc} + (1-p)F^{cp}$ with $p = 0.04$. The inset shows the same data in a log-lin scale.

in different samples and corresponding surface adsorbed charge densities are presented in Table 7.1.

The Debye length, which is a typical screening length of an electrostatic interaction, is well determined and in our experiments on commercial cyanobiphenyls, we have found that it is around 100 nm. Assuming that only singly charged ions are present in the liquid crystals that were studied, this gives an estimate of the concentration of the charged ions or molecules in the LC samples under investigation, which is in the range $1 \times 10^{21} - 5 \times 10^{21}$ m^{-3}, or $2 - 7\,\mu$M.

The surface charge densities, determined from our experiments, are less accurate, but give an important estimate. The typical value of $\sigma = 2 \times 10^{-5}$ Asm^{-2} gives a mean separation between singly charged adsorbed ionic impurities of $a = 50$ nm, which is a relatively large separation. However, since our surfaces are covered with a hydrophobic DMOAP layer, it is expected that the surface charge is low, which is in agreement with the measurements. The real adsorbed surface charge density is most probably higher, since close to the surface, the counterion density is normally higher than predicted by the Poisson-Boltzmann equation, due to the counterions, immobilized at the surface [42].

The quality of the fits to (7.6) is surprisingly good, even though, as we have mentioned above, (7.6) is not expected to be valid at small surface separations. This is due to the fact that in this approximation, the potential in the gap is a sum of potentials emerging from both surfaces which does

Table 7.1. Surface charge densities δ, surface potentials U, Debye lengths λ_D and ion number densities ρ_∞. Only singly charged ions were considered

System	$\sigma\,[10^{-5}\,\mathrm{As/m^2}]$	$U\,[\mathrm{mV}]$	$\lambda_D\,[\mathrm{nm}]$	$\rho_\infty\,[10^{21}\,1/\mathrm{m^3}]$
sapphire-8CB-BK7	$10.3(1\pm0.07)$	$80(1\pm0.07)$	$73(1\pm0.07)$	$2.8(1\pm0.14)$
LaSF-8CB-BK7	$9(1\pm0.1)$	$59(1\pm0.1)$	$61(1\pm0.08)$	$4.1(1\pm0.16)$
LaSF-5CB-BK7	$6.3(1\pm0.1)$	$65(1\pm0.1)$	$97(1\pm0.1)$	$1.6(1\pm0.2)$

not hold at small separations. At small separations, the surface potential and surface charge density are no longer simply coupled and the solution of the PB equation for the constant surface charge differs from the solution for the constant surface potential. In the case of a constant potential, the surface charge density has to decrease, which is called *charge regulation.*

In general, the electrostatic interaction falls in between the limiting values for constant surface charge density and constant surface potential [42]. In order to investigate this, we have numerically solved the PB equation for both limiting cases of constant charge and constant potential. The theoretical results are shown in Fig. 7.14, together with the measured force. The surface charge was set to $\sigma = 10^{-4}\,\mathrm{Asm^{-2}}$ in the first case and the surface potential was set to $\Psi_0 = 80\,\mathrm{mV}$ in the second case. Both curves coincide in the limit of large separations and deviate from each other only at small separations. The force, calculated for a constant surface potential F^{cp} fits the measurement better than the force for constant surface charge F^{cc}. The best fit is obtained by a model interaction, proposed by Behrens and Borkovec [55]: $F = pF^{cc} + (1-p)F^{cp}$ with the factor $p = 0.04$. It is obvious that the constant surface potential approximation is better suited for our case, indicating that there is a strong charge regulation as the distance between the surfaces changes. Any variation of surface potential strongly impacts the surface charge density. As the surfaces are brought closer together, the surface charge considerably reduces, which indicates that either the surface charges desorb as the surfaces are closer or counterions combine with adsorbed ions on the surface.

Here, we would like to add some comments on the applicability of a mean field model for the electrostatic interaction that we have used. First, the concept of the Debye screening length is only applicable, if the number of ions adsorbed on the surface is very small compared to the total number of ions in the sample [59]. In our case, we can estimate the thickness of the sample to be 1 mm, which gives the number of ions per unit area in the bulk sample $\approx 2\times10^{18}\ \mathrm{m^{-2}}$, which has to be compared with the number of surface adsorbed ions $\approx 7\times10^{14}\ \mathrm{m^{-2}}$. As discussed by Nazarenko et. al. [59] and Kühnau et. al. [60], the use of the Debye length is justified, when the separation between the charged surfaces is greater than a few times $\lambda_S = \epsilon\epsilon_0 k_B T/4\pi\sigma e_0$. λ_S is the characteristic surface screening length and describes the thickness of the region, where the electric field drops more rapidly with separation than farther away from the surface, where the drop of the electric field is appropriately described by an exponential decay with the Debye length. This is satisfied in our case, because $\lambda_S \approx 2\,\mathrm{nm}$. Another limitation for the use of (7.6) is the demand that the surface potential is lower than $\approx 25\,\mathrm{mV}$ [42]. This is not entirely true in our case, but since the assumption of equally and homogeneously charged surfaces is also not justified, we expect that the error contribution due to the higher surface potential is not too large.

It is also very important to emphasize that the electrostatic repulsion has not been observed in all experiments that were performed. At this moment it is very hard to find a perfect correlation between the process of sample preparation and the presence of an electrostatic force. One thing that we are aware of, is that the electrostatic force has been observed mainly in the experiments (but not all), where the surfaces have been cleaned in an oxygen RF plasma. Probably, the plasma cleaning process has revealed and activated several binding sites for ions on the surface. We should also add that the electrostatic force data that were used to determine the parameters of (7.6) were measured between two different surfaces. The reported surface charge densities are then most likely a geometric mean of the charge densities on both surfaces.

At the end we would like to estimate, if the detected surface charge densities and corresponding surface electric field could contribute significantly to the surface anchoring energy of nematic liquid crystal molecules at the DMOAP interface. There are two different ways in which the mesoscopic liquid crystalline orientational order couples to the electric field that emanates from the surface [45,61]. The first one is the direct dielectric coupling, where the electric field couples with the induced dipole moments of polarizable liquid crystal molecules. It can be described as a dielectric free energy contribution $f^{diel} = -\epsilon_a \epsilon_0 (\boldsymbol{n} \cdot \boldsymbol{E})^2/2$, where $\epsilon_a = \epsilon_{\|} - \epsilon_{\perp}$ is the dielectric anisotropy of the oriented liquid crystal and $\boldsymbol{n}$ is the nematic director. The second contribution is the coupling between the flexoelectric polarization $\boldsymbol{P}_{fl}$, associated with the splay and bend deformation of the nematic director $\boldsymbol{n}$, and the electric field, $f^{flexo} = -\boldsymbol{P}_{fl} \cdot \boldsymbol{E} = -\left[e_1 \boldsymbol{n}(\nabla \cdot \boldsymbol{n}) + e_3(\nabla \times \boldsymbol{n} \times \boldsymbol{n})\right] \cdot \boldsymbol{E}$. Here e_1 and e_3 are the flexoelectric coefficient corresponding to the splay or bend distortions, respectively. The flexoelectric effect is common to polar LC molecules with specific shape asymmetries or LC molecules that possesses non-zero electric quadrupole moments [62].

It has been shown by Barbero et. al. [61], that the direct dielectric coupling is dominant if $\epsilon_a > 2e^2/3\epsilon_0 K$. Here K is the effective elastic constant in the one constant approximation and $e = e_1 + e_3$. Taking typical values for cyanobiphenyls, $K = 5.5\,\mathrm{pN}$ and $e = 6 \times 10^{-12}\,\mathrm{Asm}^{-1}$, the value of ϵ_a has to be larger than 0.49. Since the dielectric anisotropy in the nematically ordered 5CB is indeed larger, $\epsilon_a \approx 10$, we can conclude that the direct dielectric coupling is dominant in this case and only its contribution to the surface coupling energy will be considered. The dielectric anisotropy of cyanobiphenyls is positive, therefore the direct dielectric coupling tends to orient the liquid crystal homeotropically. Its contribution to the linear (ordering) anchoring energy w_1, defined by Poniewierski and Sluckin [63], is approximately $f^{diel}\lambda_D$. The value of the electric field on the surface is $\sigma/\epsilon\epsilon_0 \approx 7 \times 10^5\,\mathrm{V/m}$ and the contribution to the surface free energy in the case when the orientation is homeotropic, is $f^{diel}\lambda_D \approx 2 \times 10^{-6}\,\mathrm{J/m^2}$. Comparing this value to the typical values of the surface anchoring energies on silane covered substrates $w_1 \approx 1 \times 10^{-4}\,\mathrm{J/m^2}$ [57], we can conclude that the electrostatic coupling is weak compared to the coupling between the dimers and alkyl chains of DMOAP.

7.4 Electric Force Microscope Observations of Electric Surface Potentials

Maria P. De Santo, Riccardo Barberi and Lev M. Blinov

Electrostatic Force Microscopy (EFM) allows to obtain information on the surface electrical properties of materials by measuring electric forces between a charged tip and the surface. It is particularly suitable for the study and manipulation of ferroelectric thin films with large surface charge. Interestingly, an EFM can also be used to study the surface properties of dielectric materials, that are polarized by the electric field of the tip. In this mode of operation, the EFM is sometimes called *Polarization Force Microscope* and can be used to study and image even air-liquid interfaces [64].

In the EFM, an electric field is created at the surface of the sample by applying a voltage to a conductive AFM tip, that is in close proximity to the surface. In the case of thin ferroelectric samples, it is possible to induce extremely strong and localized electric fields and hence to produce domains of electric polarization on a (sub)micrometer scale inside the film. By means of the EFM, the evolution of such domains can be studied both in organic and inorganic ferroelectric films. The interaction between the tip and the sample as a function of the material properties and system geometry can be described within a phenomenological model, described at the end of this contribution.

The aim of these studies is to understand and control the electrical properties of surfaces on a micrometric scale, that might be extremely useful for the application of these materials in the field of liquid crystal display (LCD) technology. In fact, it is well known that the LC anchoring properties depend not only on the substrate morphology but also on its electrical properties [68]. The electric polarization in *nematic* and *other non-polar* liquid crystals has essentially three origins: flexoelectricity [65], orderelectricity [66, 67] (related to the gradient in the order parameter) and the polarization of the substrate that confines the material. The substrate polarization is an external polarization, i.e. not depending on the LC and can be controlled in the case of a ferroelectric substrate using EFM, as will be shown in the following. The surface polarization can be used to modulate the other ones, that are intrinsic to the system, either by annulling or amplifying their effects.

It has been known for a long time that the surface ordering of a nematic (or other non-polar) liquid crystal is influenced by the ferroelectric domains of the anchoring substrate. In a work by M. Glogarova at al. [69], it is shown how the properties of a liquid crystal cell can be modulated and stabilized using a ferroelectric material as an anchoring substrate. These results motivated us to consider that the EFM technique could be efficiently used to create surfaces with variable anchoring conditions on a micrometric scale.

In this contribution, it is also shown how the EFM technique can be used to investigate in detail the secondary effects of the rubbing process on dielectric films used to align liquid crystals.

7.4.1 Principle of Operation of Electric Force Microscope

In an AFM, a non-conductive tip made of a dielectric material is used to sense forces between the tip and the surface. These forces may be of various origin, such as van der Waals, hard contact, or structural forces, if there is a liquid-like medium between the tip and the surface. An AFM can be easily adapted to measure variations in the surface electric potential and changes of the capacitance due to the changing gap between the tip and the sample. An EFM uses an *electrically charged conductive tip* to sense the Coulomb force between the charged tip and the charges in the sample. Depending on the electric properties of the substrate, the charges in the sample can be either (i) free charges in a conductive sample, (ii) charges due to the spontaneously polarized ferroelectric material, or (iii) electric-field induced (mirror) charges in a dielectric material. Of course, dielectric surfaces can also be additionally charged by depositing extra charge onto the surface.

Similar to the AFM, an EFM can be used either in non-contact mode (NC EFM) or contact mode (also DC EFM, direct contact EFM). In the latter case, simultaneous acquisition of both topographical and electrical signals is possible.

In the case of NC EFM, the Coulomb force on the conductive and charged tip is kept constant by monitoring the deflection of the cantilever and controlling the height of the sample by the feedback circuitry of the AFM. In most simple cases, where the surface charge density is uniform, the image that is obtained by scanning across the surface, is a topographical image. In the case of a non-homogeneous charge distribution, the topographical image is influenced by charge (or dielectric constant) inhomogeneities.

The charge on the EFM tip can also be modulated by applying a time-oscillating voltage to the tip. In this case, a slightly modified system is used to detect the time-varying deflection of the cantilever. This is done by adding

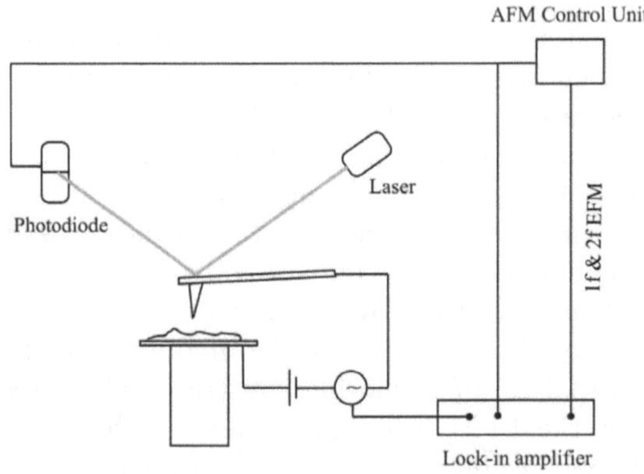

Fig. 7.15. Scheme of an Electrostatic Force Microscope.

an AC supply and a lock-in amplifier, to the DC supply, shown in Fig. 7.15, [70, 71].

In all EFM applications the tip must be conductive (metal coated or doped silicon), whereas the sample can be either conductive or dielectric. In the case of conductive samples, the electrical contacts on the supporting body of the tip and on the sample are produced using silver paste to achieve good electric conductivity in the most delicate parts of the set-up. In the case of dielectric or ferroelectric samples, a counter electrode underneath the samples is usually used.

During the scanning in the contact mode, an alternating voltage at a frequency f, usually of the order of tenths of kHz, is applied between the tip and the sample. f is well below the frequency used to obtain topography (usually hundreds of kHz) in order to minimize the superposition between this signal and the electrostatic one. Using an access module to the relevant internal signals of the AFM, the photodiode signal due to the cantilever oscillations is acquired and sent to the external lock-in amplifier that selects the signals oscillating at the fundamental and second-harmonic frequencies, f and $2f$, respectively. These signals are sent back to the acquisition card of the microscope. The topographical signal is processed independently by the electronics of the AFM. It is therefore possible to acquire simultaneously three signals for each point on the sample surface: topography, first harmonic $1f$ and second harmonic $2f$. They can be presented pixel by pixel, thus forming three different images of the surface simultaneously.

In the following we analyze the first and second harmonic signals. The electrostatic force on the tip is

$$F = -\frac{\delta U}{\delta z} = -\frac{V^2}{2}\frac{\delta C}{\delta z} \tag{7.7}$$

where U is the electrostatic energy, z is in the direction normal to the surface, C is the capacitance between the tip and the sample and the potential V of the sample with respect to the tip is

$$V = -V_S + V_{DC} + V_{AC}\sin(\omega t)\ . \tag{7.8}$$

Here, $\omega = 2\pi f$. The surface potential equals $V_S{=}A_s - A_t$, where A_s and A_t are the contact potentials of the sample and the tip, respectively. V_S (and so V) is related to the local properties of the sample, therefore it is subjected to variations during the scanning of the tip across the surface. Substituting (7.8) in (7.7) we find an expression for the electric force F

$$F = -\frac{1}{2}\frac{\delta C}{\delta z}\left[(V_{DC} - V_S)^2 + \frac{1}{2}V_{AC}^2\right] - \frac{\delta C}{\delta z}(V_{DC} - V_S)\,V_{AC}\sin(\omega t) + \frac{1}{4}\frac{\delta C}{\delta z}V_{AC}^2\cos(2\omega t) = F_0 + F_1\sin(\omega t) + F_2\cos(2\omega t)\ . \tag{7.9}$$

The force harmonics F_1 and F_2 are both proportional to the capacitance gradient $\delta C/\delta z$. More importantly, the first harmonic of the electric force F_1 is directly proportional to the surface potential V_S. At a given position, this surface potential can be measured directly by varying V_{DC}, so that the first harmonic of the electric force F_1 is equal to zero. In this way, the *DC* voltage, necessary to compensate the first harmonic, is equal to the surface electric potential V_S. This is the principle of the Kelvin Probe Microscopy. Furthermore, by dividing the first harmonic F_1 by the second harmonic F_2

$$\left|\frac{F_1}{F_2}\right|\frac{V_{AC}}{4} = |V_{DC} - V_S| \tag{7.10}$$

the capacitance is factored-out. By acquiring images of the first harmonic of the force ($1f$ EFM) at different V_{DC}, we can therefore distinguish surface variations in V_S, induced by changes of the capacitance (the term δC/δz). A rough idea of the influence of the capacitance variations on the $1f$ EFM image can be deduced from a direct comparison between the $1f$ EFM and $2f$ EFM images.

In DC EFM, the topography is obtained with the AFM operating in the contact AFM mode, while the electric properties of the surface are obtained by monitoring the vibration of the cantilever at frequency f [72].

To understand the principle of operation of this mode, let's suppose we have a polarized ferroelectric film, which is also piezoelectric. Once the film has been polarized, its thickness D will vary when a voltage is applied between the top and the bottom surface of the film. Due to an oscillating applied voltage, the thickness of the film will also oscillate at the same frequency. The response to the applied voltage depends on the polarization state of the film. When the applied electric field is parallel to the spontaneous polarization, the thickness of the piezoelectric film increases, while for a small electric field opposite to the polarization, the thickness D decreases. If the top surface is in contact with the EFM tip and the sample has been polarized, the tip will therefore oscillate, when an oscillating voltage is applied to the tip, following the time-variations of the film thickness.

The advantage of this technique is a strong contrast in the EFM image and no correlation with the topographical image, that is determined by the repulsive forces due to hard contact of the tip. There are some controversies in the interpretation of EFM images, as both electrostatic and electro-mechanic effects are considered to be responsible for the contrast in the electrical images, both in contact and in non-contact mode. Some authors suggest that the interpretation of DC EFM images considering the piezoelectric response of the sample, is not realistic, because the variation of the thickness D is considered too small to be revealed. This is based on the fact that Coulomb forces, due to the presence of charges on the surface, are negligible with respect to the repulsive force between the tip and the sample. The charges induced on the tip, modulated by the AC voltage, produce an oscillatory force as a result of the Coulomb interaction with the surface charge density. Then the amplitude of the vibration of the cantilever at the frequency of the external voltage allows one to obtain information on the surface charge density of the sample [73] [74].

On the other hand, others suggest that the piezoelectric effect should be evident also in NC EFM due to the variations of the tip-sample separation at the frequency of the applied voltage. Then the key for the interpretation of the images should be related to the physical properties of the cantilever. According to this idea, depending on the elastic constant of the cantilever, one effect could prevail over the other, the piezoelectric effect could be dominant with respect to the modulation of the electrostatic force or vice versa [75]. In our case, we will consider only a qualitative interpretation of the EFM images.

7.4.2 Inorganic Ferroelectric Films

If a small voltage (several volts) is applied between an AFM tip, that behaves as a moving electrode, and the electrode, placed below a ferroelectric film, a very strong and localized electric field causes a polarization on a micrometric scale in the bulk of the film. EFM allows the modification of the ferroelectric domains inside the material and their visualization [76, 78]. Moreover, since it is possible to move the tip along the sample surface, domains can be reoriented following given patterns. In the following we will refer to the operation that leads to the domain reorientation as "writing" operation and to the imaging as "reading" operation.

EFM is an extremely promising technique for the investigation of ferroelectric materials and, among them, PZT ($PbZr_{1-x}Ti_xO_3$) is particularly interesting for applications in electronic and optoelectronic devices. Thin films of PZT grown on Pt/MgO and on Pt/Si were imaged and modified in EFM contact and non contact mode. The images in Fig. 7.16 were obtained on PZT films grown on Pt/Si by the sol-gel technique and they were acquired

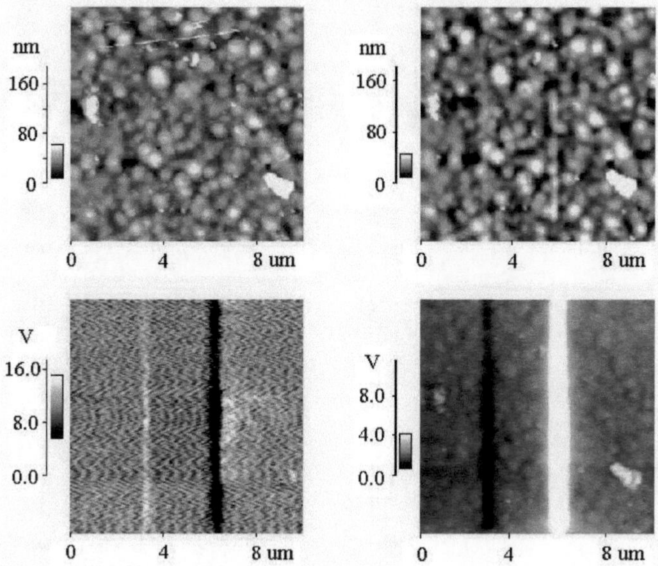

Fig. 7.16. Topography (top), and corresponding DC EFM (bottom left) and NC EFM (bottom right) images of a PZT surface. The two lines in each image were written by applying -15V (left) and +15V (right) on the tip.

at room temperature in air. The images on the left represent the topography and the DC EFM signals, the images on the right represent topography and the NC EFM signals acquired on the same surface area.

Lines were "written" by applying a DC voltage of −15V for the line on the left and +15V for the one on the right, using a coated $PtIr_5$ tip. The NC EFM image shows lines with a high contrast and of opposite phase to the DC EFM signal. The line on the left in the NC EFM image shows that the surface potential has increased with respect to the background in the positively polarized region, were the dipoles should be oriented towards the bottom of the sample. In other words, the negative charge of the dipoles should be localized on the surface, but the surface potential is positive (the white line in the EFM image). This could be explained assuming that during the "writing" operation, something like the "corona effect" occurs. Therefore the surface charge has the same sign as the charge on the tip and the presence of this charge, that stays localized on the polarized region, is considered to be responsible for the inversion of the contrast in the surface potential.

Figure 7.17 shows similar NC and DC EFM images of a thin dielectric film (Mel63, Niopik, Moscow).

Again the images on the left represent the topography (top) and the DC EFM signal (bottom), while the ones on the right show the topography and

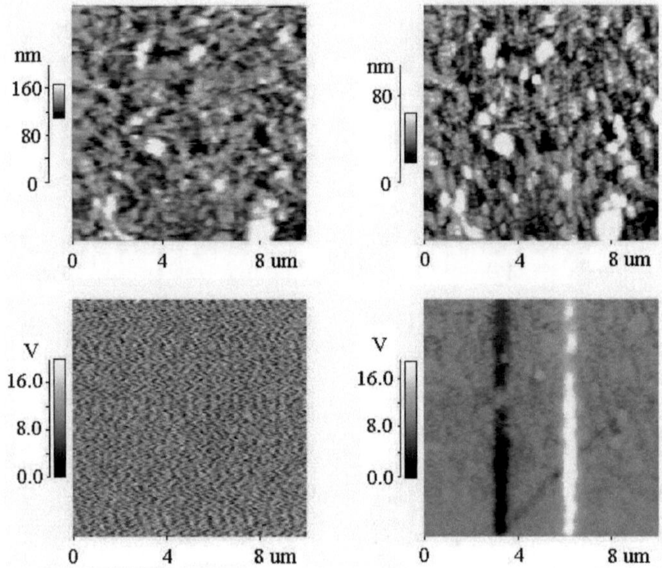

Fig. 7.17. Topography (top), and corresponding DC EFM (bottom left) and NC EFM (bottom right) images of the surface of a dielectric film. The two lines, shown on each of the bottom images were written by applying -15V (left) and +15V (right) on the tip.

the NC EFM signal. For both situations, lines were written by applying −15V (on the left) and +15V (on the right) on the tip. As expected, since Mel63 is not a ferroelectric material, the NC EFM signal originates from the charge, deposited on the surface, whereas no signal is present in the DC EFM image. We can conclude that with our set-up and the kind of tip we use, the NC EFM image gives a map of the surface potential, whereas the DC EFM image gives a map of the piezoelectric response of the material. We can also notice that the polarized areas in the NC EFM image are wider than in the DC EFM image, which can be due to charge diffusion on the film surface. However, this widening of the line in NC EFM mode can also be related to the reduced spatial resolution due to larger tip-surface separation.

In Fig. 7.18, the time evolution of the NC EFM signal coming from a point on a line written at +8V is shown. Similar results were obtained by Chen and his coworkers on PZT samples, deposited by the sol-gel technique [79]. Initially, the surface potential decays rapidly, then it slows down and stabilizes at the end. In general, the surface charge diffuses according to Ohm's law, and the surface resistivity regulates the diffusion of the charges. Moreover, the attractive force coming from the oriented dipoles of the substrate has a great effect on the charge diffusion on the piezoelectric film surface. In this

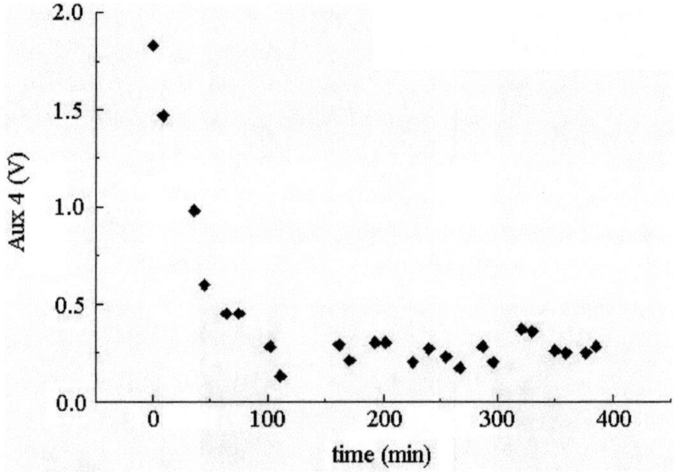

Fig. 7.18. The time evolution of the NC EFM signal measured on a line written by applying +8V on the tip.

case, the surface potential of a polarized area is the sum of a contribution due to the surface charges and a contribution due to dipole reorientation.

We have also verified the possibility to erase and overwrite existing lines. We have written two vertical lines applying ±8V and then recorded two horizontal lines with the same voltages overlapping the first ones. From our analysis we found that in the intersection between the lines written with opposite voltages, the signal coming from the surface has the same intensity as the signal coming from the non polarized areas and this is true for all the intersections. At the same time, the NC EFM signal is more intense in the intersections between the lines recorded with the same voltage than the one coming from a single line.

7.4.3 Organic Ferroelectric Films

In recent years ferroelectric polymeric materials have become more and more interesting due to the low costs in production and no particular difficulties in preparation. Polymers are easy to be modified in shape, dimensions and flexibility. Moreover they do not need to be grown epitaxially, and have good adhesion properties on a large number of substrates. In this class of materials, the polyvinylidene fluoride with trifluoroethylene (PVDF-TrFE) is a well known material with excellent ferroelectric characteristics, obtained with no special preparation treatments that can maintain ferroelectricity even when the film thickness is reduced to, 3–4 nm [80]. Using EFM, we have analyzed thin films of PVDF-TrFE deposited by the Langmuir-Blodgett technique on silicon substrates covered with Al electrodes. The investigated films were 70:30 mol%, where the 30% of the VDF (-CH2-CF2) units were substituted

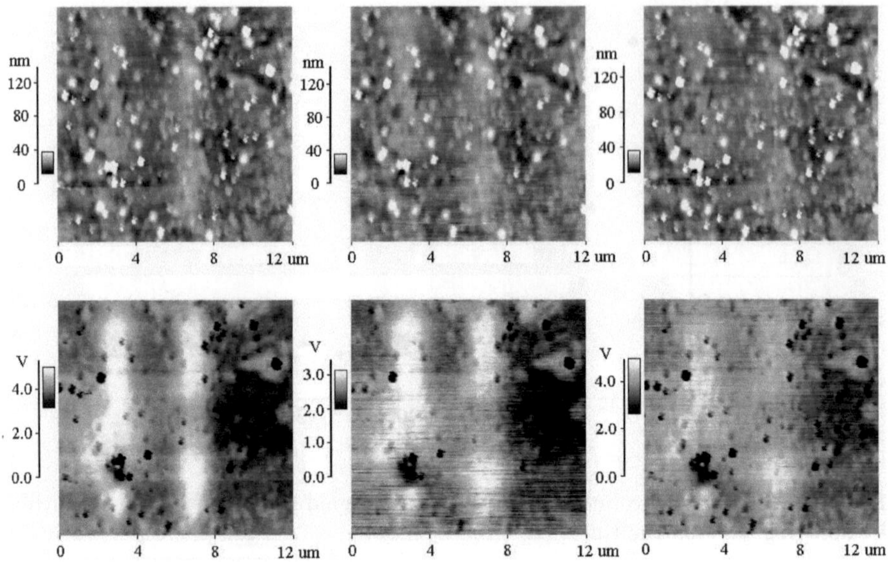

Fig. 7.19. Topography (top) and corresponding NC EFM images (bottom). The lines were written by applying -10V on the thin film of PVDF-TrFE. Images were acquired respectively 5, 20 and 35 minutes after "writing".

by TrFE ones (-CHF-CF2). The polar properties of the material are related to the dipole moments formed by the positively charged hydrogen atoms and the negatively charged fluorine atoms. The thickness of the investigated films was about 6 nm.

Images in Fig. 7.19 show the time evolution of the NC EFM signal coming from the polymeric PVDF-TrFE film for two lines, written by applying -10V between the sample and the tip. The images on the top are related to the topography and the bottom ones at the the corresponding NC EFM signal.

From the topographic images it is evident that the "writing" operation has had some effects on the surface of the polymeric film, that becomes slightly deformed due to the large force, applied by the apex of the tip during "writing". It is possible to reduce and even avoid the presence of permanent deformations on the sample surface. The images were acquired within one hour, the mean lifetime of such domains is of about one hour and a half. If we write lines with voltages of opposite polarity, the mean lifetime of the domains is less than half an hour. This asymmetric behavior of LB films with respect to the polarity of the applied voltage is probably due to the presence of a thin polarized layer of the polymeric material close the substrate, similar to what happens for spin coated films [81,82]. When the poling is parallel to the polarization of the layer induced by the substrate, it is possible to record lines with a better resolution.

7.4.4 A Simple Model of Electric Force Microscopy

In order to better understand the mechanism of "writing" and imaging of the electric surface potential using an EFM, we have developed a simple electrostatic model.

Let us first consider the sign of the surface charge. Experimentally, the sign of the surface charge, induced by the writing process, is determined by comparing the measured field on the polarized domains and the field coming from the surface, when an additional voltage of ±1V is applied between the tip and the sample. In all the pictures shown, the white lines correspond to a field directed from the surface of the film towards the tip during the reading process. The experiments show that we observe a surface charge that has a sign equal to the sign of the potential on the tip during the polarization process. We get EFM images with an opposite contrast only in very few cases.

There are in principle two electrostatic mechanisms for surface charging. In the first case, charge could be deposited from the tip on the sample surface during the writing process. For a ferroelectric, the charge can be deposited for voltages lower than the threshold for polarization inversion. Moreover, the density of this charge decays quite rapidly, sometimes in a few seconds. In the case of charge deposition, the sign of the deposited charge always coincides with the sign of the charge on the tip.

In the second case, when the applied field between the tip and the bottom electrode is strong enough to overcome the coercive field, the ferroelectric domains will have the spontaneous polarization P_S oriented along the field. In this case, we expect that the film surface has a charge with a sign opposite to the one on the tip. This is opposite to what we see in our experiments and cannot be explained by the first mechanism. The process of polarization switching has a threshold character, the polarized areas survive for a long time and the polarization inversion is observed independently using the piezoelectric technique.

In some cases, the explanation for this unusual surface charging has been suggested in terms of "overscreening", the phenomenon we have when external charges (for example charges injected from the bottom electrode or coming from the tip via the "corona" effect) arrive at the surface to compensate and sometimes to overcompensate the charge due to polarization. In general "overscreening" and "underscreening" effects can explain all the situations but they are not based on very convincing physical arguments.

We have elaborated a simple model to explain this situation [83] [84]. Suppose we have a ferroelectric thin film of thickness l and d is the distance between the tip and the sample, S* is the surface area of the tip front end, and U is the applied voltage between the tip and the sample. To model our system, we assume that the tip is covered with a thin layer of silver and that all the electric connections in our system are done through aluminium contacts. P_S is the spontaneous polarization of the material, σ is the surface charge and ΔA is the term related to the *contact potentials* at each interface of our system.

If we apply an oscillating voltage between the tip and the sample, we can measure the amplitude of the tip oscillations at the first and second harmonic of the applied voltage, that are proportional to the corresponding components of the electric force.

$$F(\omega) = -\frac{2\varepsilon_0 S^*}{(d + l/\varepsilon_f)^2}\left(\frac{(P_S - \sigma)\, l}{\varepsilon_f \varepsilon_0} - \Delta A\right) U_0 \sin \omega t \tag{7.11}$$

$$F(2\omega) = -\frac{\varepsilon_0 S^*}{2(d + l/\varepsilon_f)^2} U_0^2 \cos 2\omega t \tag{7.12}$$

The second harmonic signal observed at 2ω is independent of the polarization and the work functions ΔA and allows us to determine the effective area S* ,if the parameters d, l and ε_f are known. The signal at the fundamental frequency ω contains the most important term that depends on the parameters of the material P_S, σ and ΔA. It can be seen that the sign of F(ω) can be either positive or negative, i.e. in phase or shifted by π with respect to the applied voltage. Usually the polarization inversion is not complete and so we can replace the spontaneous polarization value P_S with the remanent polarization value P_r. When a film is very thick, we can ignore ΔA and, consequently, the F(ω) phase will depend on the sign of (P_r-σ), i.e. on the polarization state. For a not or partially screened polarization, the F(ω) signal is in phase with the applied voltage for positive values of P_r and out of phase for negative values of P_r. In contrast, for a thin film the phase of F(ω) is controlled by the ΔA term, more appropriately by a variation of the work functions at the interface.

7.4.5 EFM Measurements on Rubbed Substrates for Liquid Crystal Alignment

EFM is also a useful tool to investigate the electrostatic effects that might be induced by the rubbing of polymeric coated substrates, which is one of the most important steps in the LC display manifacturing process. Due to friction, the rubbing process induces charges on the surfaces that can influence the LC alignment and damage the TFT circuitry [85]. We have investigated the effect of rubbing on three polymers: Poly Methyl Methacrylate (PMMA, [-$CH_2C(CH_3)(CO_2CH_3)$-]$_n$), Poly Vinyl Alcohol (PVA, [-$CH_2CH(OH)$-]$_n$) and Polyimide (PI, PI2555). The rubbing was performed manually with a soft velvet cloth, before the EFM measurements. We used different clothes to avoid surface contaminations of different polymers.

The PMMA surface was first imaged before rubbing, as shown in Fig. 7.20. The image on the left shows the topography, while the one on the right shows the NC EFM signal.

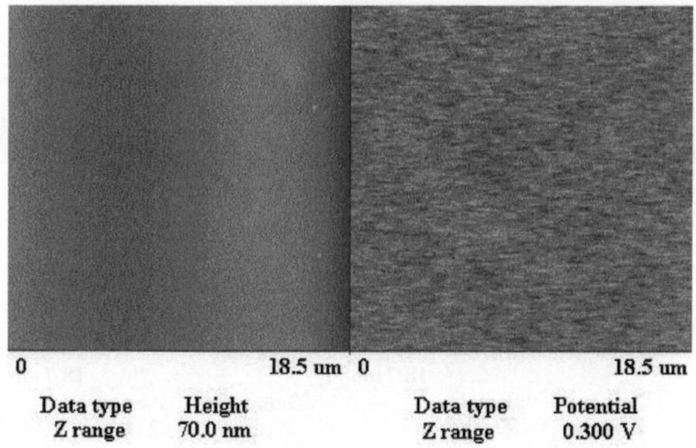

Fig. 7.20. Topography (left) and NC EFM image (right) of the PMMA surface, not rubbed [86].

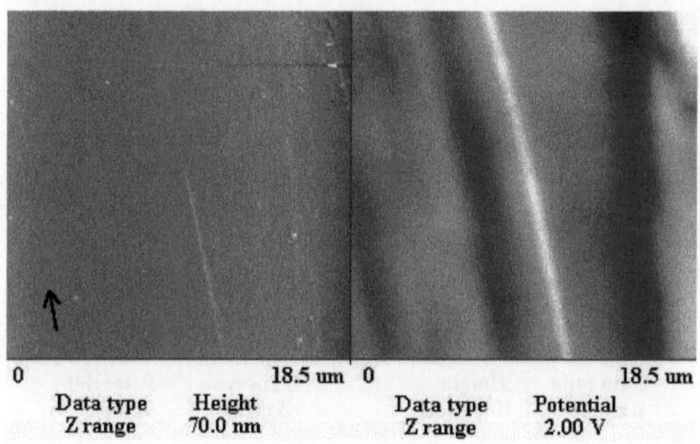

Fig. 7.21. Topography (left) and NC EFM image (right) of the PMMA surface rubbed once. Arrow indicates rubbing direction [86].

No traces of topographical deformations or electrostatic charge domains are visible. After rubbing the sample, we observed variations in both the surface topography and charged domains oriented along the rubbing direction (Fig. 7.21).

The rubbing process was repeated on the same sample for ten times, and the result is shown in Fig. 7.22. As can be noticed, the electrostatic signal coming from the surface increases and so does the surface deformation. Following the time evolution of the charge domains, we found that the signal was almost constant for five days, except for a very small reduction. The same measurements were performed also on PI and PVA.

On the PI surface, the charge domains are only slightly visible, while for the PVA no NC EFM signal is present (Figs. 7.23, 7.24). The NC EFM scale for PMMA is 2V while for PI it is 0.1V. For PVA this can be explained by considering the chemical structure of the polymers: the OH groups in PVA are

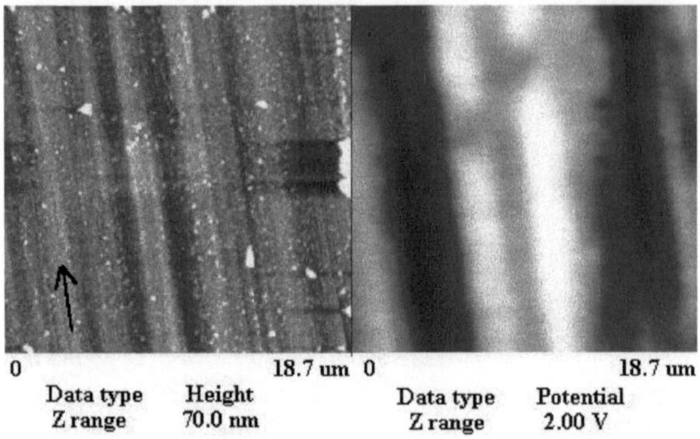

Fig. 7.22. Topography (left) and NC EFM image (right) of the PMMA surface rubbed ten times [86].

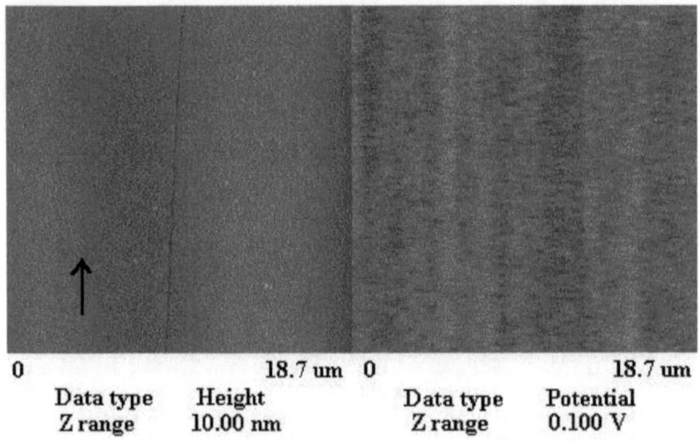

Fig. 7.23. Topography (left) and NC EFM image (right) of the PI surface rubbed once [86].

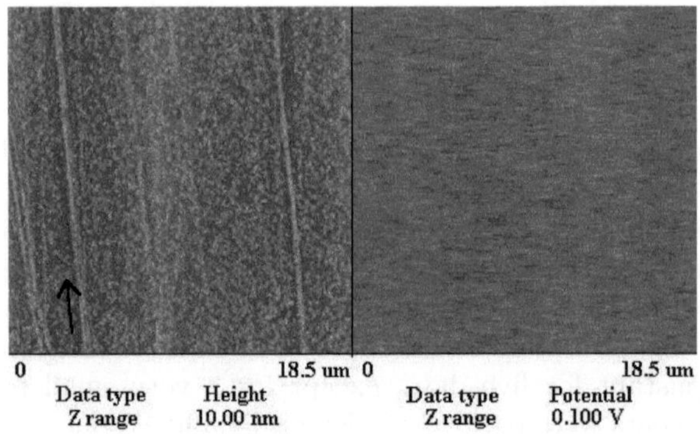

Fig. 7.24. Topography (left) and NC EFM image (right) of the PVA surface rubbed once [86].

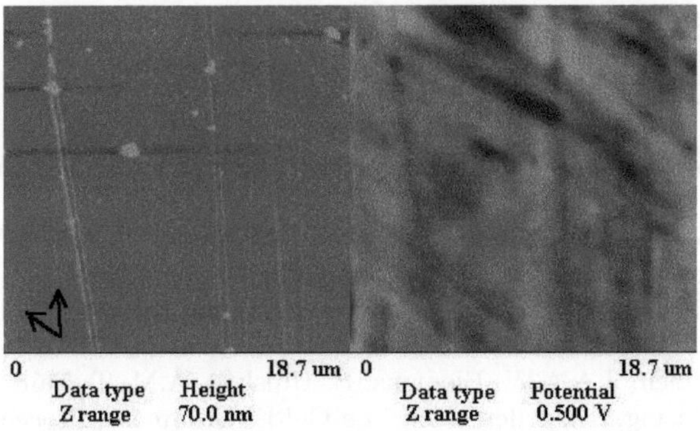

Fig. 7.25. Topography (left) and NC EFM image (right) of the PMMA rubbed once in one direction and then once again in a direction at an angle of about 45 degrees with respect to the first one [86].

good charge conductors, therefore the charges formed by the rubbing process can leak out from the surface. PMMA does not contain OH groups, so the charges stay localized at the surface for a long time. Further investigations have to be performed on PMMA and PI to better understand the dependence of their surface potential on their chemical structure.

The effect of rubbing the sample in two different directions was also investigated. The images in Fig. 7.25 show a PMMA surface rubbed once in one direction and then once in another direction forming about 45 degrees with respect to with the first one. The results show that the initial orientation of the charge domains induced by the first rubbing direction are not erased.

7.4.6 Conclusions

We have shown how Electrostatic Force Microscopy can be an extremely useful tool to investigate and to modify the electric properties of sample surfaces on a microscopic and even nanoscopic scale and we have presented a phenomenological model to help relating the experimental data to the material properties. Ferroelectric domains can locally be reoriented and their time evolution can be followed, as was shown for PZT. We have also demonstrated how the ferroelectric polymer PVDF-TrFe could be locally modified which can be used to locally vary the optical properties of a LC cell. Finally, we have demonstrated that rubbing polymer substrates can indeed result in electrostatic charging, in particular for PMMA and PI, while no charging is found for PVA.

References

1. G. Binnig, H. Rohrer, Ch. Gertber and E. Weibel, Phys.Rev. Lett. **49**, 57 (1982).
2. S. Magonov and M.-H. Whangbo, Surface Analysis with STM and AFM, VCH, New York (1996).
3. R. Wiesendanger H.-J. Güntherodt (Eds.), Scanning Tunneling Microscopy I,II and III, Springer, Heidelberg, 1992 and 1993.
4. J.W. Gerritsen, E.J.G. Boon, G. Janssens, H. van Kempen, Appl. Phys. Lett., **A66**, S78-S82 (1998).
5. J.W. Gerritsen, J.A.A.W. Elemans, B. Hulsken, A.M. Travaille, H. van Kempen, Th. Rasing, S. Speller, STM in a Gel Environment, Proceedings STM'03, p. 365, AIP Conference Proceedings 696 (2003)
6. The analog I-O board is presently a ATMIO16X from National Instruments. In a later version we use ATMIO16XE10 boards which can do all the timing without computer intervention.
7. Similar to the SCALA system of OMICRON VAKUUM-PHYSIK GmbH.
8. J. Foster and J. Frommer, Nature **60**, 1418 (1988).
9. M.Hara, Y. Iwakabe, K. Tochigi, H. Sasabe, A.F. Garito and A. Yamada, Nature **344**, 228 (1990).
10. D.P.E. Smith, J. Vac. Sci. Techn. **B9**, (1991).
11. H.Nejoh, Appl. Phys. Lett. **57**, 390 7 (1990).
12. S. Magonov, private communications.
13. D.P.F. Smith, H. Hörber, Ch. Gerber and G. Binnig, Science **245**, 43 (1989).
14. A.J. Groscek, Proc.R.Soc.London A324, 473 (1970).
15. L.L. Soethout, J.W. Gerritsen, P.P.M.C. Groeneveld, B.J. Nelissen, H. van Kempen, Journal of Microscopy, **152**, 251 1988.
16. measured by Grietje Walraven, Jorgen Willemsen, and Rob Janssen on 1st year SPM class (2003).
17. measured by Iris Silkens during STM class (2001).
18. measured by high school students Richard Sanders and Joris van Bergen (2001).
19. S.L. Brandow, J.A. Harrison, D.P. Dilella, R.J. Colton, S. Pfeiffer and R. Shashidar, Liquid Crystals **13**, 163 (1993).
20. D.C. Parks, N.A. Clark, D.M. Wlaba and P.D. Beale, Phys.Rev. Lett **70**, 607 (1993).
21. D.M. Walba, F. Stevens, D.C. Parks, N.A. Clark and M.D. Wand, Science **267**, 1144 (1995).
22. M. Ruetschi, P. Grutter, J. Funfschilling, H. J. Guentherod, Science **265**, 512 (1994).
23. O. M. Leung and M. C. Goh, Science **255**, 64 (1992).
24. E. Hamada, and R. Kaneko, Ultramicroscopy **42**, 1446 (1992).
25. X. Jin, W. N. Unertl, Appl. Phys. Lett. **61**, 657 (1992).
26. M. Ruetschi, J. Funfschilling, H. J. Guntherod, J. Appl. Phys. **80**, 3155 (1996).
27. A. J. Pidduck, S. D. Haslam, G. P. Bryan-Brown, R. Bannister, I. D. Kitely, Appl. Phys. Lett. **71**, 2907 (1997).
28. B. Wen, and Ch. Rosenblatt, J. Appl. Phys. **89**, 4747 (2001).
29. B. Wen, J. H. Kim, H. Yokoyama, Ch. Rosenblatt, Phys.Rev.E **66**, 041502 (2002).
30. B. Wen, R. G. Petschek, Ch. Rosenblatt, Appl. Opt. **41**, 1246 (2002).

31. B. Wen, M. P. Mahajan, Ch. Rosenblatt, Appl. Phys. Lett. **76**, 1240 (2000).
32. A. Rastegar, M. Škarabot, B. Blij, Th. Rasing, J. Appl. Phys. **89**, 960 (2001).
33. M. Škarabot, S. Kralj, A. Rastegar, Th. Rasing, J. Appl. Phys. **94**, 6508 (2003).
34. M. Behdani, A. Rastegar, S. H. Keshmiri, S. I. Missat, E. Vlieg, Th. Rasing, Appl. Phys. Lett.**80**, 4635 (2002).
35. J. H. Kim, M. Yoneya, J. Yamamoto, H. Yokoyama, Appl. Phys. Lett. **78**, 3055 (2001).
36. M. Yoneya, J. H. Kim, H. Yokoyama, Appl. Phys. Lett. **80**, 374 (2002).
37. J. H. Kim, M. Yoneya, H. Yokoyama, Appl. Phys. Lett. **83**, 3602 (2003).
38. J. H. Kim, M. Yoneya, H. Yokoyama, Nature **420**, 159 (2002).
39. B. Zhang, F. K. Lee, O. K. C. Tsui, P. Sheng, Phys. Rev. Lett. **91**, 215501 (2003).
40. M. Behdani, S. H. Keshmiri, S. Soria, M. A. Bader, J. Ihlemann, G. Marowsky, Th. Rasing, Appl. Phys. Lett. **82**, 2553 (2003).
41. A. J. Pidduck, G. P. Bryan-Brown, S. Haslam, R. Bannister, I. Kitely, T. J. McMaster, L. Boogaard, J. Vac. Sci.Techn. A **14**, 1723 (1996).
42. J. Israelachvili, *Intermolecular and Surface Forces.* Academic Press, 1992.
43. W. B. Russel, D. A. Seville, and W. R. Schowalter, *Colloidal dispersions.* Cambridge University Press, 1995.
44. H. De Vleeschouwer, F. Bougrioua, and M. Pauwels, Mol. Cryst. Liq. Cryst. **29**, 360 (2001).
45. G. Barbero and G. Durand, J. Phys. (France) **51**, 281 (1990).
46. G. Barbero and G. Durand, J. Appl. Phys. **68**, 5549 (1990).
47. A. L. Alexe-Ionescu, R. Barberi, J. J. Bonvent, and M. Giocondo, Phys. Rev. E **54**, 529 (1996).
48. G. Barbero, A. K. Zvezdin, and L. R. Evangelista, Phys. Rev. E, **59**, 1846 1999.
49. M. A. Osipov, T. J. Sluckin, and S. J. Cox, Phys. Rev. E **55**, 464 (1997).
50. L. M. Blinov and A. A. Sonin, JETP **60** 272 1984.
51. H. Stark, Physics Reports, **351**, 387 (2001).
52. K. Kočevar and I. Muševič, Phys. Rev. E **65**, 030703 (2002).
53. S. Marčelja, Langmuir **16**, 6081 (2000).
54. E. Matijevič and N. P. Ryde, Journal of adhesion, **51**, 1 (1995).
55. S. H. Behrens and M. Borkovec, *J. Phys. Chem. B*, **103**, 2918, (1999).
56. B. V. Derjaguin, Kolloid Zeits. **69**, 155 (1934).
57. K. Kočevar and I. Muševič, Phys. Rev. E, **64**, 051711 (2001).
58. K. Kočevar, A. Borštnik, I. Muševič, and S. Žumer, Phys. Rev. Lett. **86**, 5914 (2001).
59. V. G. Nazarenko, V. M. Pergamenshchik, O. V. Kovalchuk, A. B. Nych, and B. I. Lev, Phys. Rev. E **60**, 5580 (1999).
60. U. Kühnau, A. G. Petrov, G. Klose, and H. Schmiedel, Phys. Rev. E, **59**, 578 (1999).
61. G. Barbero, L. R. Evangelista, and N. V. Madhusudana. Eur. Phys. J. B **1**, 327 (1998).
62. P. R. Maheswara Murthy, V. A. Raghunathan, and N. V. Madhusudana, Liq. Cryst. **14**, 483 (1993) and references therein.
63. A. Poniewierski and T. J. Sluckin, Mol. Cryst. Liq. Cryst. **126**, 143 (1985).
64. L.Xu, M.Salmeron, S.Bardon, Physical Review Letters **84**, 1519 (2000).

65. R. Barberi, G. Barbero, Z. Gabbasova and A. Zvezdin, J. Phys. France **3**, 147 (1993).
66. G. Barbero, L. Dozov, J.F. Palierni and G. Durand, Phys. Rev. Lett. **56**, 19 (1987).
67. G. Barbero and G. Durand J. of Appl. Phys. **69**, 6968 (1991).
68. G.Barbero, L.R.Evangelista, N.V.Madhusudana, The European Physical Journal **B 1**, 327 (1998).
69. M.Glogarova, V.Janovec, N.A.Tikhomrova, Journal de Physique, Colloque C3, Supplement au n°4, Tome **40**, c3-c5 (1979).
70. Q.Xu, J.W.P.Hsu, Journal of Applied Physics **85**, 2465 (1999).
71. R.M.Nyffenegger, R.M.Penner, R.Schierle, Applied Physics Letters **71**, 1878 (1997).
72. G.Zavala, J.H.Fendler, S.Trolier-McKinstry, Journal of Applied Physics **81**, 7480 (1997).
73. J.W.Hong, D.S.Kahng, J.C.Shin, H.J.Kim, Z.G.Khim, Journal of Vacuum Science and Technology **B 16**, 2942 (1998).
74. J.W.Hong, S.I.Park, Z.G.Khim, Review of Scientific Instruments **70**, 1735 (1999).
75. M.Labardi, V.Likodimos, M.Allegrini, Physical Review **B 61**, 14390 (2000).
76. T.Tybell, C.H.Ahn, J.M.Triscone, Applied Physics Letters **72**, 1454 (1998).
77. C.Durkan, M.E.Welland, D.P.Chu, P.Migliorato, Phys.Rev. B **60**, 16198 (1999).
78. T.Hikada, T.Maruyama, M.Saitoh, N.Mikoshiba, M.Shimizu, T.Shiosaki, L.A.Wills, R.Hiskes, S.A.Dicarolis, J.Amano, Applied Physics Letters **68**, 2358 (1996).
79. X.Q.Chen, H.Yamada, T.Horuichi, K.Matsushige, S.Watanabe, M.Kawai, P.S.Weiss, Journal of Vacuum Science and Technology B **17**, 1930 (1999).
80. A.V.Bune, V.M.Fridkin, S.Ducharme, L.M.Blinov, S.P.Palto, A.V.Sorokin, S.G.Yudin, A.Zlatkin, Nature **391**, 874 (1998).
81. X.Chen, H.Yamada, T.Horiuchi, K.Matsushige, Japanese Journal of Applied Physics **38**, 3932 (1999).
82. X.Q.Chen, H.Yamada, Y.Terai, T.Horiuchi, K.Matsushige, P.Weiss, Thin Solid Films **353**, 259 (1999).
83. L.M.Blinov, R.Barberi, S.P.Palto, M.P.De Santo, S.G.Yudin, Journal of Applied Physics **89**, 3960 (2001).
84. M.P.De Santo, R.Barberi, L.M.Blinov, S.P.Palto, S.G.Yudin, Molecular Materials **12**, 329 (2000).
85. J.van Haaren, Nature **392**, 331 (1998).
86. I.H.Bechtold, M.P.De Santo, J.J.Bonvent, E.A.Oliveira, R.Barberi, Th.Rasing, Liquid Crystals **30**, 591 (2003).

8 Introduction to Micro- and Macroscopic Descriptions of Nematic Liquid Crystalline Films: Structural and Fluctuation Forces

Andreja Šarlah and Slobodan Žumer

In this Chapter we introduce some of the theoretical approaches for studying thin nematic liquid crystalline systems, both on the microscopic and macroscopic level. In the former, one models the microscopic interactions between the constituing molecules, leaves the system to evolve, and then determines its macroscopic properties. If the obtained macroscopic behaviour is in agreement with the experimental evidence the modeled interaction is considered appropriate. On the other hand, the macroscopic description takes into account the universal properties of systems in the vicinity of phase and structural transitions. This means that they are based on the fact that in the vicinity of phase changes the macroscopic properties of the system do not depend on the details of the microscopic interactions but on the symmetry properties and dimensionality of the system in question. Most of our attention is focused on the effects of confinement on to liquid crystalline order. Finally, we will be interested in the resulting disjoining pressure. The evidences in experiments will be briefly mentioned.

8.1 Introduction

In the past few decades the technological possibilities and interests have boosted research in systems in highly restricted geometries in almost every field of physics – recently down to lengthscales close to or even below the molecular level. In the field of liquid crystals, the importance of electro-optical applications which incorporate ordered liquid materials [1–3] has focused the research on LC systems with high surface-to-volume ratio [4]. In order to provide mechanically stable applications, liquid crystals are dispersed in polymers, stabilized by a polymer network, fill the cavities in porous materials, etc. [5,6]. The major technological interest concerns the scattering, reflective and bistable displays, optical switches, and others.

Liquid crystals confined on a submicron scale exhibit not only important applicative properties but also a variety of interesting basic physical phenomena. The host materials offer microconfinement characterized by a curved, irregular, or even fractal internal geometry. Nevertheless, the confining geometries all share a common feature: the curved boundary and/or the antagonistic boundary conditions result in frustration which leads to a number of transitions between the equilibrium structures, the control variable being either temperature or the size of the system. Because of the high surface-to-volume ratio the effects of the confining surfaces are very important, although this does not necessarily imply a strongly deformed order parameter [7–10]. However, even a weak distortion of the ordering can result in a rather strong force induced by the liquid crystal, because in this regime the fluctuation-mediated contribution to the force can be strongly amplified [11–14]. Therefore, in the last decades a special attention has been paid to the role of the confining surfaces in the ordering and dynamic properties of LC systems.

In this Chapter the basic approaches used to describe nematic liquid crystalline (NLC) systems in slab geometries under the effect of confinement are introduced. We review both, the microscopic and macroscopic approaches, however, the emphasis is on the latter. We also show the correspondence between the approaches on different levels. Special attention is devoted to effects of the confinement on the LC order and consequently to the interactions arising from that. More precise descriptions of the techniques and also more detailed results have been already published elsewhere [9–12,15–18]. In the following Section we first shortly review the microscopic origin of order and define the appropriate order parameter. Then we review the basic microscopic and macroscopic theoretical approaches to describe LC systems. In the third Section we describe in short the effect of confinement in two different types of NLC systems. The fourth Section is devoted to macroscopic interactions between confining walls, especially the ones characteristic for ordered systems. We conclude the Chapter with the discussion on the observability of structural and fluctuation forces in NLC systems.

8.2 Microscopic versus Macroscopic Theoretical Aspects

In condensed matter one is not interested in the state of each single molecule on the microscopic level but in the collective state and behaviour of the system as a whole, which is reflected in its macroscopic appearance. In the isotropic phase LC the medium possesses full symmetry. On the contrary, in the nematic phase the symmetry is lowered because the molecules exhibit collective orientational order. The two phases are schematically depicted in Fig. 8.1. To express the order of the system quantitatively, we need to define the *order parameter*. This has to vanish in the isotropic phase and has to become nonzero in the ordered nematic phase.

The (nematic) liquid crystal consists of elongated molecules characterized by unit vectors along the long axis of the molecule, $\hat{a}$. The molecule is assumed to have complete cylindrical symmetry about $\hat{a}$ which is justified by fast unbiased molecular rotations about this axis. The orientation of a single molecule is thus determined by two angles with respect to the laboratory frame (x, y, z), the polar angle θ and the azimuthal angle ϕ, where the

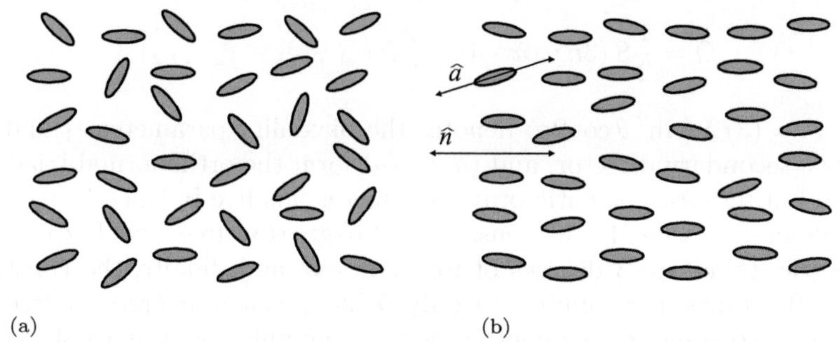

Fig. 8.1. A sketch of (a) isotropic and (b) nematic order of rod-like molecules. $\hat{n}$ is a nematic director and $\hat{a}$ denotes the long axis of a given molecule.

z axis coincides with the average direction of molecules, i.e., the *director* $\hat{n}$. The state of the alignment of molecules is described by a distribution function $f(\theta, \phi)$ which gives the probability of finding molecules in a solid angle $\mathrm{d}\Omega$, $f(\theta,\phi)\mathrm{d}\Omega/\Omega$. In undeformed bulk nematic liquid crystals, the uniaxial symmetry does not allow any ϕ dependence of the distribution function, and so the latter can be expanded in a series $f(\theta) = \sum_{m=0}^{\infty} f_m P_m(\cos\theta)$, where $P_m(x)$ are Legendre polynomials and $f_m = (2m+1)\Omega^{-1} \int \mathrm{d}\Omega f(\theta) P_m(\cos\theta)$. From this and because f is normalized, $f_0 = 1$. Due to the equivalence of $\hat{n}$ and $-\hat{n}$ yielding cylindrical symmetry, $f_m \equiv 0$ for odd m; from the same argument f_m's are nonzero for even m. The first nonzero parameter is $f_2 = \frac{5}{\Omega} \int \mathrm{d}\Omega f(\theta) P_2(\cos\theta) = 5 \left\langle \frac{1}{2}\left(3\cos^2\theta - 1\right)\right\rangle \equiv 5S$, where $S \in [-\frac{1}{2}, 1]$ is a *scalar order parameter* describing the degree of order of molecules about the average orientation. $S = 1$ corresponds to perfect nematic order where all molecules are parallel to the director, $S = 0$ in the case of uniform distribution of molecules, and the minimum value $S = -\frac{1}{2}$ corresponds to a situation where all molecules lie in the plane perpendicular to the director and there is no preferred direction in this plane. Up to the first nontrivial term, the distribution function reads

$$
\begin{aligned}
f(\theta) &= 1 + 5SP_2(\cos\theta) = 1 + 5S\frac{1}{2}\left(3(\hat{n}\cdot\hat{a})^2 - 1\right) \\
&= 1 + 5S\frac{1}{2}\left(3n_i n_j - \delta_{i,j}\right) a_i a_j = 1 + 5(\mathsf{Q})_{ij}\, a_i a_j, \qquad (8.1)
\end{aligned}
$$

where

$$
\mathsf{Q} = \frac{1}{2} S \left(3\hat{n} \otimes \hat{n} - \mathsf{I}\right) \qquad (8.2)
$$

is the *tensorial order parameter* of the nematic liquid crystal. It represents the quadrupolar moment of the distribution, i.e., deviations from the perfect spherical distribution. I is the second rank unit tensor.

Due to external fields the cylindrical symmetry of the order can be lost and the tensorial order parameter becomes somewhat more complicated,

$$\mathsf{Q} = \frac{1}{2} S \left(3\hat{n} \otimes \hat{n} - \mathsf{I}\right) + \frac{1}{2} P \left(\hat{e}_1 \otimes \hat{e}_1 - \hat{e}_2 \otimes \hat{e}_2\right) \tag{8.3}$$

where $P = (3/2)\langle \sin^2 \theta \cos 2\phi \rangle$ denotes the biaxiality parameter $\hat{e}_1$ and represents a secondary director, and $(\hat{n}, \hat{e}_1, \hat{e}_2)$ form the orthonormal triad.

The full tensorial nematic order parameter has five independent degrees of freedom: The second rank tensor has 9 degrees of freedom. Because it is symmetric, there are 3 degrees of freedom less, and, finally, the constraint $\mathrm{tr}\,\mathsf{Q} = 0$ reduces the number to only 5 independent degrees of freedom. There are two standard parametrizations of the different degrees of freedom of the nematic order: (i) two angles determining the orientation of the director, the scalar order parameter, the angle specifying the orientation of the secondary director, and the parameter of biaxiality, as in (8.2) and (8.3), and (ii) parametrization with respect to the 5 base tensors of the symmetrical traceless tensor [19, 20]

$$\begin{aligned} \mathsf{T}_0 &= \frac{3\hat{n} \otimes \hat{n} - \mathsf{I}}{\sqrt{6}}, \\ \mathsf{T}_1 &= \frac{\hat{e}_1 \otimes \hat{e}_1 - \hat{e}_2 \otimes \hat{e}_2}{\sqrt{2}}, \quad \mathsf{T}_{-1} = \frac{\hat{e}_1 \otimes \hat{e}_2 + \hat{e}_2 \otimes \hat{e}_1}{\sqrt{2}}, \\ \mathsf{T}_2 &= \frac{\hat{e}_1 \otimes \hat{n} + \hat{n} \otimes \hat{e}_1}{\sqrt{2}}, \quad \mathsf{T}_{-2} = \frac{\hat{e}_2 \otimes \hat{n} + \hat{n} \otimes \hat{e}_2}{\sqrt{2}}. \end{aligned} \tag{8.4}$$

All the above tensors are traceless and orthogonal with respect to the metric $\mathsf{T}_n : \mathsf{T}_m = \mathrm{tr}\,(\mathsf{T}_n \mathsf{T}_m) = \delta_{n,m}$. In this parametrization the order parameter reads

$$\mathsf{Q} = \sum_{m=-2}^{2} q_m \mathsf{T}_m, \tag{8.5}$$

where $q_m = \mathrm{tr}\,(\mathsf{Q}\mathsf{T}_m)$. The multiplicative constants are set so that the amplitude q_0 represents the scalar order parameter, parameters $q_{\pm 1}$ are nonzero if the order is biaxial, and parameters $q_{\pm 2}$ represent deviations in the orientation of the director with respect to the assumed director $\hat{n}$. Parametrization (i) is usually used when describing a liquid crystal deep in the ordered phase where the order is determined mostly by the orientation of the director (director description). Parametrization (ii) is usefull in the vicinity of phase changes where other degrees of freedom play an equally important role.

The order that is represented by a tensorial term in the distribution function is reflected in its influence on the macroscopic tensorial physical quantities, such as the magnetic susceptibility,

$$\underline{\chi} = \frac{2}{3}\chi_a \left[\frac{1}{2} S \left(3\hat{n} \otimes \hat{n} - \mathsf{I}\right)\right] + \chi_b \left[\frac{1}{2} P \left(\hat{e}_1 \otimes \hat{e}_1 - \hat{e}_2 \otimes \hat{e}_2\right)\right] + \chi_i \mathsf{I}. \tag{8.6}$$

Here, $\chi_a = \tilde{\chi}_3 - (\tilde{\chi}_1 + \tilde{\chi}_2)/2$ is the largest anisotropy of the magnetic susceptibility and $\tilde{\chi}_\alpha = \chi_\alpha / S$ are susceptibilities of the perfectly ordered nematic,

$\chi_b = \tilde{\chi}_1 - \tilde{\chi}_2$ is nonzero in biaxially ordered nematic ($\tilde{\chi}_\alpha = \chi_\alpha/P$), and $\chi_i = (\chi_1+\chi_2+\chi_3)/3$ is the average magnetic susceptibility, i.e., its isotropic part. $\chi_3 > \chi_1, \chi_2$ are the eigenvalues of the magnetic susceptibility, χ_3 corresponding to the director and χ_1, χ_2 correspond to $\hat{e}_1, \hat{e}_2$, respectively. The dielectric susceptibility and the corresponding refractive index are affected in the same way.

In other, more ordered LC phases, the reduced symmetry yields additional degrees of freedom of the order parameter, such as a density modulation in smectic and columnar phases.

8.2.1 Microscopic Models

Microscopic theoretical models have become more important with the increasing computer power. Though the number of particles included in simulations is still far smaller than the number of particles forming real systems, it is big enough to mimic the macroscopic behavior of studied systems.

Computer simulations serve as a bridge between microscopic and macroscopic length and time scales but also as a bridge between theory and experiment. In computer simulations we provide a guess for the interactions between molecules and probe it by comparing predicted macroscopic physical properties and observables of the system with its actual properties.

There are two basic types of computer simulations, Molecular Dynamics (MD) and Monte Carlo (MC) simulations. The former method consists of a brute-force solution of Newton's equations of motion, therefore, it corresponds to what happens in "real life" — it generates configurations time step after time step in their natural time sequence. On the other hand, the latter method can be thought of as a prescription for sampling configurations from a statistical ensemble; on achieving the equilibrium the system goes from one state to the next, not necessarily in a proper order [21, 22]. We will explain in some more detail a Lebwohl-Lasher (LL) lattice "spin" model of liquid crystals which gives good results for both bulk and confined liquid crystals [21–23].

In the framework of the LL lattice model the uniaxial nematic molecules are represented by particles, usually refered to as *spins*, that can rotate freely, but are arranged in a simple cubic lattice with spacing l. There are no translational degrees of freedom in the system, but despite this, the model can be seen to reproduce the orientational order well enough. Equivalently, particles may also be regarded as close-packed molecular clusters, maintaining their short-range orientational order in the relatively narrow temperature interval in the nematic phase, as well as across the isotropic–nematic transition [24]. A typical magnitude of the correlation length in the vicinity of the isotropic–nematic phase transition $\xi_{NI} \sim 10$ nm yields approximately 50 LC molecules per "spin".

The orientation of the spin located at the ith lattice site is denoted by a three-dimensional unit vector $\hat{a}_i$ corresponding to the director within the cluster. The interaction between the spins i and j is then modelled by a second-rank LL potential [23],

$$U_{ij} = -\epsilon_{ij} P_2(\hat{a}_i \cdot \hat{a}_j), \tag{8.7}$$

where $\epsilon_{ij} = \epsilon > 0$ for nearest-neighbor spins and it is zero otherwise, and P_2 is the second-rank Legendre polynomial. In this sense the LL model is similar to the Heisenberg model used for modeling magnetic systems, yet accounting also for the head-tail symmetry encountered in nematics.

Simulating bulk systems, periodic boundary conditions are usually applied [23, 25]. On the other hand, in confined systems a certain fraction of nematic spins (the *"ghost" spins*) is used to fix the boundary conditions. In a slab geometry, the confinement is introduced as layers of fixed spins whose orientation corresponds to the direction induced by the walls. Both nematic-nematic and nematic-ghost interactions are modeled by the interaction law (8.7), but the interaction strengths ϵ_{ij} are not necesarily equal for both interactions. In case they are, the layer distance can be recognized as the extrapolation length — the length on which the order is extrapolated to the one induced by the confinement.

Performing MC simulations, it was shown that the LL model reproduces a weakly first-order isotropic–nematic transition in bulk systems — a large sample with periodic boundary conditions [23,25] — while in confined systems of sufficiently small size the phase transition is suppressed [26, 27], which is in agreement with experimental data. The bulk isotropic–nematic transition occurs at the reduced MC temperature $T_{NI}^{MC} = 1.1232 \pm 0.0006$ [25], where $T^{MC} \equiv k_B T/\epsilon$. It should be stressed that MC simulations are particularly useful in studying ordering of nematics severely confined to complex geometries where a phenomelogical description is not very effective. One example — the hybrid cell — will later serve as a bridge between phenomenological description and simulations.

For now we have only tackled the microscopic point of view of the computer simulations. However, at the end one is interested in the macroscopic behavior of the studied system. To obtain the macroscopic physical observables, components of the order parameter tensor, $Q_{\alpha\beta} = 1/2\langle 3a_\alpha a_\beta - \delta_{\alpha\beta}\rangle$, are calculated as an average over a number of MC configurations. The order parameter tensor is determined with respect to a fixed eigenframe and then it is diagonalized to obtain the orientation of the director and the magnitude of order parameters. In addition one is interested in the total energy of the system and its correlation functions. Other physical observables can be calculated from those.

8.2.2 Macroscopic Models: Phenomenological Landau–de Gennes Theory

Let us first recall the general properties of phase transitions and the basic assumptions within the Landau description of phase transitions [28]. The term *phase transition* denotes a change in the medium which is accompanied by a discontinuity of some of the thermodynamic potentials and by a change of a certain physical quantity, e.g., density, macroscopic magnetization or polarization, magnetic, electric and optical properties, etc. If the entropy of the system is a continuous function of thermodynamic variables at the transition, then the transition is of second order or *continuous*, whereas it is of first order or *discontinuous* if the entropy is discontinuous. In the latter, latent heat is absorbed by the system when going from the low to the high temperature phase, $\mathcal{Q}_L = T_c \Delta S$, where ΔS is the difference of the entropies of the two coexisting phases and T_c is the phase transtition temperature.

In Landau theory, the information about the change of physical quantities is gathered in the order parameter $\mathrm{Q}_0 = V^{-1} \int \mathrm{d}^d r \mathrm{Q}(\boldsymbol{r})$ which is a macroscopic quantity that neglects spatial and temporal fluctuations. The basic concept of the description of phase transitions is the introduction of the Landau free energy, $\mathcal{F} = \int \mathrm{d}^d r f$, which takes into account the symmetry of the system through a power series expansion in terms of the scalar invariants of the order parameter, whereas the equation of state of the system reads

$$\frac{\partial f}{\partial \mathrm{Q}_0} = h. \tag{8.8}$$

Here h is the "external field" conjugate to the order parameter Q_0. When there are no "external fields", the nondeformed equilibrium state is the one that minimizes f. If for $h = 0$ there exists a nonzero solution of (8.8) the system can be spontaneously ordered provided that the free energy of that state is lower than the free energy of the disordered state with $\mathrm{Q}_0 = 0$.

The Landau theory assumes that the order parameter is small in the vicinity of the transition, so that only the lowest terms required by symmetry and preventing the free energy from diverging are kept in the expansion. In the case of nematic liquid crystals, the order parameter is a tensor and its scalar invariant is its trace. Thus, the Landau free energy reads

$$f = f_0 + \frac{1}{2} a \operatorname{tr} \mathrm{Q}_0^2 - \frac{1}{3} B \operatorname{tr} \mathrm{Q}_0^3 + \frac{1}{4} C \operatorname{tr} \mathrm{Q}_0^4 + \mathcal{O}(\operatorname{tr} \mathrm{Q}_0^5), \tag{8.9}$$

where f_0 is the free energy density of the disordered phase. The linear term vanishes by definition (8.2), and the 4th order invariants $(\operatorname{tr} \mathrm{Q}_0)^2$ and $\operatorname{tr} \mathrm{Q}_0^4$ are linearly dependent for symmetric tensors. The absence of the first order term is in accordance with the existence of the stable high temperature disordered phase. The third order term is included in the expansion because $\pm\mathrm{Q}_0$ have different physical meaning; from (8.6) we can see that for usual LC materials with positive magnetic anisotropy, $\mathrm{Q}_0 > 0$ corresponds to a

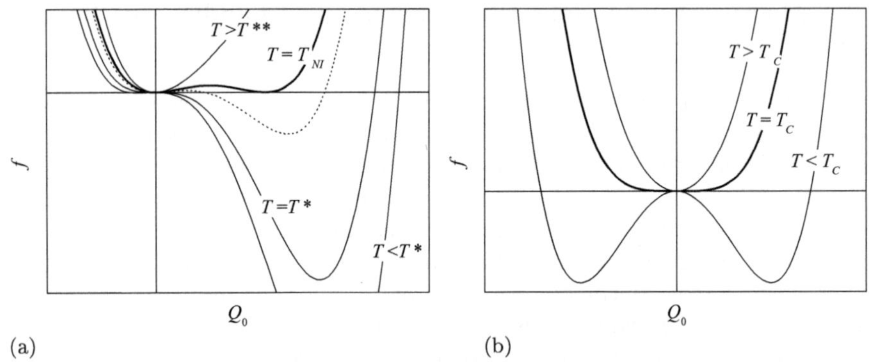

Fig. 8.2. (a) A sketch of the Landau free energy of a typical liquid crystalline material exhibiting a nematic–isotropic transition. T_{NI} is the transition temperature and T^* and T^{**} are the temperatures below/above which the corresponding metastable phase can not exist anymore. The dotted line coresponds to the temperature in the nematic phase where the isotopic phase is metastable, thus, a local minimum of the free energy at $Q_0 = 0$. As a comparison, the free energy of a system with a continuous phase transition is depicted in (b). Here the transition occurs at T_c and no metastable solutions are possible.

nematic phase in which molecules orient along the director whereas $\mathrm{Q}_0 < 0$ is associated with a nematic phase in which molecules orient in the plane perpendicular to the director. Coefficients a, B, and C are determined from the best fit to experimental results and are in general temperature dependent. Coefficient $C > 0$ is needed for f to have a local minimum for $\mathrm{Q}_0 < \infty$ whereas the parameter a must change sign at the temperature below which the solution $\mathrm{Q}_0 = 0$ is not possible anymore so that the solution $\mathrm{Q}_0 \neq 0$ becomes stable. In the expansion $a = A(T - T^*) + \ldots$ only the first term is taken into account. The transition is discontinuous. Usually, parameters B and C can be regarded as being constants. By definition, T^* is the temperature below which the disordered phase can not exist; it is refered to as *supercooling temperature.* The transition temperature is shifted from T^*,

$$T_c = T^* + \frac{B^2}{27AC}. \tag{8.10}$$

Therefore, in the interval $T^* < T < T_c$ the disordered phase is metastable whereas it is stable above T_c. The ordered phase is stable below T_c and it is metastable in the temperature interval $T_c < T < T^{**} = T^* + B^2/24AC$. Here, T^{**} is the so-called *superheating temperature.* Typically, the isotropic phase can be undercooled by ~ 1 K and the nematic phase can be overheated by ~ 0.1 K. A schematic plot of the free energy in the vicinity of the nematic–isotropic transition is depicted in Fig. 8.2.

Although the Landau description was established for continuous phase transitions it is also used for describing discontinuous transitions, though special care should be taken because of a jump in the order parameter. In the case of LCs, the phase transitions are only weakly discontinuous and the description within the Landau theory is useful.

Due to the effect of external fields, the order can vary in space and gradient terms have to be added to the Landau expansion (8.9). Usually, only the terms up to the quadratic order are considered. There are many symmetry allowed invariants related to gradients of the tensorial order parameter [29]. However, in the vicinity of the phase transition, one is not interested in elastic deformations of the nematic director but rather in spatial variations of the degree of nematic order. Therefore, the pretransitional nematic system is described adequately within the usual one-elastic-constant approximation,

$$f_{el} = \frac{1}{2} L \, \nabla \mathsf{Q} \vdots \nabla \mathsf{Q}, \tag{8.11}$$

where L is typically in the order of 10^{-11} N. Deep in the nematic phase, variations of the scalar order parameter do not play a significant role and the main contribution to the elasticity of the system is due to elastic deformations of the director field which contribute to the free energy density [30]

$$f_{el}^{Frank} = \frac{1}{2} \left\{ K_{11}(\nabla \cdot \hat{n})^2 + K_{22}[\hat{n} \cdot (\nabla \times \hat{n})]^2 + K_{33}[\hat{n} \times (\nabla \times \hat{n})]^2 \right\}, \tag{8.12}$$

corresponding to three basic deformations of the director field: splay, twist, and bend, respectively. The stiffness of the nematic with respect to a given deformation is determined by the parameters K_{ij}; typical magnitudes being 10 pN [31]. The one-elastic-constant approximation used in (8.11) corresponds to the case $K_{ii} = K$. In the uniaxial nematic LC, parameters L and K relate through $K = 9LS^2/2$, where S is the scalar order parameter.

Within the mean-field theory there are no spontaneous elastic deformations since any deformation increases the free energy. However, when a nematic LC is subject to interactions with the confining walls the homogeneous order can be perturbed. On the microscopic level, the molecules of the walls and of the liquid crystal attract each other via a short-range van der Waals interaction. In the macroscopic description this is modeled with a contact quadruple–quadruple interaction, known as the Rapini–Papoular model [32, 33], which to the lowest order reads

$$f_S = \frac{1}{2} w_2 \, \mathrm{tr} \, (\mathsf{Q} - \mathsf{Q}_{S_i})^2 \delta(z - z_S). \tag{8.13}$$

Here w_2 is the strength of the interaction and Q_S is the preferred value of the tensor order parameter at the substrate, located at $z = z_S$. In the case of uniaxial nematic order the anchoring strength w_2 can be related to the anchoring strength W for the bare director description as $W = 3w_2S^2$.

(Within the bare director description $f_S = \frac{1}{2}W \sin^2\alpha$; $\alpha = \angle(\hat{n}, \hat{k})$ and $\hat{k}$ is the easy axis of the confining wall.) Often, the anchoring strength is measured in terms of the extrapolation length which denotes the length over which the director field would relax to the one preferred by the substrate. It is a measure for the relevance of the competing elastic distortions vs. violating the substrate induced order, $\lambda = K/W$. The criterion whether the anchoring is strong or weak, is the ratio between the extrapolation length and typical dimension of the system d; $\lambda/d \to 0$ corresponds to strong anchoring whereas $\lambda/d \gg 1$ is associated with weak anchoring. We should note here, that the lowest order term alone does not reproduce the temperature dependence of the extrapolation length observed in some experiments [34].

It is useful to rewrite the quantities into a dimensionless form. Usually, coordinates are measured in terms of the film thickness d (or another typical dimension in the case of a non-planar geometry) and the correlation length $\xi_{NI} = \zeta d = \sqrt{27CL/B^2} \approx 10$ nm. The order parameter is rescaled in units of the scalar order parameter of the nematic phase at the phase transition temperature, $S_c = 2B/3\sqrt{6}C \sim 0.2 - 0.6$, and the temperature is controlled by $\theta = (T - T^*)/(T_{NI} - T^*)$; the reduced temperatures $\theta = 1$, 0, and 9/8 correspond to the bulk phase transition temperature, and to the supercooling and superheating limits, respectively.

8.2.3 Fluctuations of the Order Parameter

Within the mean-field theory the order is a macroscopic quantity and depends only on external parameters such as temperature, pressure, and external fields. However, due to finite temperatures, the order is characterized by spatial and temporal deviations from the average value, caused by *thermal fluctuations*. What is important when considering fluctuations is their correlation length which determines how big are the islands characterized by different order. When the correlation length of the fluctuations is small compared to the typical dimensions of the system, the fluctuations do not mask its average behavior. On the other hand, highly correlated fluctuations change the order on large scales and perturb the macroscopic appearance. Let us write the order parameter as a sum of the mean-field order parameter and a small part corresponding to fluctuations,

$$\mathsf{Q}(\boldsymbol{r}, t) = \mathsf{Q}_0 + \delta\mathsf{Q}(\boldsymbol{r}, t), \tag{8.14}$$

where $\langle \mathsf{Q}(\boldsymbol{r}, t)\rangle = \mathsf{Q}_0$ and $\delta\mathsf{Q}(\boldsymbol{r}, t) = \sum_m b_m(\boldsymbol{r}, t)\mathsf{T}_m$. With this, the spatial correlation function of the fluctuations, $\Gamma(\boldsymbol{r}) = \langle \mathsf{Q}(\boldsymbol{r})\mathsf{Q}(0)\rangle - \langle \mathsf{Q}(0)\rangle^2$, reads

$$\Gamma(\boldsymbol{r}) = \langle \delta\mathsf{Q}(\boldsymbol{r})\delta\mathsf{Q}(0)\rangle. \tag{8.15}$$

In the absence of external fields, the system is translationally invariant and the functions of $\boldsymbol{r}$ can be expanded in a Fourier series, $f(\boldsymbol{r}) = \sum_{\boldsymbol{q}} \tilde{f}(\boldsymbol{q})\, \mathrm{e}^{-i\boldsymbol{q}\cdot\boldsymbol{r}}$, so that $\tilde{\Gamma}(\boldsymbol{q}) = \langle|\delta\tilde{\mathrm{Q}}(\boldsymbol{q})|^2\rangle$. The amplitudes $\delta\tilde{\mathrm{Q}}(\boldsymbol{q})$ are derived from the free energy of the system which can be, by applying the Fourier series expansion, rewritten in a form $\mathcal{F} = \mathcal{F}(\mathrm{Q}_0) + \sum_m \mathcal{H}_m$, where $\mathcal{H}_m$ is a Hamiltonian associated with a given type of fluctuations,

$$\mathcal{H}_m = \frac{LV}{2} \sum_{\boldsymbol{q}} \left(\xi_m^{-2} + q^2\right) |b_{\boldsymbol{q}}|^2. \tag{8.16}$$

Here, q is a fluctuation wavevector and ξ_m is the correlation length of a given fluctuation mode. Since the fluctuations are assumed to be small, the free energy is written out only up to the quadratic terms in fluctuations. Considering the equipartition theorem and (8.16), the Fourier transform of a correlation function reads $\tilde{\Gamma}_m(\boldsymbol{q}) \propto k_BT/(\xi_m^2 + q^2)$, or in direct space

$$\Gamma_m(\boldsymbol{r}) \propto \frac{k_BT}{4\pi r}\,\mathrm{e}^{-r/\xi_m}. \tag{8.17}$$

For finite ξ_m, correlations are weak and decrease exponentially whereas for infinite correlation length the correlations are long-range and decrease inversely with distance.

There are two specific types of fluctuation modes: Goldstone and soft modes. When the correlation length of fluctuations diverges, the free energy associated with a long wavelength mode ($\boldsymbol{q} \to 0$) is very small. *Soft modes* are modes whose relaxation rate drops to zero at the transition whereas it is nonzero otherwise. The $\boldsymbol{q} = 0$ fluctuation mode of the order parameter characteristic for a given continuous transition belongs to the category of soft modes. On the other hand, the relaxation rate of a *Goldstone mode* drops to zero at the transition and stays critical within the entire range of a given ordered phase. At the phase transition from the disordered to the ordered phase, the symmetry of the system is typically lowered. This spontaneous breaking of the symmetry is accompanied by a multiple degeneration of the ground state. The system is brought from one ground state to the other by symmetry operations of the high-symmetry phase. If the broken symmetry is continuous, a fluctuation mode occurs whose deformation of the system represents a continuous change from one ground state to another and so on. In the thermodynamic limit with $\boldsymbol{q} \to 0$, this excitation does not increase the free energy of the system. The fluctuation mode associated with such deformation is called a Goldstone mode [35]. A sketch of a typical spectrum of a system with a Goldstone and soft mode is plotted in Fig. 8.3b.

Correlation Lengths of the Nematic Order Parameter

In bulk nematic liquid crystals, the average equilibrium order is uniaxial with the order parameter $\mathrm{Q}_0 = a_0\mathrm{T}_0$ and $a_0 = S$. Thus, the correlation lengths of the 5 independent degrees of freedom of the nematic order read

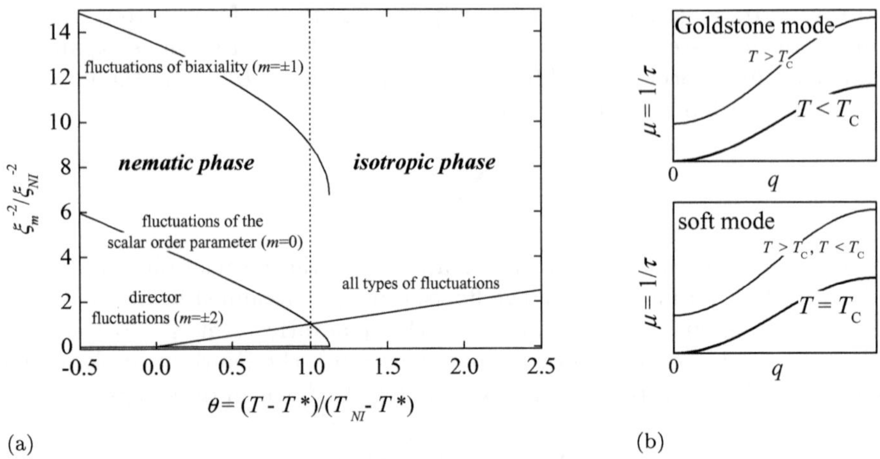

Fig. 8.3. (a) Temperature dependence of correlation lengths of 5 degrees of freedom of the nematic order in the isotropic and nematic phase. Continuations of the lines across the dotted vertical correspond to the correlation lengths in the appropriate metastable phase. Correlation lengths determine the relaxation rates of fluctuations as $\mu_i = 1/\tau_i \propto \xi_i^{-2} + q^2$, where q is the corresponding wavevector. (b) Sketch of a typical relaxation spectra for a system with a Goldstone and soft mode, respectively.

$$\begin{aligned} \xi_0^{-2}/\xi_{NI}^{-2} &= \theta - 6a_0 + 6a_0^2, \\ \xi_{\pm1}^{-2}/\xi_{NI}^{-2} &= \theta + 6a_0 + 2a_0^2, \\ \xi_{\pm2}^{-2}/\xi_{NI}^{-2} &= \theta + 3a_0 + 2a_0^2. \end{aligned} \tag{8.18}$$

Due to the uniaxial symmetry of the nematic phase the two types of fluctuations of biaxiality are degenerated and so are the two types of director fluctuations.

In the isotropic phase, there is no order and all directions are equivalent, therefore, all types of fluctuations must have the same correlation length,

$$\xi_I^{-2}/\xi_{NI}^{-2} = \theta. \tag{8.19}$$

Because the isotropic–nematic transition is discontinuous, the correlation length does not diverge at the transition, however, it diverges at the supercooling limit of the isotropic phase. In the nematic phase $a_0 = 3/4\,(1 + \sqrt{1 - 8\theta/9})$ and

$$\begin{aligned} \xi_{N,0}^{-2}/\xi_{NI}^{-2} &= \frac{9}{4}\sqrt{1 - 8\theta/9}\left(1 + \sqrt{1 - 8\theta/9}\right), \\ \xi_{N,\pm1}^{-2}/\xi_{NI}^{-2} &= \frac{27}{4}\left(1 + \sqrt{1 - 8\theta/9}\right), \\ \xi_{N,\pm2}^{-2}/\xi_{NI}^{-2} &= 0. \end{aligned} \tag{8.20}$$

The correlation length of the director fluctuations is infinite in the whole range of the stable nematic phase and the director excitation with the infinite wavelength is the Goldstone mode. Fluctuations of other degrees of freedom of the nematic order are much more energy consuming. The correlation length of fluctuations of the scalar order parameter diverges at the superheating temperature $\theta^{**} = 9/8$. The hardest type of fluctuations in the uniaxial nematic phase are biaxiality fluctuations since they oppose the established symmetry of the phase. The temperature dependence of correlation lengths of different types of fluctuations are depicted in Fig. 8.3 (a).

The temporal dependence of the fluctuations can be described by the overdamped equation of motion — the Landau–Khalatnikov equation [36, 37]. It can be understood as follows: The equilibrium configuration of the order parameter is determined by the minimum of the free energy, $\delta\mathcal{F} = \int dV \delta f(\eta, \nabla\eta) = \int dV (\delta f/\delta\eta)\delta\eta$, which is satisfied for $\delta f/\delta\eta = 0$. Here, $\delta/\delta\eta = \partial/\partial\eta - \nabla\cdot(\partial/\partial\nabla\eta)$. If the system is out of equilibrium, $\delta f/\delta\eta$ acts like a generalized elastic force which is balanced by viscous forces. When the macroscopic velocity can be neglected the viscosity is related solely to the rate of change of the order parameter $\dot{\mathsf{Q}}$,

$$\frac{\partial \mathsf{Q}}{\partial t} = -\gamma \frac{\delta f}{\delta \mathsf{Q}}, \tag{8.21}$$

where γ^{-1} is a generalized viscosity associated with the relaxation of the order parameter. For the nematic liquid crystal with a tensorial order parameter the generalized viscosity is in principle a tensor. In the ordered phase, one can expect that due to the anisotropy of the order the viscosities differ with respect to directions along different eigenvectors. In nonhomogeneously ordered nematics the generalized viscosity can be also spatially dependent. However, in order to simplify the description and since the anisotropies are usually small, the generalized viscosity is assumed to be isotropic with the effective value equal to the average viscosity, $\underline{\gamma}^{-1} = \gamma^{-1}\mathsf{I}$. The time dependence of the fluctuation modes which are described by a relaxation equation is the exponential decay, $b_i \propto e^{-\mu_i t}$, where the relaxation rate μ_i of a given fluctuation mode is the eigen-value of the operator $\hat{\mu}_i = \zeta^2(\xi_i^{-2} + \nabla^2)$.

To complete this Section we have to mention the hydrodynamics of a nematic liquid crystal. In the equations described above the macroscopic flow was neglected. Such an approximation is usually justified since we describe only systems with zero total momenum. However, one should bear in mind that due to the coupling between the hydrodynamic and orientational degrees of freedom there are local flows accompanying the orientational changes, the effect known as *backflow* [38,39]. In some cases the backflow effects can alter substantially the phenomena in question.

8.3 Confined Nematogenic Systems

In bulk the order of the molecules of a nematic liquid crystal depends only on temperature and external electric or magnetic fields. More physically and technologically interesting are however confined liquid crystals which fill the pores in porous media, holes in a polymer matrix or the space between the two confining substrates in the slab geometry of LC displays. In confined LC systems, the order of the molecules does not only depend on the interaction between the constituting molecules but also (mainly) on the lack of neighboring molecules (interactions) near the surface. The interaction with the confining substrate can result in higher or lower order than expected in the bulk, in a spatial dependence and deformation of the order, and in partial positional order. With respect to the extent of the deformation caused by the confinement, the systems fall in one of two categories: (i) systems with the deformation localized to the vicinity of the confining walls (*wetting effects*) and (ii) systems in which the *frustration* caused by competing boundary conditions is reflected in deformations spread all over the system. Furthermore, the confinement affects also the pretransitional dynamics, both through changing the equilibrium order and through setting the boundary conditions for the fluctuations themselves.

8.3.1 Heterophase Ordering: Wetting Effects

Heterophase nematogenic systems are relatively simple systems of confined liquid crystals in which there is no competition between antagonistic fields inducing nematic order in different directions, such as the surface, magnetic, etc., fields, but they are only subject to surface-induced nematic order. Especially interesting is the case, in which the surface potential is such as to induce a sufficiently large orientational ordering as compared with the bulk isotropic phase — a *paranematic system* [17, 40]. Then, a mesoscopic layer of the nematic phase intervenes at the substrate–isotropic phase interface as the isotropic–nematic transition is approached from above. The described situation is known as (orientational) *wetting*; the wetting being either partial or complete. In the case of complete wetting, the thickness of the surface-induced ordered layer diverges at the isotropic–nematic phase transition which, thus, becomes continuous. In the case of partial wetting, the thickness of the surface-induced layer is saturated before the isotropic–nematic phase transition occurs and the transition remains discontinuous. By changing the aligning power of the substrate, i.e., changing the anchoring strength and/or changing the value of the induced nematic order, the complete wetting can change to partial (or vice versa) between which the *wetting transition* occurs. The wetting behavior can be quite complex, however, complete wetting is generally related to substrates with large ordering power, i.e., inducing the nematic order with $S_s > S_c$ and with a high anchoring energy. The surface interaction may also have a disordering effect if, for example, the inner surface of the host material is rough [9, 41]. In this case a reduction of the degree of order in the boundary layer is expected below the phase transition temperature, and the substrate induces wetting by the isotropic phase. The complete wetting regime can only be obtained in the limit of $S_s = 0$. Both wetting situations are represented in Fig. 8.4.

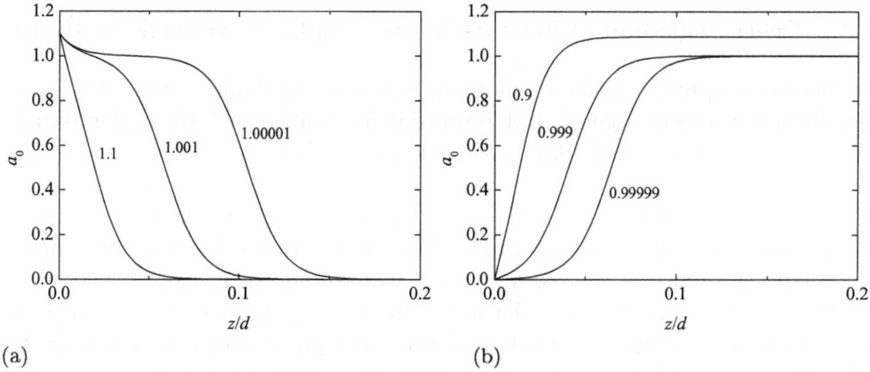

Fig. 8.4. Portrait of profiles of the scalar order parameter for various values of the reduced temperature $\theta = (T - T^*)/(T_{NI} - T^*)$ in the vicinity of the isotropic–nematic transition for (a) ordering and (b) disordering substrates. The substrate induced order is fixed to $1.1 S_c$ and 0, respectively.

In confined systems in which walls stimulate the nematic order, the transition from the nematic to paranematic phase takes place above T_{NI}. Conversely, in the case of disordering walls, the transition to the high-temperature isotropic phase is shifted below T_{NI}. The phenomenon of a formation of a nematic bridge accross the NLC system above T_{NI} is known as capillary condensation [14,40,62]. It is due to the high cost of the order parameter gradient terms in the free energy. The magnitude of the shift depends on the size of the sample and on the parameters of the surface interaction, and is practically negligible in micron-size cavities. In highly constrained surface-aligned (or surface molten) nematic systems, the transition between the isotropic and nematic phase can be lost, whereas the nematic (isotropic) order grows gradually on lowering (increasing) the temperature. This occurs after the first-order-transition line of capillary phase transition ends in the critical point (d_C, T_C).

After the interest in wetting transitions was initiated by the study of equilibrium order by Sheng in 1976, in the last few years a renewed interest was focused on the structural force among the confining substrates in a study by Borštnik and Žumer [43] and on pretransitional collective dynamics in studies by Ziherl et al. [9, 17]. Theoretical studies were accompanied by a series of experimental studies by means of NMR, [44], optical ellipsometry [41, 42, 45], and recently with the force microscopy experiments [46, 47, 62]. Similar effects as in the case of surface-aligned nematic order are observed also for the case of smectic order — wetting by a smectic phase — which is associated with both the vicinity of the transition to the smectic A phase and with the fact that the presence of a wall breaks the continuous translational symmetry resulting in a smectic-like order close to the confining substrate.

8.3.2 Pretransitional Dynamics in Heterophase Nematic Systems

As already emphasized, the phenomenological mean-field Landau–de Gennes description neglects spatial and temporal fluctuations of the order parameters These are, in particular in the vicinity of the phase transition highly pronounced and they can profoundly change the (pre)transitional behavior. Thus as a first estimate, one can study the relaxation dynamics of harmonic deviations from the mean-field order. This gives a better insight in the mechanisms of the phase change and in the nature of the transition.

Because the equilibrium order in heterophase systems is characterized by only one nonzero degree of freedom of the order parameter tensor, the fluctuation modes of all five degrees of freedom are uncoupled. Due to the uniaxial symmetry of the phase the two biaxial modes are degenerate and so are the two director modes. If a nematic layer is bounded by walls characterized by a strong surface interaction and a bulk-like value of the preferred degree of order, the fluctuation modes β_i's are sine waves, and their relaxation rates may be cast into

$$\lambda_i = \mu_i - \xi_{NI}^2(q_x^2 + q_y^2) = \xi_{N,i}^{-2}/\xi_{NI}^{-2} + \zeta^2 \left[(n+1)\,\pi\right]^2 . \tag{8.22}$$

This is the same as in bulk except that, due to the finite dimension in the z direction, the wavevector q_z can only take discrete values, $q_{z,n} = \zeta(n+1)\pi$, where n is the number of nodes of the sine function between the two substrates. q_x and q_y are the in-plane components of the wavevector.

In general, the surface-induced degree of order differs form the bulk value and the profile of the degree of order is inhomogeneous. Thus, the generalized correlation lengths of the fluctuation modes are spatially dependent and the fluctuation eigenmodes in the two wetting geometries can only be determined numerically.

Fluctuations of the Degree of Order

The primary effect of wetting is related to the existence of a slow mode characterized by a soft dispersion of its relaxation rate, whereas the upper part of the spectrum remains more or less the same as in a homophase system (see insets of Fig. 8.5). The elementary mode of fluctuations of the degree of order is localized at the phase boundary between the wetting layer and the bulk phase and it corresponds to fluctuations of the thickness of the central part of the slab. The next mode, which is also localized at the nematic-isotropic interface, represents fluctuations of the position of the core. The relaxation rates of these two modes are the same as long as the two wetting layers are effectively uncoupled.

In the complete wetting regime, the relaxation rate of the elementary excitations of the degree of order exhibits a linear critical temperature dependence typical for soft modes. The slowdown of the relaxation rates of the

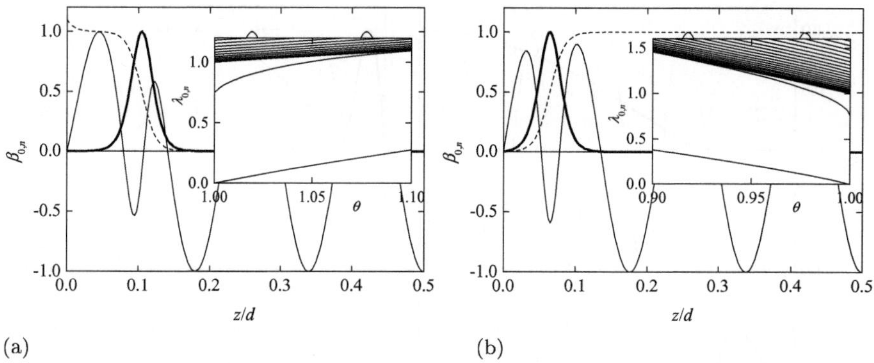

Fig. 8.5. Profile of the lowest order parameter mode (thick line) and of one of the highly excited modes for a LC heterophase system in contact with (a) ordering and (b) disordering substrates in the proximity of the phase transition. Dashed line denotes the spatial variation of the mean-field scalar order parameter. In all cases $T \to T_{NI}$. Inset: The corresponding spectrum of the relaxation rates characterized by a soft lowest order parameter mode.

surface-induced soft modes at the phase transition temperature is a clear signature of the continuity of the transition. In a finite system, however, a wetting-driven phase transition can never be truly continuous, but in samples of thickness > 100 nm this effect is detectable only if the temperature resolution of the experimental method is better than ~ 0.01 K. The behaviour of the wetting-induced elementary mode does depend strongly on the magnitude of the anchoring strength: if the wetting is partial instead of complete, the localized modes' relaxation rates do not drop to zero, but remain finite at T_{NI}. The transition stays discontinuous, however, the corresponding latent heat may be reduced considerably compared to the bulk isotropic–nematic transition.

Biaxial Fluctuations

Biaxial modes are the hardest type of fluctuations in a uniaxial nematic phase. At the phase transition temperature, the lower limit of the relaxation rates of the biaxial fluctuations in the nematic phase is 9 times larger than in the isotropic phase. This considerable difference in the energy levels of biaxial modes in the two phases is reflected in their spectra in the two wetting geometries. In the case of a nematic phase confined by a disordering wall, the lowest modes are bounded to the isotropic wetting layer (see Fig. 8.6 top) and even the high modes differ from the sine-like bulk modes. In a paranematic phase induced by an ordering substrate, biaxial fluctuations are, conversely, expelled from the ordered boundary layer, so that the allowed wavelengths

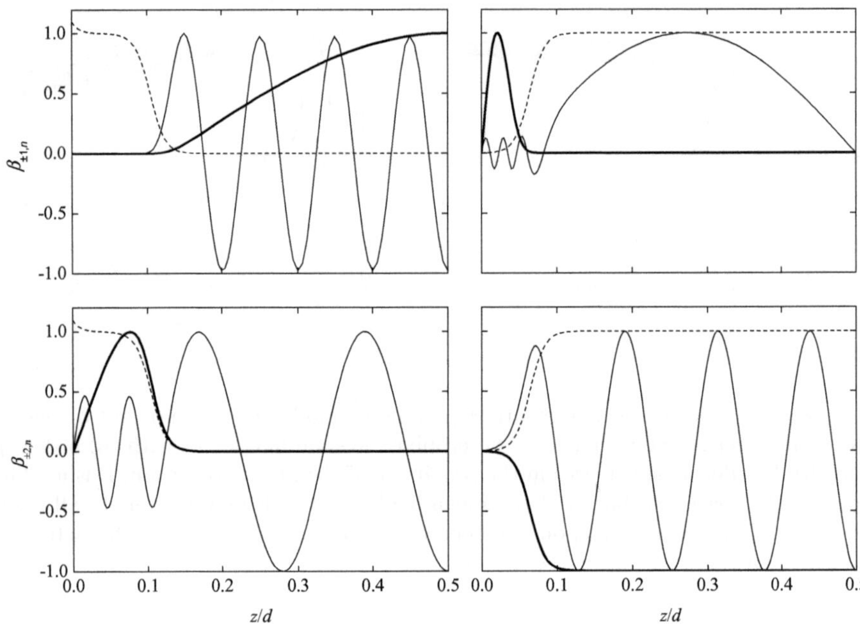

(a) (b)

Fig. 8.6. Profiles of the lowest (thick line) and of one of the highly excited modes of biaxial ($\beta_{\pm1,n}$) and director ($\beta_{\pm2,n}$) fluctuations in the LC heterophase system in contact with (a) ordering and (b) disordering substrates. Dashed lines correspond to the mean-field scalar order parameter. In all cases $T \to T_{NI}$.

of the normal modes are determined by the thickness of the central isotropic part and not by the actual thickness of the sample.

Director Fluctuations

Director modes are, as opposed to biaxial fluctuations, excited very easily in the nematic phase, where their Hamiltonian is purely elastic, whereas in the isotropic phase they are characterized by a finite correlation length. This implies that their wetting-induced behavior should be quite the inverse of that of biaxial modes. Thus, in the disordering geometry, the director modes are forced out of the substrate-induced isotropic boundary layer into the nematic core (see Fig. 8.6 bottom). The lowest mode is a Goldstone mode. In the paranematic phase a few lowest director modes are confined to the nematic boundary layer, whereas the upper ones extend over the whole sample and are more or less the same as in the perfectly isotropic phase.

8.3.3 Ordering in a Frustrated Nematic: Hybrid Nematic Cell

Here we present a simple illustration of the equilibrium ordering and pretransitional dynamics in highly constrained systems when nondirector degrees of freedom are particularly relevant. Although the origin of the high frustration in systems can be different – (i) the nematic liquid crystal is confined with substrates that have been prepared in a way to induce homeotropic nematic order at one surface and the other in a lateral direction; (ii) the nematic liquid crystal is deposited on a solid substrate inducing in-plane ordering whereas on the other sides it has a free LC–air interface which usually induces rather strong homeotropic anchoring; (iii) the frustrated situation is a result of geometrical constraints such as in a cylindrical cavity with homeotropic anchoring at the wall whereas the geometry of the cavity prefers orientation along the symmetry axis (see Fig. 8.7) – its effects on the LC order and pretransitional dynamics are similar [10,48]. Therefore one can study the basic effects of high frustration within the analysis of a simple planar system confined by two parallel substrates inducing uniaxial nematic order in mutually perpendicular directions; one in a particular direction in the plane of the confining substrate (say parallel to the x axis) and the other in a homeotropic (parallel to the z axis). The thickness of the nematic film is much smaller than the lateral dimensions of the system. Such a system is then called a hybrid nematic cell.

A number of both, theoretical and experimental studies of hybrid nematic systems, have been stimulated by their possible technological applications [49] and by the physical phenomena related to the frustration. The thickness dependence of the nematic-isotropic phase transition temperature and the stability of ordered structures in a hybrid nematic film were studied both theoretically [50] and experimentally using a quasi-elastic scattering method

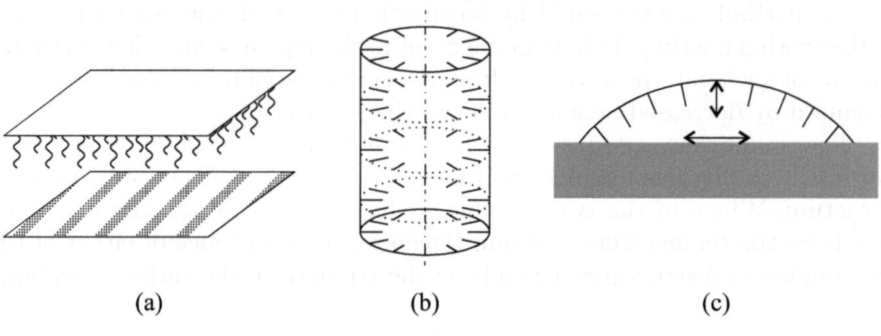

Fig. 8.7. Sketch of different systems characterized by hybrid frustration: (a) Confining substrates repared in a way, so that one induces homeotropic anchoring and the other homogeneous planar anchoring. (b) Hybrid frustration due to opposing geometry induced direction of order and the direction induced by anchoring. (c) Liquid crystal in a contact with a solid substrate and with a free LC–air interface.

[51]. Further studies have tackled the director dynamics [8] and the geometrically caused frustration [52,53]. Theoretical studies of the highly frustrated sub-micron hybrid systems where the order becomes biaxial [10, 54, 55] have recently been demonstrated also by MC simulations [16] but experimental evidence is still lacking.

Usually, the elastic distortions in such films are studied within the Frank elastic theory, where the nematic order is assumed to be uniaxial with the director field continuously bent from one substrate to the other. The bent-director configuration can only exist if the film is thicker than the critical thickness, $d_c \equiv K|1/W_H - 1/W_P|$ [50]. In thinner films the director field is uniform with the nematic director in the direction of the easy axis of the substrate with stronger anchoring. In hybrid films with equal anchoring strenghts the order should always be distorted. However, the director field is uniform below a finite critical film thickness, whereas the boundary conditions are fulfilled with the eigenvalue or director exchange [10, 54]. The other interesting consequence of high frustration is a geometry induced biaxial ordering of a uniaxial nematic liquid crystal.

To understand the complex behavior of the nematic order in the hybrid nematic cell let us first discuss its free energy. The bulk-like terms (8.9) control the melting and growing of the nematic order. The elastic part of the free energy (8.11) corresponds either to variations of the uniaxial and biaxial parameters of the order around the director or to elastic deformations of the director field. The larger the nematic order the bigger the resistance of the system against elastic deformations of the director field; the elastic energy related to deformations of the director scales as S^2. The third contribution to the free energy of a confined nematic LC system is the free energy (8.13) associated with (dis)obeying the substrate-induced order. The increase of the free energy due to the confinement induced deformation of the director field can be partially compensated by localization of the elastic deformation together with a melting of the nematic order in the region of high deformations. On the other hand, the deviation from the surface-preferred order can be also balanced by decreased nematic order at the surface.

The equilibrium ordering of a hybrid nematic film can exhibit either a distorted (hybridly bent) or undistorted (hybrid biaxial and uniform) director structure. Which of the two possible configurations will actually occur depends on the temperature and film thickness. The existence of either of the two undistorted structures depends on the strength of the surface coupling.

Bent-director Structure

In the bent-director structure the biaxiality can be neglected and the scalar order parameter corresponds to its bulk value at the temperature $\theta_{\text{eff}} = \theta + (3\pi^2/4)\zeta^2$; due to elastic deformation of the director field the temperature of the hybrid system effectively increases [10]. The increase of the effective temperature results in a smaller degree of order along the director with respect to the degree of order in the bulk where there are no elastic deformations. However, the difference is negligible in micron-size cells; for a typical liquid crystal $T_{\mathrm{eff}} - T \approx 0.5$ mK. Since the scalar order parameter varies with the distance from one of the substrates, the director tilt angle is not changing linearly as it would in the case of the uniform scalar order parameter. However, the difference is very small and, as expected, decreases further with increasing film thickness and when the boundary value of the scalar order parameter is getting closer to the value $S_{\mathrm{b}} = S_{\mathrm{bulk}}(\theta_{\mathrm{eff}})$.

Biaxial Configuration

In a hybrid film with one easy axis of the confining substrates in the direction of the x axis and the other parallel to the z axis, the director lies in the plane (x, z). The configuration can be described by two amplitudes, let us choose Q_{xx} and Q_{zz}, which refer to the scalar order parameter with respect to the director $\hat{n} = \hat{e}_x$ and $\hat{n} = \hat{e}_z$, respectively.

As shown in Fig. 8.8 for the case with equaly strong but orthogonal anchorings, on the average, near the first surface the liquid-crystal molecules are oriented parallel to the x axis while they are parallel to the z axis close to the other substrate. In the vicinity of the surfaces the order is uniaxial, however, with increasing distance from the substrates it becomes slightly biaxial. In the case of equal anchoring strengths, the order parameter profiles are symmetric with respect to the middle of the film, the *exchange region*,

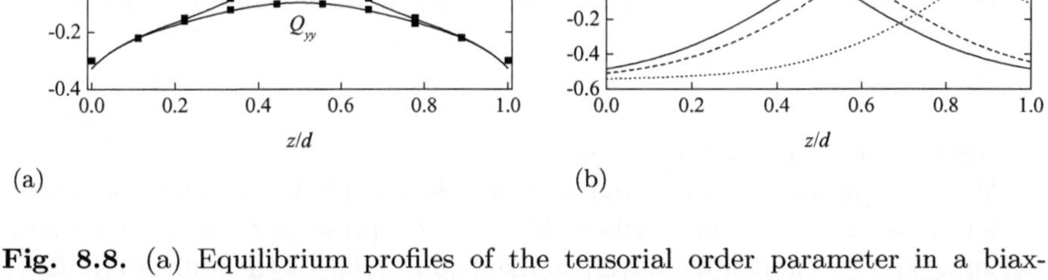

Fig. 8.8. (a) Equilibrium profiles of the tensorial order parameter in a biaxial structure. Symbols denote the corresponding results from the MC simulations. (b) Equilibrium profiles for different strengths of the homeotropic anchoring, $G = 1.2 \times 10^{-3}$ J/m^2 (solid line), 1.1×10^{-3} J/m^2 (dashed line), 0.6×10^{-3} J/m^2 (dotted line); strength of the planar anchoring is in all cases 1.2×10^{-3} J/m^2. Consequently the exchange region moves toward the homeotropic substrate. In the limiting case, the director field is uniform and parallel to the x axis.

whereas for the unequal anchoring strengths this region moves towards the substrate with the weaker anchoring [see Fig. 8.8b]. Between the maxima of the biaxiality profile the molecular ordering is characterized by a director $\hat{n} = \hat{e}_y$ and a negative scalar order parameter.

The biaxially ordered configuration would occur in highly constrained nematic liquid crystals, i.e., systems with a high surface-to-volume ratio and strong surface anchoring ($G > 10^{-3}$ J/m^2). In such systems the surface wetting layers may be in contact with each other, thus, the structure they form becomes progressively ordered on approaching the phase transition temperature. Because of the continuous growth of the ordered biaxial structure there is no nematic-isotropic phase transition. However, there is a structural transition to the low-temperature bent-director field configuration.

Structural Transition Between Bent-director Structure and Biaxial Structure

Near the bulk isotropic–nematic phase transition temperature, $T_{NI} - T = 0.1$ K, and in a hybrid film of a typical LC material (such as 5CB) the nematic order is distorted if the film is thicker than $d_t \approx 47$ nm, and it is biaxial otherwise. The metastable biaxial structure ceases to exist if the film thickness is larger than $d_s \approx 71$ nm. As the temperature is decreased both values decrease and so does the difference between them. The same structural transition can be realized if the film thickness is held constant and the temperature is varied. The (dis)continuity of the structural transition can be changed if the temperature and film thickness are low enough [10, 54, 55]. The two different regimes of the transition are separated with a tricritical point below which the transition is continuous. Our estimation for its upper limit is $T_{NI} - T_{\mathrm{TP}} = 0.28$ K and $d_{\mathrm{TP}} = 34$ nm for 5CB [10].

8.3.4 Monte Carlo Simulations of a Hybrid Cell

Experimental evidence of the biaxial structure is rather limited since its existence is very delicately tuned by the anchoring properties of the confining substrates and since it is only realized in a very narrow temperature interval. Here we present results obtained by computer simulations which mimic the microscopic behavior of the system.

We have performed our simulation within the LL lattice spin model, on lattices of $30 \times 30 \times h'$ spins, where $h' = h + 2$ represents h layers of nematic LC spins and two additional layers of fixed spins [16]. At the four lateral faces of the simulation sample we have employed periodic boundary conditions to mimic the bulk-like conditions. The standard Metropolis procedure [56] has been used to update the lattice. The state of a system was monitored by the tensorial nematic order parameter calculated with respect to the fixed frame spanned by the orthonormal triad $(\hat{e}_x, \hat{e}_y, \hat{e}_z)$; $\mathsf{Q} = \langle (3\hat{u}_i \otimes \hat{u}_i - \mathsf{I})/2 \rangle$.

The diagonal components of the tensor Q represent the degree of order with respect to the axes x, y, and z. The off-diagonal components, $Q_{\alpha\beta}$, represent the bending of the director field in the plane (α, β). The parameter coupling the two preferred directions x and z,

$$Q_{xz} = \frac{3}{2}\langle \sin 2\theta_i \cos \phi_i \rangle, \tag{8.23}$$

was used to distinguish between the biaxial and bent-director structure. In the case of a non-zero Q_{xz} the tensor order parameter was diagonalized. A typical plot of $Q_{\alpha\beta}$ in the biaxial structure is depicted in Fig. 8.8a.

Deep in the nematic phase the director's tilt angle uniformly decreases from 90° at the first layer of fixed spins to 0° at the other. The structure corresponds to a continuously changing director field in the bent-director structure. At temperatures above a certain temperature $T_C^{MC} < T_{NI}^{MC}$, the director's tilt profile becomes a step-like function of the position in the cell. In each half of the cell the director lies along the easy direction of the closest layer of fixed spins. The tensor order parameter is diagonal within the fixed coordinate frame. A plot of director's tilt angle (with respect to the z axis) as a function of temperature is shown in Fig. 8.9 (a).

Different properties of both structures can be most easily monitored in the middle of the cell. Eigen-values of the tensor order parameter and the director's tilt angle in the two layers closest to the middle of the cell are depicted in the inset of Fig. 8.9. At the point where the step-like profile changes to the continuous regime the slopes of the growth of the scalar order parameters change, indicating that the regions of the two structures are separated

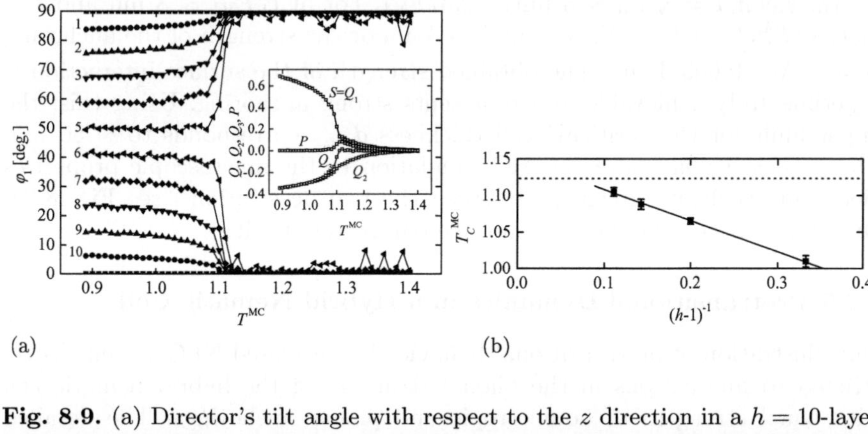

Fig. 8.9. (a) Director's tilt angle with respect to the z direction in a $h = 10$-layer hybrid nematic cell as a function of MC temperature. Different lines correspond to the tilt angle in distinct layers; 1 denotes the LC layer closest to the in-plane aligning substrate and 10 the layer nearest to the homeotropic substrate. Inset: Temperature variation of ordering parameters in the two layers closest to the middle of the cell. (b) Temperature of the structural transition as a function of a cell thickness.

by a structural transition. The same can be observed also in other layers, however, less pronounced. Another indication of the structural transition between the two ordered states is given by the inspection of the energy of the system and the corresponding heat capacity which shows a well defined peak at the temperature of the change of the director's profile [16].

The temperature of the structural transition changes with the cell thickness, see Fig. 8.9b. The thicker the cell the closer is the transition temperature to the bulk transition temperature. In thin enough cells there is a finite structural transition temperature shift whose origin is two-fold: (i) the wetting properties of substrates which induce high nematic order increase the transition temperature; in the case of non-frustrating boundary conditions the Clausius–Clapeyron type of equation yields $\Delta T \propto 1/d$ [57], (ii) the effect of the elastic deformation of the director field causes a negative transition temperature shift; $\Delta T \propto -1/d^2$, for the case of homogeneous degree of order [10]. In the case of a structural transition between the biaxial and bent-director structure the phase shift is negative, however, it exhibits an effective $1/d$ dependence. The sign of the temperature shift reflects the promoting effect of elastic deformations, whereas the different thickness dependence results from the non-homogeneous degree of order. By extrapolation we have estimated the upper cell thickness in which the biaxial structure can be found, $d_{max} = 15.6\,l$. Above this cell thickness the transition to the bent-director structure occurs before the two wetting layers get in the contact.

To relate the results from the MC simulations and the phenomenological description we have to find a suitable meaning of the “spin”. Close to the isotropic–nematic phase transition the comparison of the two corresponding energies associated with a correlated volume of LC molecules, $\epsilon \sim A(T_{NI} - T^*)V_0$, yields $l = \sqrt[3]{V_0} \sim 3$ nm $= 0.37\,\xi_{NI}$. For 5CB $\xi_{NI} \sim 8$ nm and $\epsilon = k_B T_{NI}/T_{NI}^{MC} \sim 0.024$ eV, so that $\lambda \sim 3$ nm or the strength of the anchoring $G = L/\lambda \sim 0.003$ J/m^2. The obtained strength of the surface interaction is experimentally achievable and represents strong anchoring. Using this, the upper limit for the “critical” cell thickness d_{max} corresponds to ~ 50 nm. The proposed relation of the MC simulation to the macroscopic properties yields an excellent agreement between the two descriptions, see Fig. 8.8a. One should note that there are no free parameters to fit.

8.3.5 Pretransitional Dynamics in a Hybrid Nematic Cell

Our illustration of pretransitional behavior in frustrated NLC systems is restricted to fluctuations in the biaxial structure of the hybrid nematic cell with equal strengths of homeotropic and planar anchorings. The detailed analysis of the pretransitional dynamics of all five degrees of freedom around the bent-director configuration is somewhat more complicated because of the nonuniformity of the base tensors.

Since the equilibrium profiles of the biaxial hybrid structure are described by two nonzero amplitudes, the corresponding fluctuation modes are coupled.

In the following, $\beta^x_{0,1}$ and $\beta^z_{0,1}$ denote the order parameter fluctuations with respect to the nematic director parallel to the x and z axis, respectively. The other three fluctuation modes are uncoupled and represent either director fluctuations (β_{-1} and low $\beta_{\pm 2}$ modes) or biaxial fluctuations, high $\beta_{\pm 2}$ modes.

Order Parameter Fluctuations

The order parameter eigenmodes are either symmetric or antisymmetric functions with respect to the middle plane of the cell. The lowest symmetric mode is associated with fluctuations of the thickness of the central director exchange region and therefore also with the fluctuations of the magnitude of biaxiality of the nematic order. The lowest antisymmetric mode corresponds to fluctuations of the position of the exchange region. The portrait of the lowest antisymmetric order parameter mode is plotted in Fig. 8.10a. The maxima of the two corresponding profiles, $\beta^x_{0,1}$ and $\beta^z_{0,1}$, are "localized" at the the position of the maximum slope of the scalar order parameter like in the case of heterophase systems, thus, being responsible for the growth of the wetting layers. The relaxation rate of the lowest mode is strongly decreased on approaching the structural transition, however, it does not show any critical behavior.

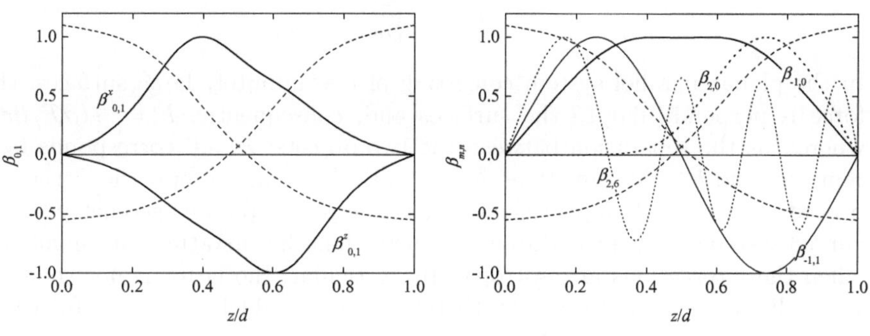

Fig. 8.10. Spatial dependence of the (a) lowest order parameter fluctuation mode and (b) director (solid) and biaxial (dotted) fluctuation modes in the biaxial structure close to the transition to the bent-director structure. Dashed lines correspond to the equilibrium profiles.

Director and Biaxial Fluctuations

Director fluctuations β_{-1} bend the nematic director in the plane of the two easy axes. The lowest mode is responsible for the bending of the director field at the transition to the one characteristic for the bent director structure (see 8.10b). Its relaxation rate exhibits a critical slowing down when the film thickness approaches the critical point.

Biaxial fluctuations $\beta_{\pm 2}$ are degenerated, whereas the eigenfunctions are just mirror images with respect to the middle plane of the cell. The few lowest modes are expelled from the part of the film where they bend the secondary director field and so they effectively represent the director fluctuations (see Fig. 8.10b). Higher modes are spread over the whole film whereas the unfavorable manner of biaxial fluctuations is compensated by the shorter wave vector of a deformation.

8.4 Forces Acting on a Thin Liquid-crystalline Film

The presence of walls causes a change of the free energy of the system. In general, this change depends on the separation between the walls which, thus, results in an attractive or repulsive force between them. Here, by system we refer to the liquid crystal in which the confining walls are immersed, so that the volume and the surface of the liquid crystal are preserved by changing the separation between walls. In an isothermal process the force is related to the free energy of a system as

$$\boldsymbol{F} = -\frac{\partial \mathcal{F}}{\partial \boldsymbol{r}}. \tag{8.24}$$

In the planar parallel system consisting of two infinitely large surfaces, the force is perpendicular to the surfaces and, consequently, $F = -(\partial \mathcal{F}/\partial d)$, where d is the separation between the two objects; $F > 0$ corresponds to a repulsive and $F < 0$ to an attractive force. Let us comment here the Derjaguin procedure [58] which applies the forces in the planar parallel geometry to the curved systems. This procedure is justified when the curvatures are small and when the interactions are isotropic. In particular, the latter is not the case when discussing liquid crystals, therefore one should be very careful when applying the Derjaguin approximation.

First observations of forces in liquid crystal were performed in the 80's by Horn et al. [59]. They studied the force between two mica plates separated by a liquid crystal 5CB using the surface force apparatus. Later on, the forces due to the structure were briefly discussed by Poniewierski and Sluckin [57] and a detailed theoretical study of forces in paranematic systems was performed by Borštnik and Žumer [43]. In the meantime, much attention was paid to the fluctuation-induced forces [11,12,60,61]. Theoretical studies were followed by renewed experimental interest, studies being performed with surface forces apparatus [46] and atomic force microscopes [62].

8.4.1 Structural Force

Confinement has, especially in the vicinity of phase and structural transitions, a great impact on the equilibrium order of a liquid crystal. The force arising from the change of the corresponding part of the free energy is denoted as *structural force.* Here, the free energy of a system is determined by means of the phenomenological mean-field theory and the structural force denotes the force associated with the equilibrium order obtained within this theory,

$$\boldsymbol{F}_{struct.} = -\frac{\partial \Delta \mathcal{F}_{MF}}{\partial \boldsymbol{r}} = -\frac{\partial \mathcal{F}_{MF}}{\partial \boldsymbol{r}} + f_{bulk}\mathcal{A}. \tag{8.25}$$

$\Delta\mathcal{F}_{MF}$ is the variation of the free energy of the system in the framework of mean-field theory with respect to the free energy of the same amount of the liquid crystal in bulk, $\mathcal{F}_{MF}^{bulk} = f_{bulk}\mathrm{V}$. In a planar geometry it is useful to express the force in terms of presure, $\Pi = F/\mathcal{A}$.

In the following we will analyse the pressure in a few structures discussed above.

Heterophase Systems: Nematic Order Variation

The structural force in such heterophase systems arises only from the deformation of the scalar order parameter. Thus, the force is short-range and attractive. Typical thickness dependence of the structural force in both nematogenic heterophase systems is presented in Fig. 8.11a.

The order in thin cells is determined by the surface interaction. If the cell thickness is increased, the corresponding amount of LC material enters the cell and its order is changed from the bulk order to the one induced by the substrates. Thus, the free energy of the system changes by $(f - f_{bulk})V$ which produces a constant force between the confining walls. The force is attractive, because in the given systems the surface-induced order corresponds to a higher free energy. In thick enough cells, the order parameter profile is characterized by a bulk-like core and surface-induced wetting layers. Now when changing the cell thickness the change of the free energy of the system is due to the minor changes in the order parameter profile while most of the material enters the core of the cell, thus, its free energy is not changed. In this regime, the structural force decays exponentially with respect to the cell thickness,

$$\Pi \propto \exp(-d/\xi), \tag{8.26}$$

where $\xi = \xi_{N,0}$ in the case of disordering confinement and $\xi = \xi_I$ in the case of a paranematic cell. The hysteresis loops in Fig. 8.11 (a) are the evidence of the metastable phases. Details of the profile of the structural force depend on details of the surface-induced interaction, whereas the existence of the hysteresis depends on the temperature — whether or not the system is above the critical point for the capillary phase transition.

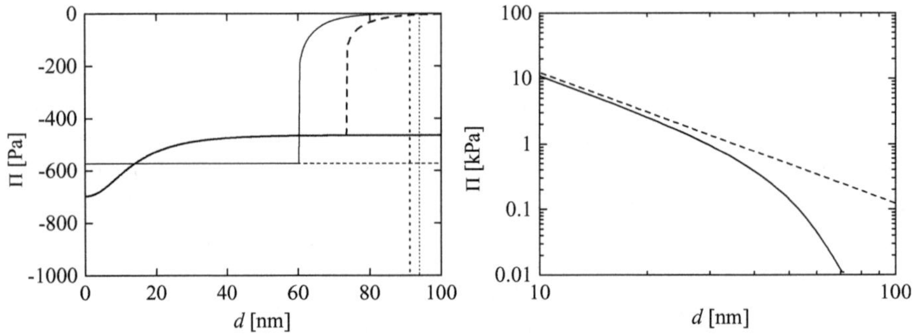

(a) (b)

Fig. 8.11. (a) Structural force per unit area in a heterophase paranematic (thick lines) and nematic system with molten boundary layers (thin lines). Solid lines correspond to the force in the nematic phase and dashed lines to the force in the isotropic phase. For the thicknesses above the corresponding verticals the isotropic (paranematic) or nematic phase is stable, respectively. The force is short-range and attractive. (b) Structural presure in the hybrid nematic system in a biaxial structure (solid line) and bent-director structure (dashed line). In both cases the interaction is long-range and repulsive.

Frustrated Systems: General Deformation of Order

The common feature of all ordered structures in a hybrid nematic cell is the repulsive character of the structural force. The repulsion is due to the antagonistic boundary conditions, which always lead to at least small deformations, and to the fact that within a certain ordered structure the frustration is stronger if the confining substrates are brought closer to each other. The magnitude of the force is tuned by the anchoring strength at both substrates.

Within the bare director picture the elastic deformation of the director field in the *bent-director structure* gives rise to strong structural repulsion. In the limit of infinitely strong surface anchorings $\mathcal{F}_{elastic} = \frac{3}{2}LS^2\mathcal{A}d\int_0^1 \mathrm{d}z(\phi')^2 = 3\pi^2 LS^2\mathcal{A}/8d$, and the structural force per unit area is given by

$$\Pi = \frac{3\pi^2 LS^2}{8d^2}. \tag{8.27}$$

In *biaxial structures*, the force is repulsive and exhibits $1/d^2$ behavior at small cell thicknesses whereas at large d's the exponent of the power law is smaller than -2 [see Fig. 8.11 (b)]. For small cell thicknesses, the elastic deformation, although of the scalar fields rather than the director field, is spread over the whole cell and the force exhibits a typical elastic dependence. The decrease of the range of the force for larger cell thicknesses is a consequence of the localization of the deformation when approaching the stability limit of this structure.

8.4.2 Pseudo-Casimir Force

The equilibrium order determined within the mean-field theory is perturbed due to thermal fluctuations which give rise to collective excitations. Except in the close vicinity of the phase/structural transitions, the thermal fluctuations of the order parameter can be assumed small, and the free energy of the fluctuations can be considered a correction to the mean-field free energy. In such a case, the fluctuations of the liquid-crystalline order are described consistently by a harmonic Hamiltonian of the form

$$\mathcal{H}[b] = \frac{L}{2}\left\{\int \left[\xi^{-2}b^2 + (\nabla b)^2\right] dV + \sum_{i=1,2} \lambda_i^{-1} \int b^2 \mathrm{d}\mathcal{A}_i\right\}. \tag{8.28}$$

The free energy corresponding to collective thermal excitations is then given by a partition function

$$\mathcal{F}_{CAS} = -k_B T \ln\left(\int \mathrm{d}b\, \mathrm{e}^{-\mathcal{H}[b]/k_B T}\right), \tag{8.29}$$

where k_B is the Boltzmann constant, T is the temperature, and the integral is over all configurations of a fluctuating field b which satisfy the boundary conditions [37].

The name pseudo-Casimir force for the fluctuation-induced interaction is due to the analogy with the Casimir effect [63]: at $T = 0$ quantum fluctuations of the electromagnetic field in a cavity yield a weak yet measurable attraction between the walls of the cavity. Because the force between the walls is determined by a derivative of the free energy of a system rather than by a derivative of its energy, a similar effect is expected above absolute zero where the interaction is not just due to quantum but also due to thermal fluctuations. In liquid crystals, the fluctuation-induced interaction is due to thermal fluctuations of the order parameter field instead of the electromagnetic field.

As a universal property of the Casimir interaction, depending on the boundary conditions, the fluctuation-induced interaction can be either repulsive or attractive [60, 61]. Its magnitude depends strongly on the surface coupling. In general, the sign of the pseudo-Casimir interaction is determined by the type of boundary conditions, provided that the system is not subject to electric or magnetic fields; $b = (\lambda/d)\, b'|_{\text{at wall}}$, where $b' = \mathrm{d}b/\mathrm{d}z$ and d is the separation between the walls. Fluctuation modes constrained by strong, $\lambda_{1,2} \ll d$, or weak, $\lambda_{1,2} \gg d$, anchoring at both substrates lead to an attractive force. In a mathematical language this corresponds to Dirichlet, $b|_{\text{at wall}} = 0$, or Neumann, $b'|_{\text{at wall}} = 0$, boundary conditions at both substrates. In contrast, the asymmetric situation with one surface enforcing a strong anchoring and the other a weak anchoring yields a repulsive force (mixed boundary conditions). The range of the interaction is determined by the correlation length of a given fluctuating field; a finite correlation length yields a short-range interaction whereas fluctuations with infinite correlation length lead to a long-range interaction.

Heterophase Systems: Order Parameter Variation

The pseudo-Casimir force in a heterophase system will be illustrated with the example of a thin nematogenic film with order-inducing wetting layers on both confining surfaces discussed before. The resulting fluctuation force can be interpreted in terms of two contributions: (i) the interaction between the substrates and the phase boundaries and (ii) the interaction between the two phase boundaries.

The interaction between a solid substrate and a phase boundary consists of three contributions corresponding to three non-degenerated fluctuation modes. The fluctuations of the degree of order and the biaxial fluctuations give rise to a short-range repulsion between the substrate and the phase boundary proportional to $\exp(-2d_W/\xi)$, where d_W is the thickness of the wetting layer and $\xi = \xi_{N,0}$ or $\xi_{N,\pm1}$ for order parameter and biaxial fluctuations, respectively. The short-range of the interaction is a consequence of finite correlations in both nematic and isotropic phase. The repulsion between the substrate and the phase boundary can be understood in terms of effectively mixed boundary conditions. The anchoring at the substrates is strong, whereas the maxima of the order fluctuation modes at the phase boundaries indicate that these modes experience effectively weak "anchoring" conditions. A similar argument applies to biaxial fluctuations. Here a weak "anchoring" condition at the phase boundary can be understood as a consequence of the fact that biaxial fluctuations are much more favorable in the isotropic phase than in the nematic phase. The main contribution to the interaction between the solid substrate and the phase boundary is induced by the director fluctuations, which are characterized by an infinite correlation length in the nematic phase; $\mathcal{F}_{CAS}^{director} \propto -k_BT/d_W^2$. This long-range interaction is attractive because in the case of director fluctuation the isotropic–nematic interface induces an effective strong anchoring.

Interaction between phase boundaries gives rise to an attractive fluctuation-induced force which is proportional to $\exp\left(-2(d-2d_W)/\xi_I\right)$. The attraction is due to the identical boundary conditions. Except in the vicinity of the metastability limit of the paranematic phase, the distance between the substrate and the phase boundary is much smaller than the distance between the two boundaries, and the interaction between the phase boundaries is very weak.

The effective interaction between solid substrates is a superposition of the two contributions discussed above. In the range of stable paranematic phase ($\theta > 1$, $d_W \ll d$) the fluctuation-induced force between the two substrates is governed by the interaction between the solid substrate and the phase boundary. It is mediated by the structural interaction which determines the functional dependence of the wetting layer thickness on the sample thickness

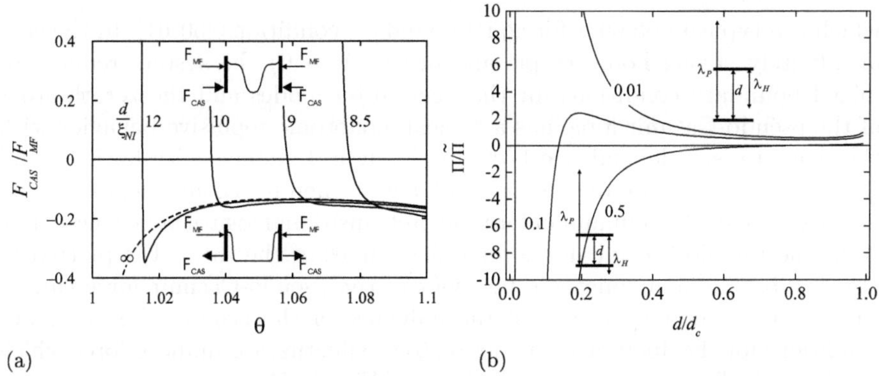

Fig. 8.12. (a) Pseudo-Casimir force in the paranematic heterophase system compared to the structural force as a function of temperature. (b) Pseudo-Casimir force in a hybrid nematic cell with uniform director field. Labels denote the ratio λ_H/λ_P.

$\partial d_W/\partial d \approx -\text{const.} \times \exp(-d/\xi_I)$ [64]. Therefore, the leading term in the substrate-to-substrate fluctuation-induced force

$$F_{CAS} \approx F_{CAS}^{director} \frac{\partial d_w}{\partial d} \propto \mathrm{e}^{-d/\xi_I} \tag{8.30}$$

is repulsive and short-range. Its range, ξ_I, is identical to the range of the structural force, whereas its magnitude is smaller but comparable to the magnitude of the structural attraction. Since the two forces have the same range, the structural force is proportionally diminished by the fluctuation-induced contribution. The Pseudo-Casimir force in the paranematic system is presented in Fig. 8.12a.

Frustrated System: Hybrid Nematic Cell

Up to now, in the frustrated nematic systems the pseudo-Casimir force has only been determined for the simplest structure with a uniform director field ($d < d_c = \lambda_H - \lambda_P$, where $\lambda_H \ll \lambda_P$ is the extrapolation length of the homeotropic substrate and λ_P the extrapolation length of the degenerate in-plane anchoring which preserves the full rotational symmetry about the substrate normal) [12]. Then, within the bare director description the correlation length in the Hamiltonian (8.28) is constant and the partition function of the fluctuation modes can be derived analytically. The derivation of the force in the bent-director and biaxial structures is more complex due to the deformed equilibrium order.

The shape of the pseudo-Casimir force in the uniform structure depends on the ratio between the two extrapolation lengths ($\Lambda = \lambda_H/\lambda_P$) whereas its magnitude is tuned by the difference between them, d_c. For finite but similar extrapolation lengths ($\Lambda \to 1$) the force is attractive and decreases as d^{-3}

which is a typical behavior for equal boundary conditions [60,61]. In the case of infinitely strong homeotropic anchoring ($\Lambda \to 0$) the system reduces to mixed boundary conditions for the fluctuation modes and the zeroth-order of the pseudo-Casimir force has a typical monotonic repulsive behavior with the characteristic dependence $1/d^3$. In real hybrid systems with both λ_P and $\lambda_H > 0$, the force is attractive at small d/d_c's, then it becomes repulsive and may reach a local maximum before the pretransitional logarithmic singularity [12]. The pseudo-Casimir force for a few values of the ratio Λ is plotted in Fig. 8.12b. The nonmonotonic behavior of the pseudo-Casimir force can be simply understood by means of the influence of the type of the boundary conditions for the fluctuating modes on to the fluctuation-induced force which is schematically represented in the inset of Fig. 8.12b.

8.5 Experimental Evidence of Structural and Pseudo-Casimir Forces in Liquid Crystals

The first observations of forces in liquid crystals were performed in the 80's by Horn et al. using the surface force apparatus (SFA) [59]. Later on, the studies were extended to the atomic force microscope (AFM) [13]. An alternative set-up with several advantages compared with the standard techniques is, however, the one for spinodal dewetting [65]. In this experiment, a thin layer of liquid is spread on a solid substrate, usually by spin-casting. Depending on the derivative of the force between the solid–liquid and the liquid–air interfaces with respect to the film thickness, the film can be either stable or it can disintegrate into an array of droplets. In the case of an unstable film, by measuring the size of the droplets and the dewetting time as a function of the initial thickness of the film, one could reconstruct the profile of the interaction between the interfaces.

Let us explain the mechanism of the spinodal dewetting in some more detail. Imagine a thin liquid film deposited on a solid substrate so that it has a free liquid–air interface. Due to thermal fluctuations of the free interface, *capillary waves*, the film is not flat but rather wrinkled. Because of the interactions in the liquid, the total pressure in the film depends on its thickness, being either higher or lower than the external pressure. The difference in pressures, $\Pi = p_0 - p$ where p_0 is the external pressure, is denoted as a *disjoining pressure*. It is either repulsive, $\Pi > 0$, or attractive, $\Pi < 0$, and it vanishes in the limit of an infinitely thick film. Thus, if the disjoining pressure is characterized by a monotonic behavior, the magnitude of the disjoining pressure is larger for smaller film thicknesses. The two possible situations for the monotonic disjoining pressure are depicted in Fig. 8.13. Imagine now a film whose interface is perturbed at a given moment. If the disjoining pressure is repulsive the repulsion is stronger in a thinner region of film. Thus, the tendency to thicken the thinner part of the film will be stronger than to thicken the thicker part and the differences in the film thickness will be smeared. On the

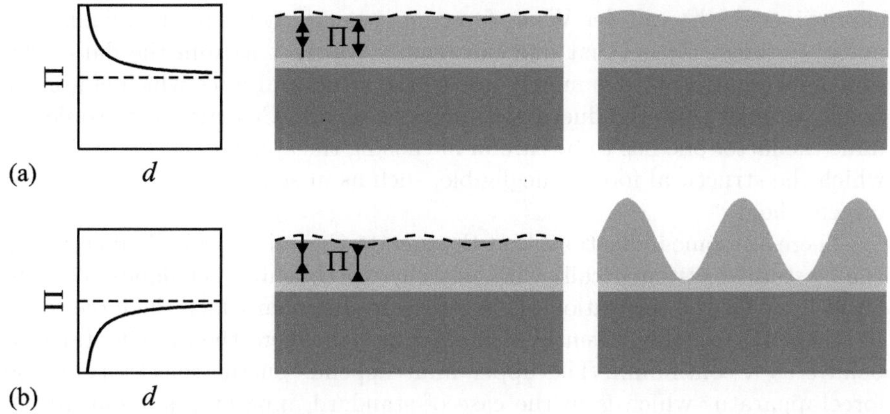

Fig. 8.13. Schematic representation of conditions for the stability of thin liquid films. Thermal fluctuations of the free liquid–air interface are (a) diminished and (b) amplified, resulting in stable film or decomposition of the film in liquid drops and dry patches, respectively. Left: schematic representation of the disjoining pressure; middle: uniform film and a sketch of a capillary wave with a disjoining pressure indicating the sign and the magnitude; right: resulting film profile.

other hand, if the disjoining pressure is attractive, the stronger attraction in the thinner regions amplifies the differences in the film thickness untill finally the thickness of the thinner regions drops to zero and dry patches occur.

In the lubrication limit and for small amplitudes of the capilary waves $[\zeta(x) = d_0 + \sum_q u_q \, \mathrm{e}^{iqx} \, \mathrm{e}^{-t/\tau_q}$ and $u_q \ll d_0]$ the relaxation times of the interfacial fluctuations can be derived analytically [66, 67]

$$\frac{1}{\tau_q} = \frac{d_0^3}{3\eta} \left[\gamma q^4 - q^2 \Pi'(d_0)\right] . \tag{8.31}$$

For $\Pi'(d_0) < 0$, $\tau_q > 0$ for every q and fluctuations decrease exponentially after they are excited. On the other hand, if $\Pi'(d_0) > 0$ a wavevector q_c can be found so that $\tau_{q<q_c} < 0$ and the given fluctuation mode is amplified resulting in a decomposition of the film. Typical size of drops which result from the spinodal dewetting of a film is given by the most amplified capillary wave $(\partial \tau_q / \partial q = 0)$ with

$$\lambda_M = \frac{2\pi}{q_M} = \sqrt{\frac{8\pi^2 \gamma}{\Pi'(d_0)}} . \tag{8.32}$$

The force measured in liquid crystals is a superposition of structural, pseudo-Casimir, and non-structural forces. The non-structural forces include electrostatic and polarization forces, such as the van der Waals force. The behavior of non-structural forces is well-understood [68]. Recently, the improved expression for the van der Waals force has been derived for the uniaxial nematic systems [15], so that one can readily subtract it from the data. The remaining contribution is mostly due to the structural force which is usually much stronger than the fluctuation-induced pseudo-Casimir force. To determine the latter one has to be careful in chosing the appropriate LC system in which the structural force is negligible, such as in structures with a uniform director field.

There are some indications that the pseudo-Casimir force in liquid crystals can be studied systematically with existing experimental techniques [69]. The lower limit for the separation of the substrate between which the force could be studied is roughly given by a molecular size where the system starts to behave as a continuum. The upper limit depends on the sensitivity of the force apparatus which is in the case of standard apparatuses about 10 pN [70]. This means that these apparatuses are precise enough to detect the pseudo-Casimir force at distances up to ≈ 40 nm [69].

In spinodal dewetting experiments, the upper limit of separation of the substrates depends on the stability of the film. There are indications that in some cases this upper limit can reach ≈ 100 nm. However, the data collected in a spinodal dewetting experiment could be less straightforward to interpret. It may well prove that the existing theory of spinodal dewetting must be extended to include non-linear effects if the observables are to be related accurately enough to the interaction between the substrates. Recently, there are some indications that the spinodal dewetting process in LC films could be smeared by other dewetting processes [71]. A more detailed study of a dewetting process initially studied by Vandenbrouck et al. [72] has shown, that the LC film becomes inhomogeneous on a short time scale only it the temperature is within a certain window below the clearing point. Above and below this window, whose width depends on the initial thickness, the films are homogeneous. The observed inhomogeneity was now identified as terraces of two different but finite heights. On long time scales LC films dewet via hole nucleation.

The first interpretation of this newly identified short-time breakup of thin nematic films was put forward by van Effenterre et al. [73] and is rooted in basic thermodynamics, i.e., in the biphase coexistence of the two (meta)stable phases taking part in the discontinuous transition. However, as pointed out in detail by Ziherl and Žumer [74], the explanation misses two critical points: First, that the isotropic–nematic coexistence can only exist in the temperature range where both phases are thermodynamically metastable or stable, and second, that the nematic director is only distorted in thick enough films (i.e., above the critical thickness $d_c = |\lambda_1 - \lambda_2|$). Therefore, the dewetting process in thin LC films is still an open question for both theoretical and experimental studies.

8.6 Conclusions

Confined LC systems have been among the main areas of LC research for a long time. Their importance for applications is determined by the fact that in many cases weak external forces are sufficient to induce structural changes and thereby altering the optical properties of the system. The main ingredient in a switching process is the competition of two or more aligning forces, typically the anchoring and the electric field. With decreasing size of the system the variety of phenomena resulting from this competition increases considerably.

In this Chapter, we have summarized results of our theoretical studies of effects of the anchoring on the equilibrium ordering and pretransitional fluctuation dynamics, both from the micro- and macroscopic point of view. We have shown two basic effects of the anchoring, the wetting with nematic or isotropic phase, and the distortion of the nematic director field. In the last case, there is a good qualitative and quantitative agreement between a macroscopic phenomenological description and a microscopic description, i.e. a MC simulation of a lattice spin model. Further studies were focused on structural, pseudo-Casimir, and van der Waals forces in confined nematic systems. Due to the effect of the confinement there is an additional — disjoining — pressure in the LC system whose magnitude and range is tuned by the extent of the deformation of the order, by the correlation lengths of the fluctuating modes, and by the dielectric properties. We have also discussed possibilities for the experimental observation of structural and pseudo-Casimir forces in confined LC systems.

To conclude, let us point out, that more detailed and target studies of confined LC systems are still required. On the theoretical side, the complete phase diagram of confined nematogens is still needed. Since most of the existing studies correspond to the simplest planar geometry, the more complex and realistic geometries should be considered as well. Here, a microscopic analysis such as MC or molecular dynamics simulations would be welcome to complement the continuum theoretical description. On the experimental side, the main focus should be to extract structural and especially pseudo-Casimir forces from the observed data.

References

1. G. P. Crawford, D. Svenšek, and S. Žumer, in *Chirality in liquid crystals*, H. S. Kitzerow and C. Bahr, Edts., Springer-Verlag, New York, (2001).
2. M. Ibn-Elhaj and M. Schadt, Nature **410**,796 (2001).
3. H. Sirringhaus, P. J. Brown, R. H. Friend, M. M. Nielsen, K. Bechgaard, B. M. W. Langeveld-Voss, A. J. H. Spiering, R. A. J. Janssen, E. W. Meijer, P. Herwig, and D. M. de Leeuw, Nature **401**, 685 (1999).
4. G. P. Crawford and S. Žumer, Edts., *Liquid Crystals in Complex Geometries*, Taylor & Francis, London, (1996).

5. S. Žumer, Phys. Rev. A **37**, 4006 (1988).
6. G. P. Crawford, J. W. Doane, and S. Žumer, in *Handbook of liquid crystal research*, P. J. Collings and J. S. Patel, Edits., Oxford University Press, Oxford, (1997).
7. X. I. Wu, W. I. Goldburg, M. X. Liu, and J. Z. Xue, Phys. Rev. Lett. **69**, 470 (1992).
8. S. Stallinga, M. M. Wittebrood, D. H. Luijendijk, and Th. Rasing, Phys. Rev. E **53**, 6085 (1996).
9. P. Ziherl, A. Šarlah, and S. Žumer, Phys. Rev. E **58**, 602 (1998).
10. A. Šarlah and S. Žumer, Phys. Rev. E **60**, 1821 (1999).
11. P. Ziherl, R. Podgornik, and S. Žumer, Phys. Rev. Lett. **82**, 1189 (1999).
12. P. Ziherl, F. K. P. Haddadan, R. Podgornik, and S. Žumer, Phys. Rev. E **61**, 5361 (2000).
13. K. Kočevar and I. Muševič, Liq. Cryst. **28**, 599 (2001).
14. A. Borštnik Bračič, K. Kočevar, I. Muševič, and S. Žumer, Phys. Rev. E **68**, 011708 (2003).
15. A. Šarlah and S. Žumer, Phys. Rev. E **64**, 051606 (2001).
16. C. Chiccoli, P. Pasini, A. Šarlah, C. Zannoni, and S. Žumer, Phys. Rev. E **67**, 050703(R) (2003).
17. P. Ziherl and S. Žumer, Phys. Rev. Lett. **78**, 682 (1997).
18. S. Žumer, A. Šarlah, P. Ziherl, and R. Podgornik, Mol. Cryst. Liq. Cryst. **358**, 83 (2001).
19. S. Hess, Z. Naturforsch. Teil A **30**, 728 (1975).
20. V. L. Pokrovskii and E. I. Kats, Zh. Eksp. Teor. Fiz. **73**, 774 (1977) [Sov. Phys. JETP **46**, 405 (1977)].
21. M. P. Allen and D. J. Tildesley, *Computer simulation of liquids*, Clarendon Press, Oxford, (1989).
22. P. Pasini and C. Zannoni, Edts., *Advances in the Computer Simulations of Liquid Crystals*, Kluwer, Dordrecht, (2000).
23. P. A. Lebwohl and G. Lasher, Phys. Rev. A **6**, 426 (1972).
24. E. Berggren, C. Zannoni, C. Chiccoli, P. Pasini, and F. Semeria, Phys. Rev. E **50**, 2929 (1994).
25. U. Fabbri and C. Zannoni, Mol. Phys. **58**, 763 (1986).
26. C. Chiccoli, P. Pasini, F. Semeria, and C. Zannoni, Phys. Lett. A **150**, 311 (1990).
27. C. Chiccoli, P. Pasini, F. Semeria, and C. Zannoni, Mol. Cryst. Liq. Cryst. **212**, 197 (1992).
28. L. D. Landau, Fiz. Z. Sowejetunion **11**, 26 (1937).
29. A. L. Alexe-Ionescu, G. Barbero, and G. Durand, J. Phys. II (Paris) **3**, 1247 (1993).
30. F. C. Frank, Discuss. Faraday Soc. **25**, 19 (1958).
31. P. G. de Gennes and J. Prost, *The Physics of Liquid Crystals*, Clarendon Press, Oxford, (1993).
32. A. Rapini and M. Papoular, J. Phys. Colloq. **40**, C3–490 (1969).
33. M. Nobili and G. Durand, Phys. Rev. A **46**, R6174 (1992).
34. A. Mertelj and M. Čopič, Phys. Rev. Lett. **81**, 5844 (1998).
35. K. Huang, em Statistical Mechanics, Wiley, New York (1987).
36. L. D. Landau and I. M. Khalatnikov, Dokl. Akad. Nauk SSSR **96**, 469 (1954).
37. P. M. Chaikin and T. C. Lubensky, *Principles of condensed matter physics*, Cambridge University Press, Cambridge, (1995).

38. D. Svenšek and S. Žumer, Continuum. Mech. Therm. **14**, 231 (2002).
39. D. Svenšek and S. Žumer, Phys. Rev. Lett. **90**, 155501 (2003).
40. P. Sheng, Phys. Rev. A **26**, 1610 (1982).
41. T. Moses and Y. R. Shen, Phys. Rev. Lett. **67**, 2033 (1991).
42. H. Hsiung, Th. Rasing and Y. R. Shen, Phys. Rev. Lett. **57**, 3065 (1986).
43. A. Borštnik and S. Žumer, Phys. Rev. E **56**, 3021 (1997).
44. A. Golemme, S. Žumer, D. W. Allender, and J. W. Doane, Phys. Rev. Lett. **61**, 2937 (1988).
45. M. I. Boamfa, M. W. Kim, J. C. Maan, and T. Rasing, Nature **421**, 149 (2003).
46. L. Moreau, P. Richetti, and P. Barois, Phys. Rev. Lett. **73**, 3556 (1994).
47. K. Kočevar, R. Blinc, and I. Muševič, Phys. Rev. E **62**, R3055 (2000).
48. N. Schopohl and T. J. Sluckin, Phys. Rev. Lett. **59**, 2582 (1987).
49. D. Andrienko, Yu. Kurioz, Yu. Reznikov, and V. Reshetnyak, Sov. Phys. JETP **85**, 1119 (1997).
50. G. Barbero and R. Barberi, J. Phys. (Paris) **44**, 609 (1983).
51. M. M. Wittebrood, Th. Rasing, S. Stallinga, and I. Muševič, Phys. Rev. Lett. **80**, 1232 (1998).
52. P. Ziherl and S. Žumer, Liq. Cryst. **21**, 871 (1996).
53. S. Kralj, E. G. Virga, and S. Žumer, Phys. Rev. E **60**, 1858 (1999).
54. P. Palffy-Muhoray, E. C. Gartland, and J. R. Kelly, Liq. Cryst. **16**, 713 (1994).
55. H. G. Galabova, N. Kothekar, and D. W. Allender, Liq. Cryst. **23**, 803 (1997).
56. N. Metropolis, A. W. Rosenbluth, M. N. Rosenbluth, A. H. Teller, and E. Teller, J. Chem. Phys. **21**, 1087 (1953).
57. A. Poniewierski and T. J. Sluckin, Liq. Cryst. **2**, 281 (1987).
58. B. V. Derjaguin, Kolloid Z. **69**, 155 (1943).
59. R. G. Horn, J. N. Israelachvili, and E. Perez, J. Phys. (Paris) bf 42, 39 (1981).
60. A. Ajdari, L. Peliti, and J. Prost, Phys. Rev. Lett. **66**, 1481 (1991).
61. H. Li and M. Kardar, Phys. Rev. Lett. **67**, 3275 (1991).
62. K. Kočevar, A. Borštnik, I. Muševič, and S. Žumer, Phys. Rev. Lett. **86**, 5914 (2001).
63. H. B. G. Casimir, Proc. Kon. Ned. Akad. Wet. **51**, 793 (1948).
64. G. Gompper, M. Hauser, and A. A. Kornyshev, J. Chem. Phys. **101**, 3378 (1994).
65. S. Herminghaus, K. Jacobs, K. Mecke, J. Bischof, A. Fery, M. Ibn-Elhaj, and S. Schlagowski, Science **282**, 916 (1998).
66. P. G. de Gennes, Rev. Mod. Phys. **57**, 827 (1985).
67. F. Brochard-Wyart and J. Daillant, Can. J. Phys. **68**, 1084 (1990).
68. J. Mahanty and B. W. Ninham, em Dispersion Forces, Academic Press, London, (1976).
69. P. Ziherl and I. Muševič, Liq. Cryst. **28**, 1057 (2001).
70. J. Israelachvili, *Intermolecular & Surface Forces*, Academic Press, London, (1985).
71. D. van Effenterre, R. Ober, M. P. Valignat, and A. M. Cazabat, Phys. Rev. Lett. **87**, 125701 (2001).
72. F. Vandenbrouck, M. P. Valignat, and A. M. Cazabat, Phys. Rev. Lett. **82**, 2693 (1999).
73. D. van Effenterre, M. P. Valignat, and D. Roux, Europhys. Lett. **62**, 526 (2003).
74. P. Ziherl and S. Žumer, Eur. Phys. J. E **12**, 361 (2003).

9 Applications

Cindy Nieuwkerk and Bianca van der Zande

In this Chapter an overview of the state of the art of liquid crystal applications is given, in particular the current market development for LCD's and the recent developments using non contact alignment techniques like photo-alignment and ion-beam alignment.

9.1 Introduction

By far the largest application area of aligned liquid crystals is in Liquid Crystal Displays (LCDs). In LCDs the LC material is sandwiched in between two glass plates with conducting layers indium-tin-oxide (ITO). Figure 9.1 shows the lay out of an LCD. Apart from the ITO covered glass plates and the LC material, the LCD consists of a stack of various optical layers such as retarders, polariser, analyser, colour filter and a backlight.

The alignment of the LC molecules is crucial for successful operation of an LCD. Not only does it control the electro-optical switching mode of the display, appropriate alignment also prevents the formation of random multidomains, disclination lines or small mismatches in LC director orientation that deteriorate the displayed image. An alignment layer imposes the orientation onto the LC molecules. Conventionally this alignment layer is rubbed by synthetic or natural fabric, e.g. polyester or velvet [1–4].

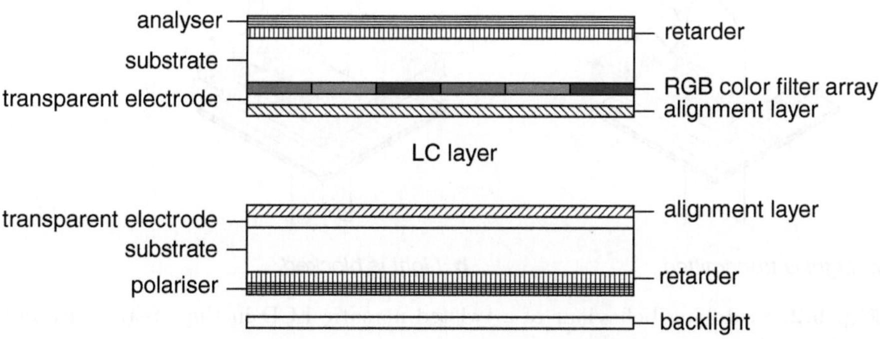

Fig. 9.1. Lay-out of a liquid crystal display.

Although rubbing is the state-of-the-art technology used by most LCD manufacturers, there is a strong demand for non-contact alignment techniques because rubbing generates dust and electrostatic damage of the electronics. Both effects can cause a yield drop during the manufacturing process. In the previous Chapters of this book various alternative alignment techniques have been described that may be useful to orient the LC molecules in the next generation LCDs.

The LC molecules orient on the alignment layer in a predefined in-plane and out-of-plane direction. The out-of-plane (polar) orientation is called the pretilt *angle*. The directional orientations and pretilt angles on both surfaces in combination with a chiral dopant added to the LC mixture determine the twist direction of the LC molecules in an LCD. Most commonly left-handed dope molecules that cause a left-handed twist sense are added to the LC mixture. Both the in-plane and out-of-plane directions are important to obtain the proper switching behaviour upon driving the display.

In LCDs various LC modes of operation are applied like twisted nematic (TN) [5], super twisted nematic (STN)[6], vertically aligned nematic (VAN) [7, 8], optically compensated birefringence (OCB) [9] and in plane switching (IPS) [10–12]. The LC mode used depends on the demands of a specific application, like the viewing angle, power consumption and manufacturing cost.

In Fig. 9.2 the switching behaviour of a twisted nematic (TN) LCD is depicted. Figure 9.2a shows a TN LCD in the off-state. The light coming from the bottom is polarised by the first polariser and due to the bire-

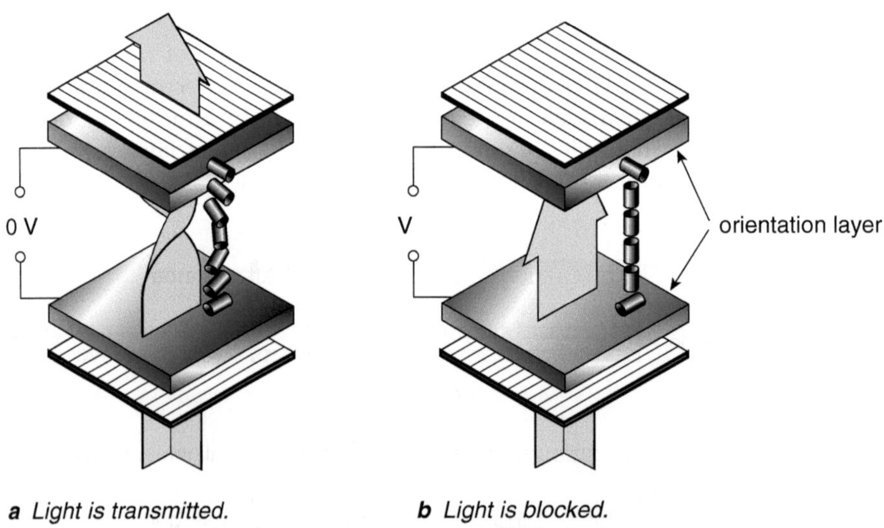

Fig. 9.2. Switching behaviour of a twisted nematic LCD in the off-state (a) and on-state (b).

fringent properties of the liquid crystal layer the polarised light traversing through the LC layer is rotated over 90 degrees. As the top analyser is crossed with respect to the bottom polariser, the light is transmitted resulting in a bright (white) state. Upon applying a voltage, the LC molecules orient normal (homeotropic) to the substrates (Fig. 9.2b). During this process the initially present twist is lost. As a consequence the polarised light is no longer rotated and the analyser blocks the light. This is the black state of the TN LCD. For proper switching it is important that upon reducing the voltage both the twist sense and tilt directions are regained in order to prevent the formation of reverse tilt and twist disclinations that will disturb the image [13].

The type of LCD used for a specific application may differ. There are three types of LCDs: transmissive, reflective and transflective LCDs. In Fig. 9.3 the three types are presented. In transmissive LCDs a backlight behind the display provides the light to display the image whereas reflective LCDs use the ambient light or a frontlight. Transflective LCDs contain both reflective and transmissive areas within each pixel and are provided with a backlight to display the image in a dark environment. In general transmissive LCDs are used in large size applications like monitors and televisions, whereas reflective and transflective LCDs are used in mobile applications, like telephones and Personal Digital Assistants (PDA's). This implies that the size of LCDs covers a wide range: from smaller than 1" for e.g. wristwatches and projection systems up to 54" for televisions.

This Chapter is organised as follows. After introducing the current market situation, the use of non-contact alignment methods like photo-alignment and ion-beam treatment are discussed, as they can be preferred alignment methods in manufacturing without dust generation and electrostatic damage to TFTs. Photoalignment is illustrated for two main application areas.

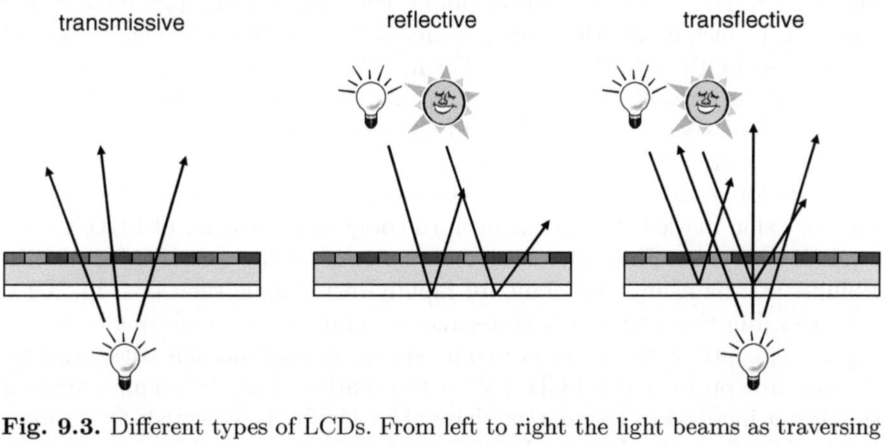

Fig. 9.3. Different types of LCDs. From left to right the light beams as traversing through a transmissive, a reflective and a transflective LCD are depicted.

The first application is the alignment of the LC molecules in an electro-optical switch, where the LC director varies with the driving voltage on the display (Sect. 9.3). The second application is the alignment of reactive LC molecules that are used for the manufacturing of optical films required to enhance the LCD performance (Sect. 9.4). In this application a photo-polymerisation process following the LC alignment permanently fixes the orientation of the reactive LC molecules. The first application of photoalignment is found to be usefull for large size displays whereas the second application is presently promoted for the use in smaller displays in the mobile market segment.

Finally, a completely different approach to noncontact alignment, namely ion beam treated amorphous carbon films, will be briefly discussed (Sect. 9.5). This approach, which resulted from earlier investigations of ion beam irradiated polymer films, has been demonstrated to deliver excellent front-of-screen quality LCD's for laptops and desk tops.

9.2 Current market situation

The market share of LCDs varies with the application area and size. Since the introduction of LCDs in small devices like wristwatches, new applications like mobile telephones and laptops have emerged. Recently the monitor market and even the television market have become very attractive areas to apply LCDs. Figure 9.4 gives an overview of the flat panel revenues by application from quarter 1 in 2001 to quarter 1 in 2004. The highest revenues are obtained in LCDs for desktop monitors and notebooks. However, the market of mobile displays is by far the largest regarding the number of sold displays.

For laptop computers currently the only display principle available is LCD due to its restrictive demands (low power, low weight, thin, high information content). In fact it was the laptop business that enabled the penetration of larger sized LCDs onto the market. For monitors the market share is increasing rapidly as a result of replacement of the Cathode Ray Tube (CRT) based monitors. The space saving aspect of LCD monitors in the office environment was the first drive for the CRT monitor replacement. Nowadays also the CRT monitors for personal use at home are being replaced in a rapid pace. For the television market, Sharp has been the only manufacturer of LCD TV for a number of years. Recently other major players in the electronic arena like Philips, LG, Samsung and Sony are rapidly introducing their own LCD TV products. Improvements in both technology and display performance (viewing angle, contrast ratio and switching speed) in combination with price reduction are pushing the LCD TV to the market. For the coming years a significant increase in the market share of LCD TVs is expected, as is shown in Fig. 9.5, where the decrease in CRT market share versus the increase in the market share of flat panel displays (including plasma display panels (PDP))

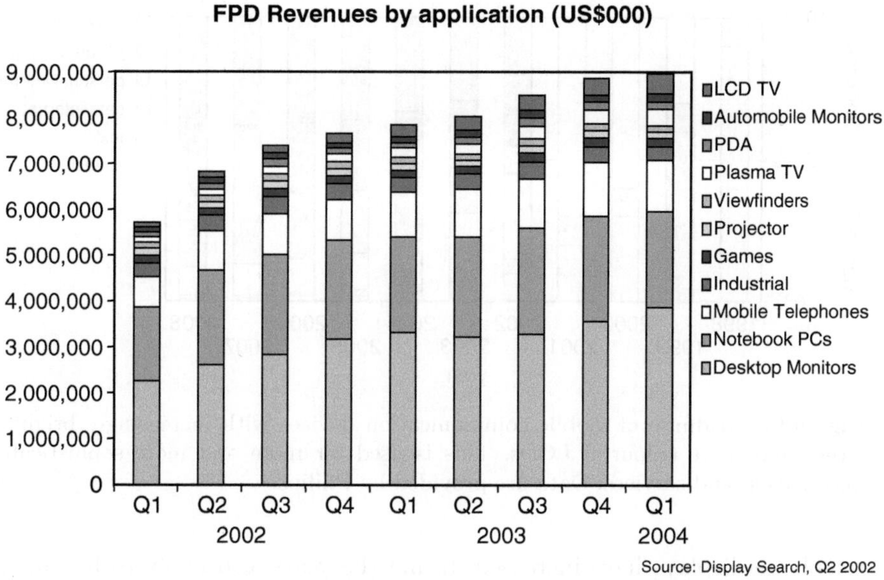

Fig. 9.4. Overview of flat panel revenues by application in the period Q1, 2002 to Q1, 2004.

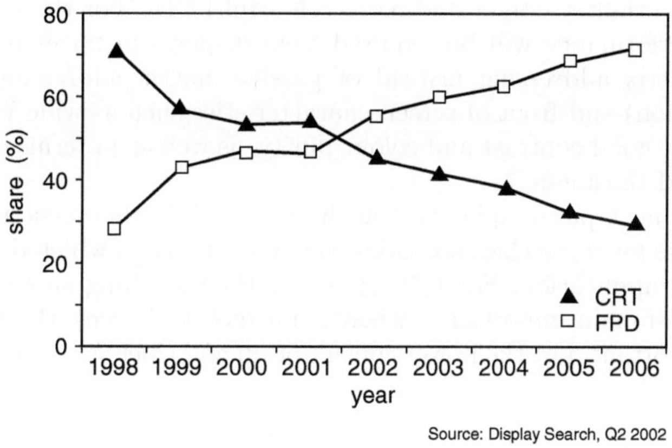

Fig. 9.5. Estimation of the market share of CRT vs flat panel displays (including PDPs) between 1998 and 2006.

is predicted. The forecast from Display Search predicts a growth in the number of LCD-TVs from 1.3 million units in 2002 to 31.1 million units to be shipped in 2007.

Although for mobile applications new display principles based on organic light emitting diodes (OLED/PLEDs) are close to market introduction, the market share of LCDs in telephone displays is still 100%. The demands on

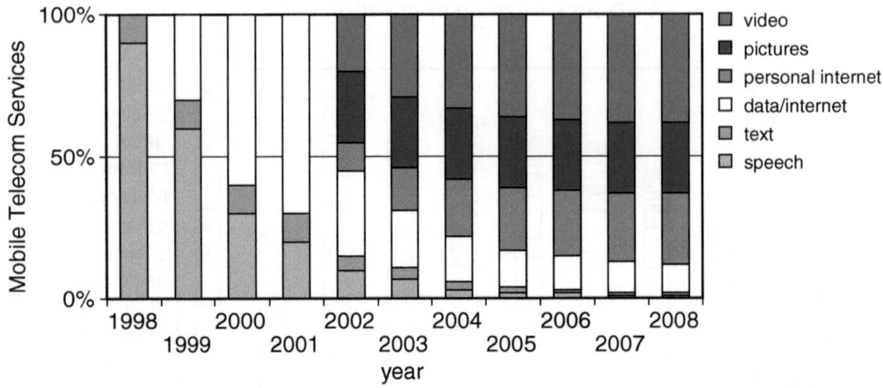

Fig. 9.6. Roadmap of mobile communication devices with increasingly brighter, larger and more colourful LCDs. This is used for more and more sophisticated information and services. Data are provided by Philips.

telephone displays have increased during the years, especially to be able to display image and video information. As a consequence, the current trend for displays used in telecommunication is to become ever more complex.

Figure 9.6 shows a roadmap for mobile communication devices with increasingly brighter, larger and more colourful LCDs. For these purposes improved performance will be required from displays in terms of driving (i.e. active matrix addressing instead of passive matrix addressing, low power consumption) and front of screen characteristics such as wide viewing angle, brightness, good contrast and colour purity, as well as in terms of robustness, weight and thickness.

In the next paragraphs we will discuss the LC alignment tools we have to optimise for these characteristics, the importance of which depends on the particular application. For television sets the switching speed and viewing angle are of main importance, whereas for mobile devices, the viewing angle is less important, but the power consumption and thickness (and thus weight) are.

9.3 Large Size LCD – Attempts Towards Wide Viewing Angle LCDs

In Sect. 9.2 the increasing penetration of LCDs in the monitor and television market was addressed. Especially for television, the viewing angle is of great importance. Here the CRT sets a reference performance that is difficult to match because of the Lambertian light emission of its phosphors providing good contrast and brightness at all angles. Transmissive LCDs are in fact nothing more than light valves in front of a backlight. Conventional LCDs use TN cells. These types of LCDs have an intrinsic dependence of the transmission on the viewing angle. TN displays generally operate in the normally white mode, which means that the display is used between crossed polarisers. The transmission in the bright state is almost independent of the viewing angle. Upon applying a voltage, the light is blocked by the outside polarizers (black state, see Fig. 9.2). The display shows a high-quality dark state in the direction of normal view (i.e. perpendicular to the glass substrates) because the light from the backlight does not experience a birefringent medium for normal view. At sufficiently high voltage any residual transmission is mostly determined by the properties of the polarisers, such as contrast ratio and colour neutrality. At off-axis angles, the transmission is however affected by the birefringence of the addressed medium. Obliquely passing light will partially leak through the pair of crossed polarisers, reducing the obtained contrast, which is the ratio between the intensities in the bright and the dark state. In fact the viewing angle dependence may be split up in two artefacts: (1) contrast degradation at larger polar angles and (2) gray scale inversion.

Figure 9.7 shows both artefacts. The right hand side photograph in Figure 7a shows the image on a TN display under normal view, the photograph on the left hand side shows the same image at an oblique angle. The contrast at oblique angle is lower than for the normal view (i.e. the black image appears light gray). In Fig. 9.7b the bottom photograph displays the image at normal view and the top photograph shows the image at an oblique angle. Within the white square it can be seen that the intentionally white image turns gray under oblique angles, whereas the black area is perceived as white. This is called gray scale inversion. As shown in Fig. 9.7 both artefacts depend on the azimuthal angle of view, which relates to the director profile of the LC molecules in the driven state.

In Fig. 9.8 the origin of contrast degradation is explained. When looking at the mid-plane director of the LCD, the image perceived from the right

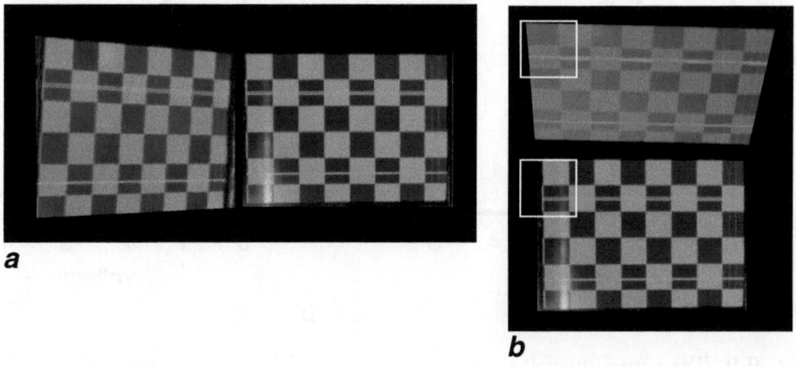

Fig. 9.7. Contrast degradation (a) and gray scale inversion (b) as observed in a twisted nematic display.

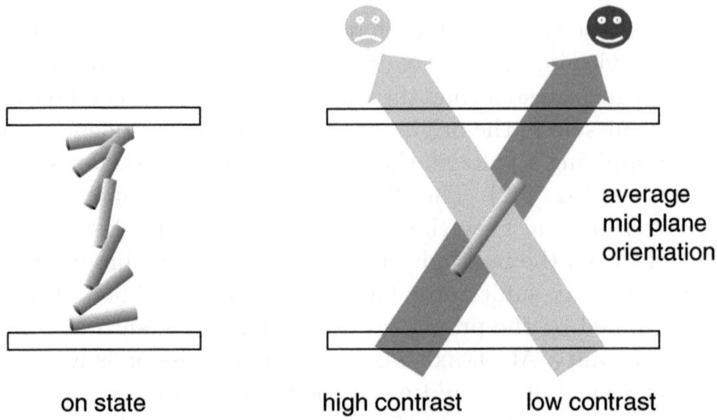

Fig. 9.8. Schematic explanation of viewing angle artefacts when looking at an LCD under oblique angles.

hand side differs from the left hand side image. This is due to a difference in the optical path length for the two directions as the birefringence is different for the two viewing directions. As a result the viewer on the right hand side the light is modulated in the "correct" manner and this viewer sees the correct image, whereas for the viewer on the left hand side a lower contrast is perceived.

Gray scale inversion is explained in Fig. 9.9. Figure 9.9a shows a transmission voltage curve for normal incidence angles. For normal incidence the

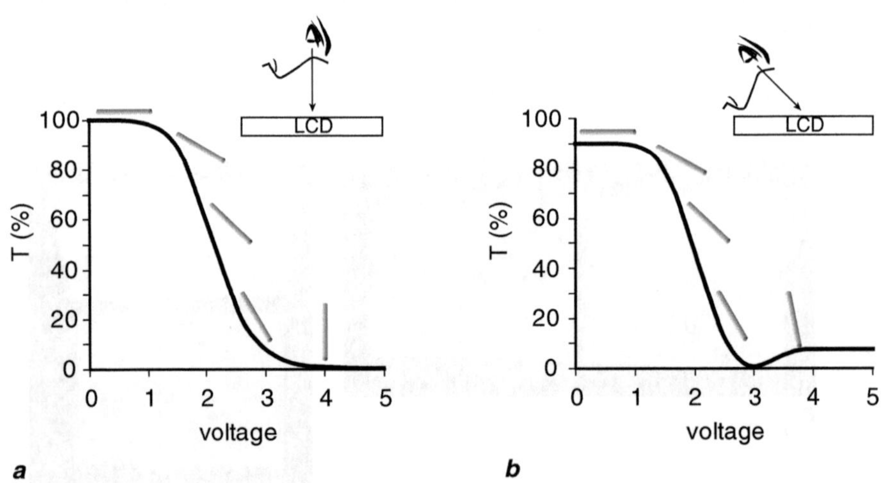

Fig. 9.9. Transmission-voltage curves for a 90° TN LCD for normal view (a) and for view at oblique angle (b). The ovals represent the mid plane director of the LCD.

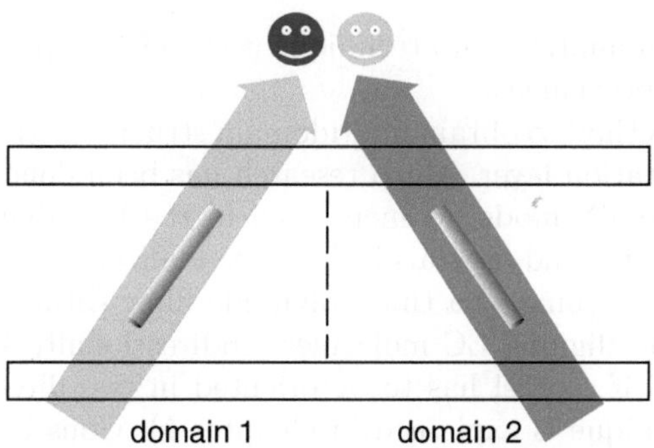

Fig. 9.10. Improvement of viewing angle performance of a TN LCD by dividing pixels into domains with opposite low and high contrast directions.

transmission of an LCD decreases monotonically as a function of increasing voltage. The mid-plane director gradually changes from a planar orientation to a homeotropic orientation upon applying a voltage, as was explained in Fig. 9.2. As a result, the image will gradually change from white to light gray, gray, dark gray and finally to black. The different appearances of the image for the various values of the applied voltages are called gray scales. The transmission voltage curve may show a minimum in transmission for oblique incidence (Fig. 9.9b). This means that driven states with voltages higher than this minimum look brighter than the minimum transmission state (with increasing applied voltage the image appears e.g. white, gray, black, gray instead of going gradually from white to black). In Figure 9.9b this is depicted by the orientation of the mid-plane molecules.

For many applications, the contrast degradation and gray scale inversion as observed at larger viewing angles are not acceptable. In general, the viewing angle of LCDs is enhanced by simultaneous optimisation of the liquid crystal mode, the use of multidomain structures and the application of one or more retardation foils [14]. In the following the focus will be on the use of multidomain structures and retardation foils. The various liquid crystal modes are discussed in references [5-12].

Multidomain structures are useful in increasing the viewing angle of LCDs. They are especially helpful in reducing the gray scale inversion, generally at the cost of contrast loss. In multidomain structures the pixels are divided into domains with different optimal orientations with respect to the viewer. In Fig. 9.10 a two-domain structure is shown for the TN mode. The pixel is divided into two domains of which the mid-plane directors are rotated with respect to each other. In this configuration the viewers' eye will integrate the light rays coming from both domains. As a result the gray scale inversion is reduced. The contrast is slightly degraded because a low and high contrast domain are mixed [15–17]. Compensation of the contrast reduction is achieved by the use of retardation foils [18–20]. The combination of retardation foils with a multidomain technique is therefore a preferred method for viewing angle improvement.

The easiest method to obtain multidomain structures in an LCD is by patterning the orientation layer. Much research has been done in this direction. Especially for the TN mode, as mentioned in the introduction, the conventional alignment of liquid crystals in LCDs is realized at a polymer interface by unidirectional rubbing of a thin polyimide film. Although it is a very effective process for aligning LC molecules unidirectionally, it becomes a very elaborate process if a pixel has to be oriented in two directions. (The photoalignment technique is explained in Chapter 1). Consecutive rubbing and photolithographic steps then need to be performed. Photoalignment [21,23] is an alternative technique to pattern orientation layers. This is a non-contact alignment technique in which the polymer orientation layer is exposed to UV light prior to contacting it with the liquid crystal. The major advantage of photoalignment over rubbing is the possibility to expose the orientation layer through a mask in such a way that several orientations can be obtained within one polymer layer.

Factors that determine the uniformity of LC alignment on a photoalignment layer include exposure dose, exposure angle, degree of polarisation and wavelength of the UV source. These parameters are specific for each set of photoalignment material / liquid crystal type and need to be optimised for each combination in order to obtain satisfying alignment.

In Fig. 9.11 examples of patterned LCDs are presented. Figure 9.11a shows a stripe pattern with line widths of 3 and 10 μm, and in Fig. 9.11b a checkerboard pattern with 10 and 100 μm features is presented. The alignment material used in these examples is Staralign 2110 (Vantico). Features down to 3 μm size are well resolved, which underlines the strength of the pat-

a *b*

Fig. 9.11. Polarisation microscope photographs of test displays composed of a patterned photoalignment layer (Staralign 2110 from Vantico) and a uniformly rubbed layer, filled with LC mixture MLC 7700 000 (Merck). The polarisers are crossed. (a) Stripe pattern with line widths of 10 and 3 μm, (b) checkerboard pattern of 10x10 and 100x100 μm.

terning capabilities of the photoalignment technology. Alternative alignment techniques that can be used for patterning have been described throughout this book. However, because the substrate size is between 1000x1100 and 1500x1800 mm^2 for the current generation of cleanrooms, photoalignment is the technology that can most easily be introduced in LCD manufacturing lines.

The retardation foils are used to enhance the front-of-screen performance of both mono- and multidomain LCDs [14]. In Fig. 9.12 the action of retardation foils is explained for a 90° TN display [19]. For a conventional LCD, the polarisers are mutually crossed, such that the LCD is transmissive (white) in the off-state and light blocking (black) in the on-state. In the black state, the LC director is almost upright throughout the display except at the interface between the LC and the alignment layer. It is this thin layer of LC molecules that causes background retardation in the LCD responsible for loss of contrast when viewed under oblique angles. In addition, due to non-sufficient driving voltage of the display drivers, the mid plane director is sometimes not completely upright, which also causes loss of contrast. To improve the contrast a retardation foil may be added. These foils are designed to optically

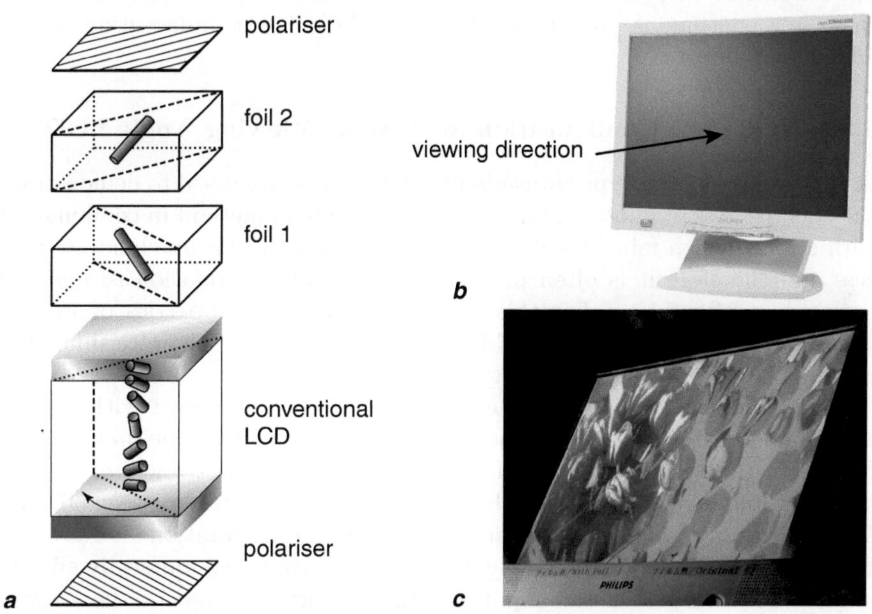

Fig. 9.12. Action of two retardation foils for a 90° TN LCD. (a) Configuration, (b) viewing direction w.r.t. display as shown in (c) example of viewing angle improvement by use of the two foils when looking to the display obliquely from the right hand side. On the left hand side of the photograph the foils are attached whereas on the right hand side the image without foils is presented.

match the director profile in a particular driven state of the LCD, e.g. a dark gray tone. As it is a static foil compensation it is only optimal for the selected single gray tone. These foils mainly act to enhance the contrast under oblique angles of view, but they cannot entirely compensate for gray scale inversion. In the example shown in Fig. 9.12 two retardation foils composed of calamitic molecules are oriented such that they correct for the non-upright LC molecules at both the bottom-side and the top-side of the LCD. To optimise the viewing angle in a proper manner, the foils need to have the correct retardation value as well as good orientation with respect to both polarisers and LCD. Figure 9.12b shows a photograph of a display with retardation foils on the left hand side and without foils on the right hand side. This photograph clearly shows the contrast improvement by the addition of retardation foils.

For each specific application a thorough consideration of the advantages and disadvantages of the chosen combination of multidomain structures and retardation foil(s) is necessary. Furthermore, there is a delicate balance to find the most suitable method to make wide viewing angle displays taking into account both performance aspects and economic considerations. The enhanced viewing angle performance should not compromise other aspects like brightness, response speed and driving voltage ranges. At the same time, there is a continuous pressure on the cost of display manufacturing and of added components.

9.3.1 New LCD Configuration for a Wide Viewing Angle LCD

In this section a prototype transmissive LCD will be discussed to demonstrate the effectiveness of photoalignment as a multidomain method in combination with a retardation foil. Prototyping of new LCD designs is an elaborate process and therefore it is often preceded by computer simulation to find the optimum configuration. For this purpose the optical properties of various combinations of multidomain LC layers and retardation foils were simulated using 2x2 Jones matrix modelling [24–26].

Single and dual domain LC configurations were combined with respectively two or one optical compensation foils [27]. For the single domain configuration the standard TN effect was taken with a twist angle of 90°. In the dual domain configurations the twist angles were varied between 60° and 90°. A schematic of the mono- and dual domain configurations is shown in Fig. 9.13. The monodomain configuration is built from two uniformly aligned substrates (with pretilt and alignment direction as indicated by the arrows) with a left-handed twist sense, and two retardation foils (Fig. 9.13a). The dual domain configuration is composed of a uniformly aligned layer on the top substrate with an orientation and pretilt angle of the LC director, as indicated by the continuous arrow, and a patterned layer on the bottom substrate with no pretilt angle and two orientations, as indicated by the dotted arrows (Fig. 9.13b).

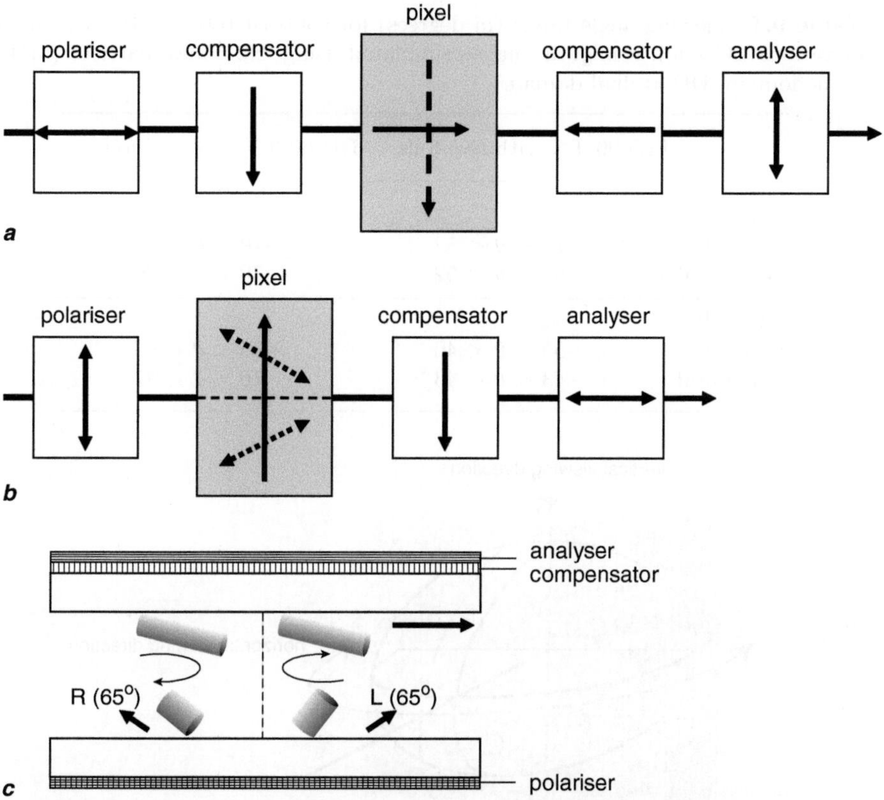

Fig. 9.13. Configurations of the studied LCDs. Top view of a monodomain 90° TN LCD (a), and dual domain 65° TN LCD (b) and cross section of a dual domain 65° TN LCD (c). The light path through the LCD is from left to right. Compensator denotes the retardation foil.

The dual domain display has a polariser with its transmission axis parallel to the LC orientation at the top substrate (continuous arrow) and an analyser with its transmission axis crossed to the polariser. It can be seen from Fig. 9.13b that the LC orientation at the bottom surface is neither parallel nor perpendicular to at least one of the polarisers. This means that the optimum contrast (obtained when a 90° TN LCD is placed between crossed polarisers) cannot be obtained. Therefore a retardation foil needs to be inserted in the light path to optimise the contrast of the display.

The simulation results are shown in Table 9.1. The viewing angle range for contrast ratios above 5 and 10 are given for horizontal and vertical viewing directions. The horizontal and vertical viewing directions are taken as depicted in Fig. 9.14.

Table 9.1. Viewing angle range (in degrees) for contrast ratios (CR) of 5 and 10 of various LC / foil configurations as simulated using 2x2 Jones matrices (MD = monodomain, DD = dual domain).

	MD 90 TN with two foils	DD 60 TN with one foil
CR $\geq$ 5		
vertical	$-70 < \theta < 52$	$-70 < \theta < 70$
horizontal	$-53 < \theta < 52$	$-70 < \theta < 70$
CR $\geq$ 10		
vertical	$-55 < \theta < 40$	$-65 < \theta < 65$
horizontal	$-43 < \theta < 43$	$-70 < \theta < 47$

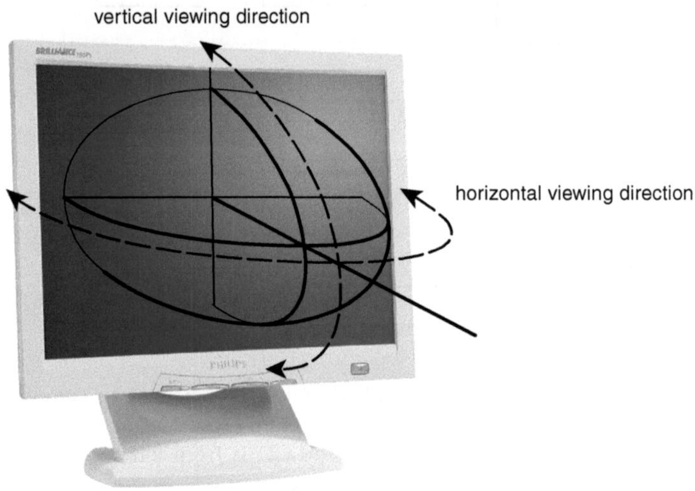

Fig. 9.14. Explanation of horizontal and vertical viewing directions.

Comparison of the monodomain TN display with two retardation foils with the dual domain configuration shows that the viewing angle range for contrast ratios of 5 and 10 is improved for the dual domain configuration. It was therefore concluded that the dual domain TN LCD with only one compensation foil should give an improved viewing angle performance compared to the foil compensated monodomain TN display.

9.3.2 Demonstration of Dual Domain LCD

To demonstrate the actual performance improvement of a dual domain LCD, a demonstrator display has been made. In this display the domains are made by patterning a photoalignment material supplied by LG Cable. This material reacts via a [2+2] cycloaddition reaction. On this alignment layer the LC material aligns perpendicular to the direction of linearly polarised UV light. The patterned layer is exposed with normally incident linearly polarised UV light in a two step sequence to obtain the two different orientations within the pixels (Fig. 9.13b). In the first step the substrate is exposed through a mask with an exposure dose sufficient to attain high quality uniform alignment within the exposed areas. In the second step the whole substrate was exposed, but as the alignment material has already reacted in the areas exposed in the first step, only the non-exposed areas obtained the new orientation direction. On the uniformly aligned layer the orientation and pretilt angle are induced by photoalignment under oblique angle.

In an LCD the twist sense of the LC material is determined by the combination of the average tilt angle through the display, the angle between the orientation directions on the two substrates and the concentration of chiral dopant in the liquid crystal. In the dual domain configuration the twisted nematic layer in the two domains exhibits an unusual opposite twist sense. To obtain the opposite twist sense within each pixel, the LC mixture may not contain a chiral dopant. As there is no pretilt on the patterned substrate and no dopant in the LC material, the twist angles in the two domains are restricted to values below 90° in order to prevent reverse twist in the display. From experimental results with various twist angles in combination with simulations, a twist angle of 65° was chosen for the dual domain configuration. As LC material a non-doped super-fluorinated LC mixture (Merck, Darmstadt) for TFT applications was used. The optical retardation in the LCD, $d\Delta n$, was 0.34 μm.

Figure 9.15 shows the viewing angle properties of a 5-inch display made in the configuration as shown in Fig. 9.13b. This display shows the improvement of the optical performance under oblique angles as compared to a normal non-compensated TN display Fig. 9.7.

In Table 9.2 the results of the measurements, of the viewing angle range and gray scale inversion of a foil compensated active matrix TN LCD and the new dual domain configuration are compared. From this Table it can be concluded that the range of viewing angles with contrast ratios above 5 and 10 is better for the dual domain display. Additionally, the viewing angle range, where the gray scale inversion occurs, has decreased considerably for the dual domain LCD.

The presented configuration is shown to have the potential to make LCDs with improved viewing angle properties. For this LCD only one of the alignment layers needs to be patterned and additionally no pretilt is required on this layer. Therefore the configuration is relatively tolerant against variations in the photoalignment process. In addition, the need for only one retardation foil makes it an interesting option in view of cost reduction.

In principle all kinds of domain configurations can be made using the photoalignment technique. Besides the LCD configuration described above,

Fig. 9.15. Viewing angle properties of a 5 inch dual domain LCD as made using the configuration shown in Fig. 9.13b.

Table 9.2. Viewing angle range and gray scale inversion range (in degrees) of various LC / foil configurations as measured on active matrix test displays (MD = monodomain, DD = dual domain). The angular range accesible for measurements was limited to angles between −70 and 70 degrees.

	MD 90 TN with two foils	DD 60 TN with one foil
$CR \geq 5$		
vertical	$-70 < \theta < 51$	$-70 < \theta < 70$
horizontal	$-70 < \theta < 70$	$-70 < \theta < 70$
$CR \geq 10$		
vertical	$-65 < \theta < 39$	$-45 < \theta < 70$
horizontal	$-55 < \theta < 55$	$-70 < \theta < 65$
Range with no gray scale inversion		
vertical	$-10 < \theta < 15$	$-70 < \theta < 45$
horizontal	no GSI	no GSI

alternative twist angles and different numbers of domains have been described in the literature. However, in display manufacturing, the photoalignment is not used as the alignment technology yet. This is mainly due to the problem of image sticking. Image sticking means that a previously displayed image is still visible in an image displayed afterwards. Currently it is not known whether the image sticking is a result of the materials used for photoalignment, the photoalignment process itself or other processes and materials used in LCD manufacturing. In applications with active switching of LC molecules, the image sticking is a serious issue. It thus remains important to continue the search for alternative alignment techniques to replace rubbing and make patterning of alignment layers possible to increase the number of parameters to tune the LCD performance to the customers' demands.

9.4 Performance Boost in Small Portable Displays Enabled by Director Control of LC Networks in Retardation Foils

In the previous section the necessity of a uniform alignment of LC molecules used as a light valve in large area dual domain LCDs was discussed. Apart from the alignment of LC molecules used as a light valve, alignment of LC molecules is also crucial for the fabrication of the optical components that are required to boost the optical performance of the LCD. In this section we will address the alignment of *LC molecules applied in thin film retardation foils*. Currently, the retardation layers are laminated on the exterior of the display. Such exterior films consist of an optical active layer, protection layers and a carrier sheet. These films add to the thickness of the display (typically >100 micron per film), which is undesirable for mobile applications, where they have to be integrated in flat and light-weighted devices such as mobile phones and palm top computers. In near future these foils will start to dominate the cell thickness as demonstrated in Fig. 9.16. The panel thickness is currently mainly determined by the thickness of the individual glass plates being about 0.5 mm. However in 2008 the thickness is expected to be reduced to only 0.2 mm. In order to prevent the optical foils being thickness determining, a new technology is developed to make the foils and with that the LCD thinner and lighter [28].

The new technology should allow for the fabrication of micron thin films with the desired optical properties that can be applied inside the LCD. Direct integration into the device is also beneficial to the robustness of the display since the exterior of the display is now formed by the (glass) substrate. Techniques that allow for micrometer thin films that can be applied inside the display are spin-coating or print-coating techniques. In addition to being wet-coatable, the used materials have to be stable against further processing such as ITO sputtering. Apart from making the device thinner, lighter and more robust, the integration of components into the display leads to a number of other advantages for the optical performance, because parallax issues are minimised and local variation in the optical properties (also referred to as patterning) is enabled by the presented technology (i.e. reactive mesogen). These additional degrees of freedom are a prerequisite to improve the current state-of-the-art LCDs further on front-of-screen performance. The

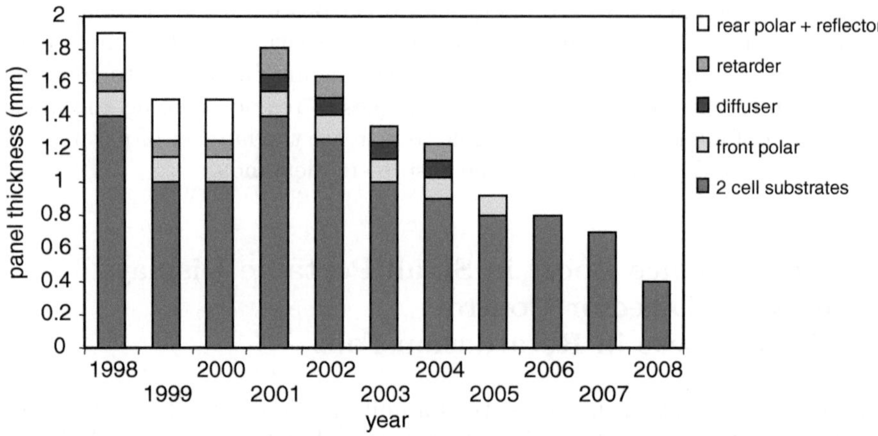

Fig. 9.16. Philips roadmap for the thickness of small LCD panels for mobile communication equipment. The LCD panel thickness is reduced by thinner glass substrates (eventually plastic) and by the replacement of external foils by in-cell optical films. The thickness of external foils contribute significantly to the total panel thickness.

reason for that is that the number of free parameters determining the optical mode is less than the number of front-of-screen performance parameters in the current display design [29]. Computer simulations have indicated that a patterned retardation foil may improve the viewing angle in transmissive displays [30], while it may improve contrast and brightness in both transflective [29] and reflective displays.

9.4.1 The Reactive Mesogen Technology Applied for Uniaxial and Biaxial Optical Fetarders

The commercially available retardation foils are formed by carefully stretching extruded plastic sheets to obtain a fixed retardation value. These sheets are glued on top of the display, between the glass and the top polariser film. The reactive mesogen film technology, in which the retardation films are applied as thin layers directly via wet-coating processes uses reactive liquid crystal molecules to form the birefringent film. Figure 9.17 presents the molecular structure of such a molecule. The reactive LC molecule consists of a stiff central core that enables the formation of a nematic phase, and a polymerizable end group, e.g. an acrylate. This acrylate group reacts via chain reaction polymerisation upon the UV exposure in the presence of a photoinitiator. The reactive end group is attached to the stiff central core by a flexible spacer to facilitate the formation of an ordered layer by the polymerisation reaction [31–34].

Figure 9.18 shows the process steps for the fabrication of a thin film retarder. A solution of a mixture of the reactive LC monomers doped with

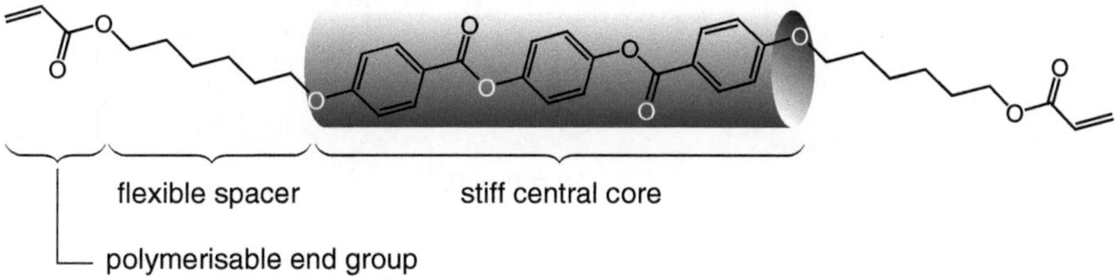

Fig. 9.17. The molecular structure of a reactive LC molecule.

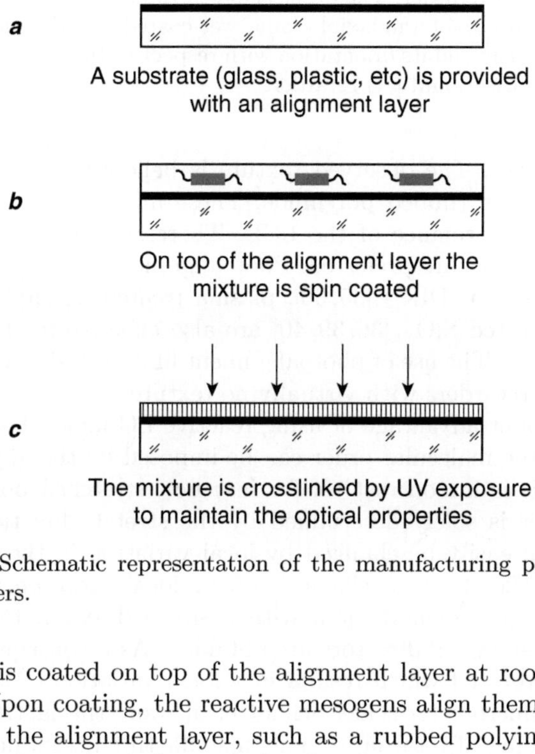

Fig. 9.18a–c. Schematic representation of the manufacturing process of coated thin film retarders.

photoinitiator is coated on top of the alignment layer at room-temperature (Fig. 9.18b). Upon coating, the reactive mesogens align themselves in a single domain on the alignment layer, such as a rubbed polyimide layer. The achieved molecular order is frozen-in by the photo-polymerisation of the reactive end groups upon UV exposure (Fig. 9.18c).

As a result, an ultra-thin birefringent layer is obtained that exhibits a good chemical, thermal and photochemical stability. By varying the coating conditions, as well as the concentration, retardation values up to 600 nm are attainable within 2 % accuracy. The high stability of the films allows for the deposition of ITO conductive layer sputtering or for the fabrication of stacks of retardation films as used in for example wide band $\frac{1}{4}$ λ retardation foils.

Figure 9.19a is a schematic representation of the molecular order on the alignment layer: a planar orientation is obtained over the entire substrate. Figure 9.19b shows an example of a uniaxial planar retarder viewed in between two crossed polarisers. In this example a rubbed polyimide is used as

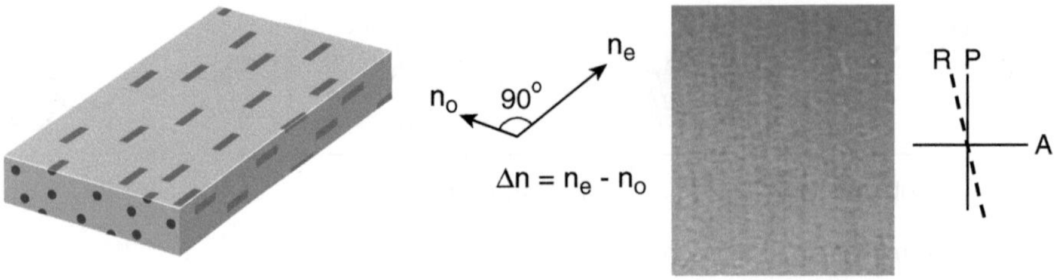

Fig. 9.19. (left) A schematic of the molecular order of the reactive LC molecules in a coated thin film retarder. The order is forced to be planar by the rubbed polyimide. (right) An image of a uniaxial planar wet-coated retarder viewed in between two crossed polarisers and its orientation with respect to the crossed polarisers. (A) Analyser, (P) polariser, and (R) retarder.

an alignment layer. The observed texture is believed to be a result of the microgrooves of the rubbed polyimide. These microgrooves can reduce the front-of-screen-performance of the LCD. Therefore alternative non-contact alignment techniques are preferred. Photoalignment [21–23], plasma treated diamond like carbon (DLC) [35, 36], plasma treated polyimide [37, 38], and slantwise evaporated SiO_x [36, 39, 40] are also effective in the alignment of reactive mesogens. The use of photoalignment films and slantwise-evaporated SiO_x results in retarders with virtually no texture.

The paramount advantage of using reactive LC molecules is that the directionality of the molecular order can be imposed by the applied boundary conditions. This is in contrast to, for instance, stretched polymer films, in which the order is solely determined by the stretch direction. Hence, the LC director can easily be changed by local variation in the boundary condition. Figure 9.20 shows a schematic of the local variation in LC director. Instead of having a monodomain with a single director, two uniform domains with different LC directors are obtained. As a consequence, the main optical axis is rotated which results in a local variation of the effective retardance on a micrometer scale. Schadt et al. were the first to demonstrate director patterning of retardation foils manufactured from liquid crystalline polymers with the aid of photoalignment [21, 41] [see Chap. 1]. The basis for the thin film optical retarder technology, as described in the following, is in-situ photo-polymerisation of liquid crystalline diacrylates [42]. The photoalignment technique is applied to manufacture a novel, high performance transflective LCD. The novel transflective LCD consists of a dual domain retardation foil with effective retardance values of $\frac{1}{4}\ \lambda$ and $0\ \lambda$ (see also Sect. 4.2).

Several mechanisms of photoalignment reactions have been reported in the literature and hence a variety of photoalignment materials are available for director patterning. For the manufacturing of a patterned retarder applied in a transflective display (Sect. 9.4.2), we have tested two commercially avail-

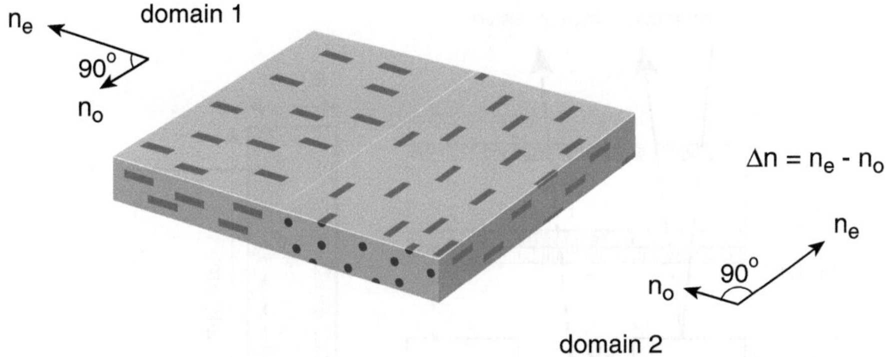

Fig. 9.20. A schematic of the molecular order of the reactive LC molecules after director patterning. The order is forced to be planar in two domains by a dual domain photoalignment layer. The domains exhibit identical $d\Delta n$ values, but the effective retardation depends on the angle between the main optical axis and the transmissive axis of the polariser.

able photoalignment materials, Staralign 2110 (Vantico) and RN1349 (Nissan Chemicals). The cycloaddition material Staralign 2110 has C=C double bonds in the side chain that can be crosslinked via [2+2]-cycloaddition reaction upon UV exposure (313 nm). The photodegradation material RN1349 is a photosensitive polyimide in which covalent bonds that are oriented with their absorption axes parallel to the electrical field vector of the UV light break (254 nm) (see Chap. 1). Patterning of the retarder is performed by two exposure steps being a mask and flood exposure of the photoalignment layer with polarised UV light prior to applying the retardation film. The sequence of the two step exposure depends on the mechanism by which the photoalignment material reacts with UV light. In case of the cycloaddition material, the first exposure is a mask exposure in which all bonds in the exposed areas are fixed. The second exposure is a flood exposure with a dose that is generally lower than the first exposure. The director orientation in the two domains will depend on the ratio between doses of the two subsequent exposures as well as on the actually applied doses and exposure times (see [28]).

9.4.2 Application in a Transflective LCD

A director patterned retarder can be used in a transflective LCD to boost the optical performance and to lower the power consumption. A transflective display is the combination between a reflective and a transmissive display.

Figure 9.21 shows a simplified cross section of a transflective LCD. It consists of two ITO covered glass plates, with the LC molecules in between, that act as the optical valve. In order to distinguish between the on- and off-state, two polarisers are required. A full colour display is created by addition of a colour filter. The reflective mode is enabled by a mirror on top of the bottom

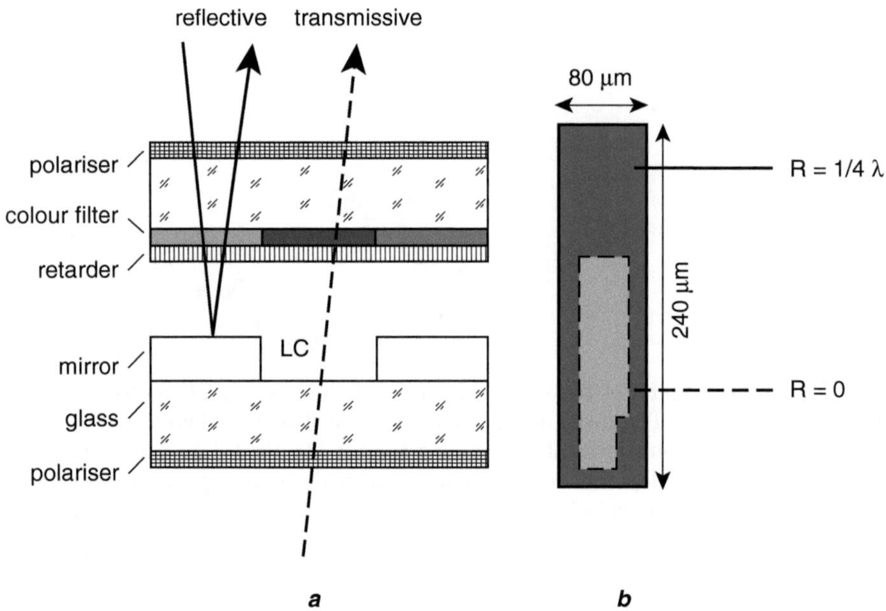

Fig. 9.21. (a) A cross section of a transflective LCD and (b) a top view of the pixel layout demonstrating the transparent hole in the mirror and the required optical activity expressed in retardation values.

glass plate. Incoming ambient light passes through the top polariser and the LC layer until it hits the mirror. The mirror reflects the light back into the LC layer and the top polariser. Hence the light passes the LC layer twice. The operation in the transmissive mode is enabled by transparent holes in the mirror. In this case light from the backlight passes through the bottom polariser, the LC layer, colour filter and top polariser before reaching the eye. Hence, in the transmissive mode the light passes only once through the LC layer. The fact that the length of the optical path in the transmissive mode differs from that in the reflective mode has the consequence that the reflective part is black in the non addressed state, whereas the transmissive part appears white. To achieve a simultaneously white picture in the transmissive and reflective part, a $\frac{1}{4}$ λ retardation foil is required in the reflective part, whereas the transmissive part does not need a retardation foil at all [29]. Figure 9.21b shows a top view of the pixel layout to clarify the area that needs a $\frac{1}{4}$ λ retardation foil.

Figure 9.22a shows an image of the patterned $\frac{1}{4}$ λ retarder made by the technology explained in this Chapter, viewed in between two crossed polarisers. The retarder is spin-coated on top of a colour filter. If the orientation of the optic axis of the $\frac{1}{4}$ λ retarder is parallel to the transmissive axis of one of the polarisers, the effective retardation is zero. An effective retarda-

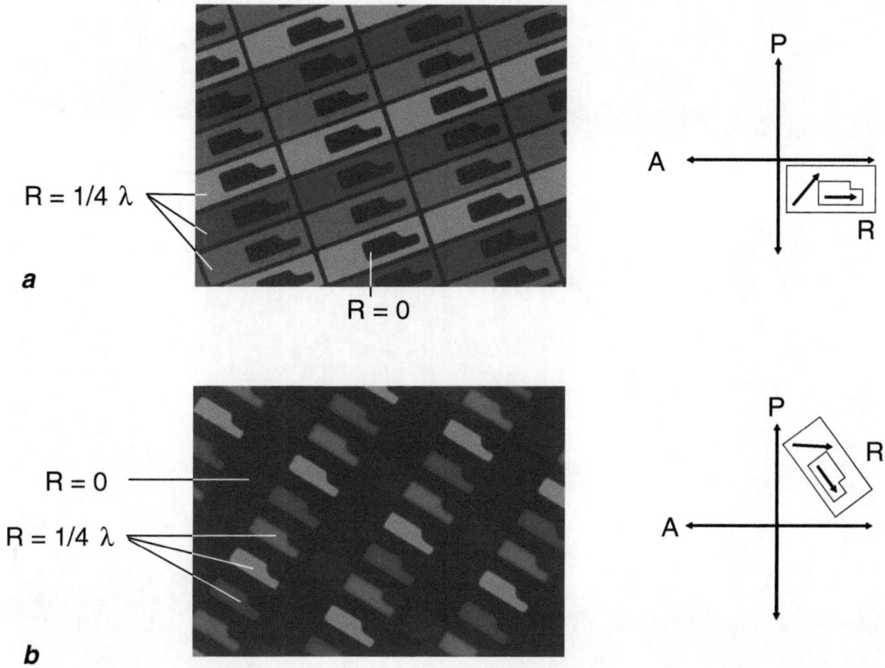

Fig. 9.22. (a) The patterned retarder as a single component coated on top of a colour filter,, viewed in between two crossed polarisers and its orientation with respect to the crossed polarisers. (A) Analyser, (P) polariser, and (R) retarder (a). This orientation represents the orientation as if it is applied in a transflective LCD.

tion of zero appears as a black state when viewed between crossed polarisers. Staralign 2110 purchased at Vantico is used as a photoalignment layer for the patterned retardation foil. An Ushio UV exposure system is used to irradiate the photoalignment layer. Figure 9.23 shows a picture of the transflective LCD in which a patterned retarder is applied. The pixels in the white part are not addressed (off-state), whereas the pixels in the black part have a voltage across (on-state) [29].

Using photoalignment as director patterning technique yields high resolution patterned retarders (down to 10 μm sized domains). However, director patterning leads to two domains having an intrinsic identical retardation value. As a consequence misalignment in the main optical axis of the retardation foil with respect to the transmissive axis of one of the polarisers will result in light leakage upon which the contrast decreases. In addition to the misalignment, the optical properties of the retardation foil are also viewing-angle-dependent in the transmissive part. Both effects strongly reduce the front-of-screen performance of the display and are therefore disadvantageous. Figure 9.22b shows an image of the retardation foil presented in Fig. 9.22a, but now 45° rotated to illustrate the sensitivity to misalignment. In this ori-

Fig. 9.23. A photograph of a transflective LCD with a patterned retarder. The black part is addressed, whereas the white part is not.

entation, the effective retardation of the reflective part becomes zero, while the transmissive part has an effective retardation of $\frac{1}{4}$ λ.

For an optimum front-of-screen performance, the transmissive part should preferably be isotropic (i.e. without preferred orientation). Isotropic domains can be created by a technology referred to as 'temperature patterning'. As mentioned earlier, the advantage of using reactive mesogens for the creation of retardation foils is that the order of the reactive mesogens before polymerisation depends on the boundary conditions. Isotropic domains are obtained by raising the temperature above the clearing point before polymerisation.

Figure 9.24 shows an example of a temperature patterned retardation foil viewed between crossed polarisers. In this example the transmissive part does not have any birefringence, which is demonstrated in Fig. 9.24b. The black

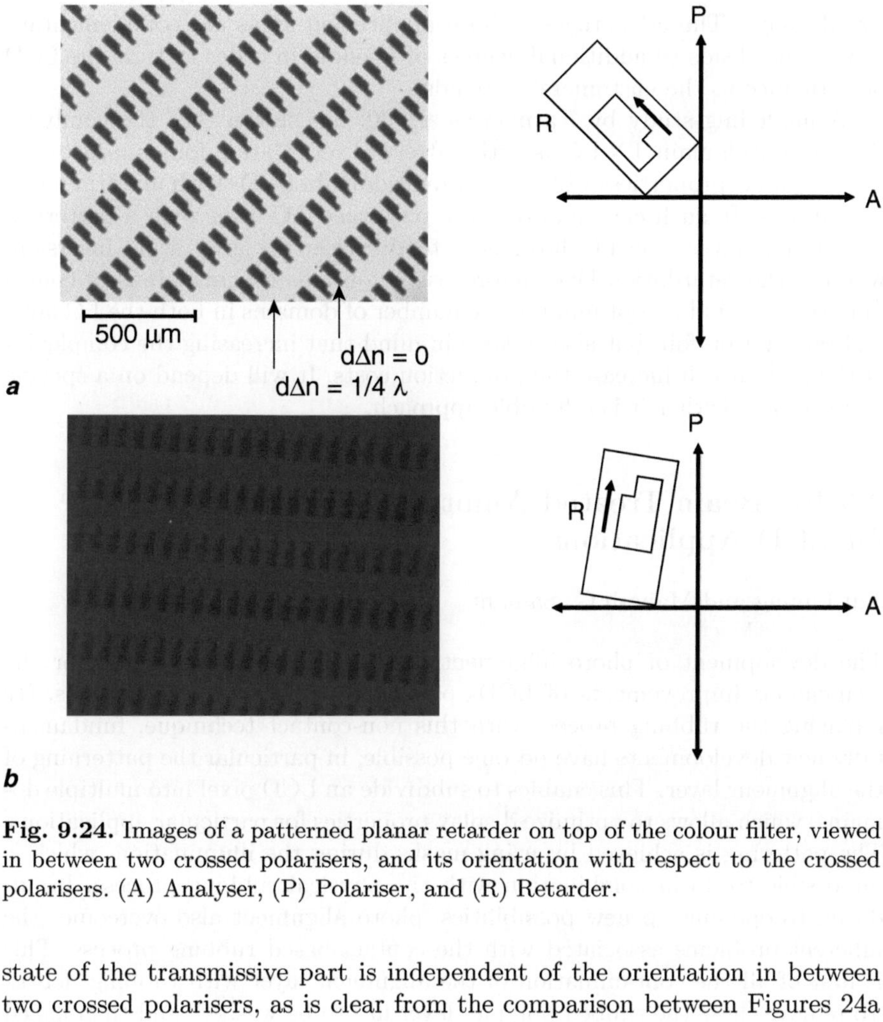

Fig. 9.24. Images of a patterned planar retarder on top of the colour filter, viewed in between two crossed polarisers, and its orientation with respect to the crossed polarisers. (A) Analyser, (P) Polariser, and (R) Retarder.

state of the transmissive part is independent of the orientation in between two crossed polarisers, as is clear from the comparison between Figures 24a and b.

9.4.3 Ultimate LCD Performance by Local Director Variations in Both LC Layer and Retardation Foils

In the previous sections we have discussed the use of the photoalignment techniques, in order to improve specific optically active parts of LCDs. We have shown that for the LC layer that forms the optical valve both in-plane and out-of-plane orientation is important to attain the correct switching behaviour. This is, however, not sufficient with respect to the front-of-screen performance, like contrast, brightness and viewing angle. In order to improve these properties of LCDs, patterned alignment for use in multidomain LCDs and retardation layers was discussed for respectively large and small size displays. The advantages of both application areas are complementary as their use leads to additional degrees of freedom in order to tune the LCD performance to the customers' demands.

A modelling study by Karman et al. [30] has shown that the combination of multidomain TN LCDs with patterned retardation foils improves the viewing angle properties of LCDs over multidomain LCDs with uniform retardation foils. In such combinations of a patterned LC layer with a patterned retardation layer, the LC layer acts to decrease the gray scale inversion, whereas the retardation layer improves the off-axis contrast. It is of course important to find an optimum in the number of domains in both the LC layer and retardation foil, but also to keep in mind that increasing the complexity of the system will increase the production costs. It will depend on a specific application whether it is a feasible approach.

9.5 Ion Beam Treated Amorphous Carbon Films for LCD Applications

Jan Lüning and Mahesh G. Samant

The development of photo alignment has played an essential role for the tremendous improvements of LCDs presented in the previous sections. By replacing the rubbing process with this non-contact technique, fundamentally new developments have become possible, in particular the patterning of the alignment layer. This enables to subdivide an LCD pixel into multiple domains, which allows to optimize display properties for particular applications. The pattering is achieved by using masks during the illumination, which is impossible to do in combination with the physical rubbing process. In addition to opening up new possibilities, photo alignment also overcomes the inherent problems associated with the contact-based rubbing process. This is first of all the contamination of the alignment layer with rubbing debris. To remove these the rubbed polymer layer has to be washed, and then baked in a furnace to remove the effects of the washing itself. Also, rubbing can lead to damage of the underlying electronics layer by local charge build up in the insulating polymer film. Furthermore, the rubbing process has serious reliability issues associated with the unpredictable local degradation of the rubbing roller.

For these and other reasons, it has been of great interest to industry to remove the rubbing process from the LCD manufacturing. Much effort went into the search and development of new, preferably non-contact, techniques capable of introducing the for liquid crystal alignment necessary anisotropy at a polymer surface. That this can be achieved with directional ion beam irradiation was discovered in the late 90's by Chaudari et al. [36]. Liquid crystal anchoring energies and pretilt angles (up to 12°), as required for LCDs were demonstrated on ion beam irradiated polyimide films. Based on the explanation of liquid crystal alignment given in Chap. 6, we now understand that preferentially breaking of directional phenyl rings and/or carbon double or triple bonds oriented perpendicular to the ion beam leads to the for liquid crystal alignment necessary charge anisotropy at the film surface. The low kinetic energy of the ions (typically in the range of 50 eV to 500 eV) ensures that only the surface layer of a few nanometer thickness is modified. For the preparation of alignment layers this is of tremendous importance as it means that the ion beam irradiation for LCDs, will not affect the underlying electronic circuit. Also, most of the process parameters (e.g., ion energy, irradiation dose and incidence angle) are not very critical, which is a desirable aspect from a manufacturing point of view.

Another part of the LCD manufacturing process that is of high interest to industry to be replaced is the polymer film itself. The polymer layer is printed in a wet deposition process for which relatively large amounts of organic materials and solvents are needed. To remove these, the deposited film has to be baked. These process steps are inherently dirty and time consuming, and it is desirable to replace them with a 'high-tech' compatible, clean, and fast process. Since all the process steps before the deposition of the alignment layer are done in a vacuum chamber, it would be advantageous, if the alignment layer could also be deposited in vacuum.

These considerations make it clear why the discovery of liquid crystal alignment on ion beam irradiated amorphous carbon, as discussed in Chapter 6, was of so much interest, as it opened the way to a completely dry and contact free processing technique. The risk of developing the required new manufacturing technology was outweighted by the prospect that simultaneously the polymer layer and the rubbing process, both costly steps of LCD production, would be replaced. The polymer layer is replaced with an only 5 to 20 nm thin film made out of a low cost material, which is deposited in a one-step process by evaporation or sputtering. Either one of these processes is not only more cost-effective than the polymer film deposition, but also more environmentally friendly. In addition, the rubbing process is replaced by a non-contact technique, which is based on a reliable, predictable process performed in the clean environment of a vacuum chamber. In fact, it even has been demonstrated that by directional sputtering of the amorphous carbon layer itself, one can combine film deposition and alignment preparation in a single process step [47]. Finally, both processes are fully compatible with the deposition of ITO, and could even be incorporated into the same vacuum vessel.

The fact that this new manufacturing technology was developed by IBM, was also due to the circumstance, that there, the deposition of amorphous carbon layers was a well-known and well-established process, since these layers are used as protective coatings on magnetic hard disks. Based on the developed know-how and the experience in depositing these as homogeneous and ultra-smooth thin coatings, there was confidence that amorphous carbon films could be deposited on large substrates as required for LCDs. It was also known how the properties of amorphous carbon films can be tailored by doping of amorphous carbon with hydrogen or nitrogen [46]. This allows to optimize, for example, the transparency and the electric resistivity of amorphous carbon for the application as alignment layer in an LCD.

Within about a year, in which further NEXAFS studies were an important tool for the optimization of the involved processes, twisted nematic LCDs (15-inch diameter) as used in laptops as well as the in-plane switching displays (22-inch) used in desktops were manufactured [36, 47]. One of the first LCDs using ion beam irradiated amorphous carbon as alignment layers is shown in Fig. 9.25. Both kinds of displays showed superior front-of-screen quality to displays produced by the rubbing process, primarily because of the absence of rubbing streaks. Not even on a microscopic level did AFM studies reveal any micro-grooving due to the directional ion beam irradiation. Also, as in the case of photo alignment, masking of the amorphous carbon layer during the ion beam irradiation [47] allows for the creation of multiple domains

Fig. 9.25. Image of an LCD using ion beam irradiated amorphous carbon as alignment layer (reproduced with permission from reference [36]).

within a single LCD pixel. As presented in the previous sections this can be exploited to improve the performance of LCDs relying on ion beam irradiated amorphous carbon as alignment layer.

Summary

Aligned LC molecules are proven to be useful in many applications ranging from the well-known LCDs, to lenses [39], shutters and components used in optical data storage [38]. For proper functioning of all these applications the control of the alignment of the LC molecules is of crucial importance.

In this Chapter we have tried to give an overview of the recent developments in the area of LC applications, in particular for LCDs. The current market situation shows that this application area is large and growing, but also puts challenges in optimization of LCD designs and functioning. Important developments in this respect are the use of non-contact alignment techniques like photo-alignment and ion-beam radiation.

Acknowledgements

Cindy Nieuwkerk and Bianca van der Zande acknowledge Marice van Deurzen, Titie Mol, Sander Roosendaal, Christel Renders and Johan Osenga for their contributions in the demo-making; Sjoerd Stallinga, Frans Leenhouts and Sander Roosendaal for the modelling work; Johan Lub, Emiel Peeters and Wim Nijssen for their work on the synthesis and processing of reactive mesogens; Marice van Deurzen, Christel Renders and Jan Steenbakkers for their assistance on the alignment of reactive mesogens; Dick Broer, Dick de Boer, Jos van Haaren, Peter van de Witte, Rob van Asselt, Gerwin Karman and Ciska Doornkamp for the fruitful discussions; and Olaf Gielkens and Hennie Alblas for their help in preparing this manuscript. Jan Lüning and Mahesh G. Samant acknowledge the manifold contributions to this work of their colleagues from the IBM Research Division in Almaden and Yorktown, USA, and the IBM Display Unit, Japan.

References

1. H. Zocher, Naturwissenschaften, **49**, 1015 (1925).
2. J. Stöhr, M. G. Samant, J. Elec. Spectrosc. Relat. Phenom. **189**, 98-99 (1999).
3. N. Ito, K. Sakamoto, R. Arafune, S. Ushioda, J. Appl. Phys. **88**, 3235 (2000).
4. J. M. Geary, J. W. Goodby, A. R. Kmetz, J. S. Patel, J. Appl. Phys, **62**, 4100 (1987).
5. M. Schadt, W. Helfrich, Appl. Phys. Lett. **18**, 127 (1971).
6. T. J. Scheffer, J. H. Nehring, Appl. Phys. Lett. **45**, 1021 (1984).
7. M. F. Schiekel, K. Fahrenschon, Appl. Phys. Lett. **19**, 391 (1971).

8. K. A. Crandall, M. R. Fish, R. G. Petschek, C. Rosenblatt, Appl. Phys. Lett. **64**, 1741 (1994).
9. P. J. Bos, K. R. Koehler-Beran, Mol. Cryst. Liq. Cryst. **113**, 329 (1984).
10. S. Kobayashi, SID International Symposium 1972, Digest of technical papers III, 68.
11. R. Kiefer, B. Weber, F. Windsched, G. Baur, IDRC '92 Digest, 547 (1992).
12. M. Oh-e, M. Ohta, S. Aranti, K. Kondo, IDRC '95 Digest, 577 (1995).
13. For general information on fabrication and driving of LCDs please refer to S. S. Kim, Information Display 2001, 8/01, 22-26, and S.S. Kim, Information Display, 9/01, 22 (2001).
14. For an overview see I -W. Wu, D. -L. Ting, C. -C. Chang, IDW'99, 383 (1999); C. -K. Wei, I. -W. Wu, IDW'98, 139 (1998).
15. N. Watanabe, M. Hirata, S. Mizushima, AMLCD Conference **97**, 45 (1997).
16. E. Hoffmann, H. Klausmann, E. Ginter, P. M. Knoll, H. Seiberle, M. Schadt, SID 98 Digest, 734 (1998).
17. M. S. Nam, J. W. Wu, Y. J. Choi, K. H. Yoon, J. H. Jung, J. Y. Kim, K. J. Kim, J. H. Kim, S. B. Kwon, SID 97 Digest, 933 (1997).
18. P. van de Witte, J. van Haaren, J. Tuijtelaars, S. Stallinga, J. Lub, Jpn. J. Appl. Phys. **38**, 748 (1999).
19. P. van de Witte, J. van Haaren, S. Stallinga Jpn. J. Appl. Phys. **39**, 101 (2000).
20. H. Mori, Y. Itoh, Y. Nishiura, T. Nakamura, Y. Shinagawa, Jpn. J. Appl. Phys. **36**, 143 (1997).
21. M. Schadt, K. Schmitt, V. Koznikov, V. Chigrinov, Jpn. J. Appl. Phys. **31**, 2155 (1992).
22. M. Schadt, H. Seiberle, A. Schuster, Nature **381**, 212 (1996).
23. H. Seiberle, K. Schmitt, M. Schadt, Euro Display 99 Digest 121 (1999).
24. R. C. Jones, J. Opt. Soc. Am. **31**, 488 (1941).
25. P. Yeh, J. Opt. Soc. Am. **72** 507 (1982).
26. A. Lien, Appl. Phys. Lett. **57**, 2767 (1990).
27. C. Nieuwkerk, T. Mol, S. Stallinga, F. Leenhouts, P. van de Witte, J. van Haaren, D. Broer, SID 00 Digest 2000, 842 and references therein.
28. B. M. I. van der Zande, A. C. Nieuwkerk, M. van Deurzen, C. A. Renders, E. Peeters, S. J. Roosendaal SID 03 Digest, 194 (2003).
29. S. Roosendaal, B. M. I. van der Zande, A. C. Nieuwkerk, C. A. Renders, J. T. M. Osenga, C. Doornkamp, E. Peeters, J. Bruinink, J. A. M. M. van Haaren, S. Takahashi SID 03 Digest, 78 (2003).
30. G. P. Karman, A. C. Nieuwkerk, B. M. I. van der Zande, E. Peeters, R. H. M. Cortie, M. H. W. A. van Deurzen, R. van Asselt, D. K. G. de Boer, D. J. Broer, Eurodisplay 02 Digest, 515 (2002).
31. R. Hikmet, J. Lub, Prog. Polym. Sci. 1165 (1996).
32. D. J. Broer, H. Finkelmann, K. Kondo, Makromol. Chem. **189**, 185 (1988).
33. D. J. Broer, J. Boven, G. N. Mol, G. Challa, Makromol. Chem. **190**, 2255 (1989).
34. D. J. Broer, R. A. M. Hikmet, G. Challa, Makromol. Chem. **190**, 3201 (1989).
35. J. van Haaren, Nature **411**, 29 (2001).
36. P. Chaudhari, J. Lacey, J. Doyle, E. Calligan, S. -C. A. Lien, A. Callegari, G. Hougham, N. D. Lang, P. S. Andry, R. John, K. -H. Yang, M. Lu, C. Cai, J. Speidell, S. Purushothaman, J. Ritsko, M. Samant, J. Stöhr, Y. Nakagawa, Y. Katoh, Y. Saitoh, K. Sakai, H. Satoh, S. Odahara, H. Nakano, J. Nakagaki, Y. Shiota, Nature **411**, 56 (2001).

37. O. Yaroshchuk, R. Kravchuk, A. Dobrovolskyy, L. Qui, O. D. Lavrentovich, SID 03 Digest, 1062 (2003).
38. O. Yaroshchuk, R. Kravchuk, A. Dobrovolskyy, S. Pavlov, Eurodisplay 02 Digest, 421 (2002).
39. J. Cognard, Mol. Cryst. Liq. Cryst. **78** suppl. 1, 1 (1982).
40. Minhua Lu, K. H. Yang, T. Nagasogi, S. J. Chey, SID 00 Digest, 446 (2000).
41. M. Schadt, H. Seiberle, A. Schuster, M. Kelly, Jpn. J. Appl. Phys. **34**, 3240 (1995).
42. D. J. Broer, in Radiation curing in polymer science and technology, vol. 3, Polymerisation mechanisms (ed. Fouassier, Jean-Pierre; Rabek, J.) Elsevier Ch. 12, 1993.
43. J-H. Kim, Y. Shi, S. Kumar, S-D. Lee, Appl. Phys. Lett. **71**, 3162 (1997).
44. H. R. Stapert, S. del Valle, E. J. K. Verstegen, B. M. I. van der Zande, J. Lub, S. Stallinga, Adv. Funct. Mater. **13**, 1 (2003).
45. A. Shishido, O. Tsutsumi, T. Ikeda, Mat. Res. Soc. Symp. Proc. **425**, 213 (1996).
46. J. C. ANGUS, Diamond and diamond-like films, THIN SOLID FILMS **216(1)**, 126 (1992).
47. J. P. Doyle, P. Chaudhari, J. L. Lacey, E. A. Galligan, S. C. Lien, A. C. Callegari, N. D. Lang, M. Lu, Y. Nakagawa, H. Nakano, N. Okazaki, S. Odahara, Y. Katoh, Y. Saitoh, K. Sakai, H. Satoh, and Y. Shiota, Ion beam alignment for liquid crystal display fabrication. Nuclear Instruments and Methods in Physics Research B **206**, 467 (2003).

Epilogue

We started the work on liquid crystal surfaces and interfaces with a question *"Why do certain surfaces align liquid crystals and others not?"*. Now, at the end of this book we can give the following answer: *"We do understand and see what are the mechanisms that are responsible for liquid crystal alignment. Although these may differ from one surface to another, there are clearly common features"*. On polymer surfaces, the predominant alignment mechanism is due to polymer chain and side group alignment in the top layer of the polymer, together with surface corrugations. By the combined output of the various methods presented, we do now understand, why rubbed polyimide shows alignment parallel to the rubbing direction, why rubbed polystirene aligns perpendicular to the rubbing direction, and what is the origin of the pretilt angle. We know, that only a statistically significant number of bonds with preferred orientation in the top layer of a polymer is enough to induce macroscopic liquid crystal alignment. There is no need for regions of crystalline or quasi-crystalline order, which is most clearly demonstrated in amorphous carbon alignment layers, formed by random bombardment of a polymer surface.

On the molecular level, we see that certain crystalline surfaces induce collective orientational and positional order in the first layer of liquid crystalline molecules. We observe, that there are several energetically degenerate domains that coexist at the interface. This is a clear signature of the *"orientational rigidity"* of an ensemble of liquid crystalline molecules that collectively adopt certain spatial orientation via the molecule-molecule interactions.

The "orientational rigidity" of liquid crystalline phases is most strikingly demonstrated in experiments on liquid crystal alignment on surface gratings, formed by laser ablation. Although the polymer surface is disordered on the microscopic scale, the geometrical, macroscopic surface features induce the alignment of a liquid crystal very efficiently. This leads to a clear answer to the old question of microscopic or macroscopic origins of alignment mechanism: both mechanisms are efficient and the interplay of both is responsible for the aligning action of a particular surface.

The surface anisotropy, which is necessary to induce a certain direction of liquid crystal alignment, can be produced in many different ways, such as rubbing, oblique evaporation, oblique ionic bombardment, surface bombardment by photons in photo-alignment techniques, nano-rubbing and patterning etc. Of particular importance are contact free methods, that allow a much larger flexibility and will likely take over the standard rubbing process in future applications.

Speaking of future, we see two directions where the liquid crystal surfaces and interfaces might evolve. The first direction is nano and micro-patterning of surfaces using light, Scanning Force Microscopes or other more efficient techniques to create surfaces with novel anchoring properties. Good examples in this direction are nano-structured surfaces, showing bistability, multistability and extraordinary large pretilt due to frustration. The second direction is towards active surfaces or command surfaces, where the surface anchoring conditions could be directed either by light or external electric field, resulting in chemical, electrochemical or conformational changes of the top molecular layer of the aligning surface. These two directions might be efficiently combined not only in displays, but also in novel photonic devices.

Acknowledgements

M.P. De Santo, R. Barberi and L.M. Blinov would like to thank Taylor and Francis, Ltd, for granting permission to reprint figures from the paper "Rubbing-induced charge domains observed by electrostatic force microscopy: effect on liquid crystal alignment" by I.H.Bechtold, M.P. De Santo, J.J.Bonvent, E.A.Oliveira, R.Barberi and Th.Rasing, Liquid Crystals, (2003), Vol. 30, No. 5, pp. 591-598.

J.Lüning and M. Samant would like to thank Nature for granting permission to reprint a figure from the paper: P. Chaudhari, J. Lacey, J. Doyle, E. Galligan, S.C.A. Lien, A. Callegari, G. Hougham, N.D. Lang, P.S. Andry, R. John, K.H. Yang, M.H. Lu, C. Cai, J. Speidell, S. Purushothaman, J. Ritsko, M.G. Samant, J. Stöhr, Y. Nakagawa, Y. Katoh, Y. Saitoh, K. Sakai, H. Satoh, S. Odahara, H. Nakano, J.Nakagaki and Y. Shiota, Nature 411, 56-59 (2001)

Index

MIX
Papier aus verantwortungsvollen Quellen
Paper from responsible sources
FSC® C105338

If you have any concerns about our products,
you can contact us on
ProductSafety@springernature.com

In case Publisher is established outside the EU,
the EU authorized representative is:
Springer Nature Customer Service Center GmbH
Europaplatz 3, 69115 Heidelberg, Germany

Printed by Libri Plureos GmbH
in Hamburg, Germany